Water Treatment Handbook

Water Treatment Handbook

Edited by **Herbert Lotus**

R CALLISTO REFERENCE

New York

Published by Callisto Reference,
106 Park Avenue, Suite 200,
New York, NY 10016, USA
www.callistoreference.com

Water Treatment Handbook
Edited by Herbert Lotus

International Standard Book Number: 978-1-63239-622-8 (Hardback)

Printed in the United States of America.

Contents

Preface

Water is a limited resource which needs effective management. Economic growth, population escalation, and ecological pollution occurring in numerous parts of the world are placing huge demands on accessible resources of fresh water and reflecting a "water crisis". Resource supervision, well-organized consumption of the water resources, and water refining are the alternatives to resolve the water crisis. Refinement techniques include conventional techniques that have lasted for more than a few centuries without major modifications, as well as, latest modern techniques. This book consists of numerous topics related to either improving the existing treatment techniques or the latest advanced techniques.

The information contained in this book is the result of intensive hard work done by researchers in this field. All due efforts have been made to make this book serve as a complete guiding source for students and researchers. The topics in this book have been comprehensively explained to help readers understand the growing trends in the field.

I would like to thank the entire group of writers who made sincere efforts in this book and my family who supported me in my efforts of working on this book. I take this opportunity to thank all those who have been a guiding force throughout my life.

Editor

Management and Modeling of Treatment Systems

Informatics, Logistics and Governance in Water Treatment Processes

Rezaul K. Chowdhury and Walid El-Shorbagy

Additional information is available at the end of the chapter

1. Introduction

Increasing demands of water for diverse uses of domestic, agricultural, and industrial consumption, both in quantity and quality, pose tremendous pressure on the need of well-articulated management approaches. Such approaches are expected to address several challenges such as limited supplies, deteriorated quality of available and produced water, conflicting interests of public stakeholders and groups, adverse environmental and ecological impacts, climate changes, etc. Large number of these approaches exist and still evolve and develop for different purpose including data management, operation and governance, conjunctive management of water and energy, asset management, and intelligent systems in water treatment process.

Planning and execution of water treatment processes involve a coordinated effort from different stakeholders. This involves collection of data (water availability data, water demand data, demographic data, water quality data, land use data etc.), identification of appropriate authority for their sources and development of effective communication to these authorities. Logistics involved in establishment of water treatment plants vary from country to country and depend on the country's socio-political systems. For any city or town, municipality or city council authority is generally responsible for execution of water treatment plants, however their approval process may involve several stakeholders. Government system of any country significantly affects selection of water treatment processes (desalination plant, surface water treatment plant, groundwater treatment and selection of alternative water sources such as stormwater, rainwater and treated greywater). Collection of treatment plants' performance monitoring data is essential for execution of similar projects. Now-a-days, climate change and their impacts on freshwater availability is a matter of great concern to water professionals. Scientific researches throughout the world anticipate reduction of water availability. Water treatment and water supply schemes require significant amount of energy (electricity) while water is essential for energy and electricity production. This paradox is widely known as "water energy nexus". It is also known as "climate, water and energy nexus". Significant progress has been made on uses of intelligent systems in real time monitoring of water quality and water treatment processes. All of these issues can generally be considered as advanced management issues in water treatment and management.

This chapter discusses a number of management approaches associated with different aspects. First, it discusses the data requirements for water treatment and water supply scheme. The management approaches are mostly tailored and oriented to achieve the scheme of Integrated Urban Water Management (IUWM). Second, the chapter discusses the governance in water treatment and water supply scheme. Water governance is defined as the political, social, economic and administrative systems developed to manage water resource and to deliver needed societal water services. Suitable water governance for some sector depends on the societal and environmental conditions prevailing in that sector. Efficient governance system leads to efficient operation of the water system and adequate benefit of the technologies applied in this system. Different classes of water governance, mainly public, public-private, and private systems, are discussed and evaluated. Third, the chapter discusses the joined management of water and energy resources practiced in many cases, known as water energy nexus. Finally, intelligent systems in water treatment process is addressed.

2. Data requirements

Because of rapidly increasing population, tremendous pressure on water quantity and quality is generally observed across all aspects of urban water cycle. Urban water cycle includes water supply, wastewater, stormwater, groundwater and aquatic ecosystems (Fletcher et al., 2008). Water treatment and water supply cannot be considered as an individual component of urban water cycle, rather this is connected to other components. For example collection, storage, treatment and distribution of stormwater for non-potable domestic consumptions (toilet flushing, gardening) may affect groundwater recharge and water quantity and quality in the downstream of streams. Therefore it is important to manage these interactions between different urban water cycle components. The term Integrated Urban Water Management (IUWM) indicates management of individual water cycle component in an integrated way. Availability of appropriate database is the prerequisite for reliable integrated management of individual water cycle components (water treatment, stormwater, wastewater etc.). Monitoring is a fundamental part for water treatment processes and monitoring data is also important for adaptive management of water projects.

Water treatment process is a component of water supply scheme. Components of water supply scheme are collection of raw water, conveyance of raw (untreated) water, supply reservoir, treatment plant, transportation of treated water, distribution system and finally water consumers. Water treatment process is linked with other components of water supply scheme. Water supply scheme is intrinsically related with other components of IUWM. Wastewater may enter into the water supply scheme through leakage on water reticulation

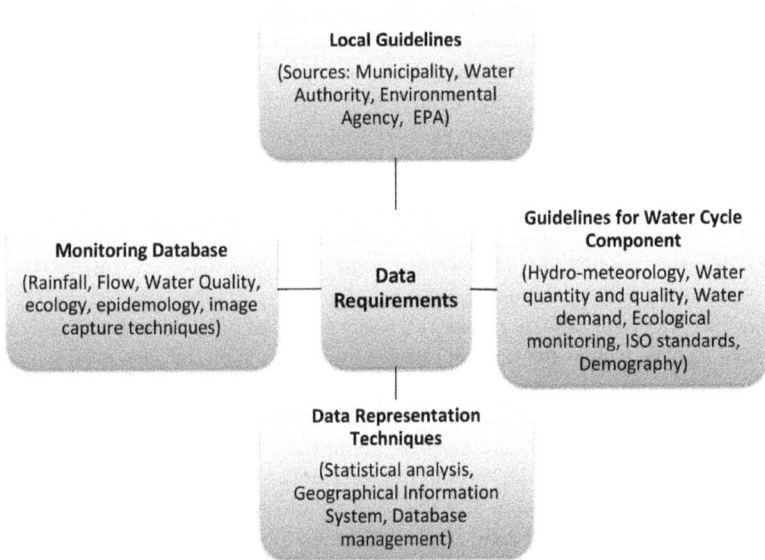

Figure 1. Data requirements guidelines for water treatment processes as a component of integrated urban water management (based on Fletcher et al., 2008)

pipes. More uses of treated water generate more greywater and wastewater. Monitoring of water quality data is an intrinsic property of water supply scheme. Different agencies may collect and store data on various individual components of IUWM. Therefore data exchange between agencies is significantly important for promoting the IUWM concepts. A guideline for data requirements for an integrated urban water scheme is shown in Figure 1 (Fletcher et al., 2008).

Uncertainty is generally involved in monitoring programs. An ideal monitoring program should consists of monitoring data, results and/or predictions from these data analyses and analysis of uncertainty involve in data and predicted results. Uncertainty arises from two sources, bias errors such as systematic error from erroneous sensor calibration and random error from natural processes and equipment measurement variations. When uncertainty level is more than allowable limit, the monitoring programs need to be redesigned.

3. Governance in water

As of now, water sector is driven by investments in technological innovations and development of infrastructures. The goal of such investments is to allow access of large number of people to water supply. However, there are many cases where infrastructures are not operated in an effective manner. In such cases, benefits from appropriate technologies are not fully utilized. This is because of lack of good governance in water management. According to Global Water Partnership (2003), governance can be defined as the range of political, social, economic and administrative systems that are in place to develop and manage water resources, and to deliver water services to different levels of society. There are two main values on which good governance rests, inclusiveness (ensures that all members of the group receive equal treatment) and accountability (ensures that those in authority answer to the group they serve if things go wrong and are credited when things go well) (Tropp, 2007).

There are four competing approaches on how water in different processes should be governed; these are [1] Water (drainage, sanitation, recycling, and reuse) seen as an economic good or commodity, with IWRM focus, [2] Water (and sanitation) seen as a human right and a social good and can be complementary to other approaches, [3] water (drainage and ecological sanitation seen as a socio-ecological good, also can be seen as a h, and human right as well as the right of other living beings and ecosystems, [4] water (and sometimes sanitation) seen as a sector. It is important to realize that for water governance of a city how the territorial dimension of water governance is dealt with and from which perspective. In particular, thre groups here should be distinguished; [1] those mainly looking at water from outside the cities, [2] those mainly looking at water from within the cities, and [3] those mainly looking at water from a multi-scalar perspective combining the global and regional scale with the city, its territory, and the neighborhoods within the cities and vice-versa (Miranda et al., 2011) .

Governance processes determine decision making about selection of water source, water storage, regulation of extraction from aquifers, regulation of discharges, and allocation between competing end users including allocations for environmental flow. For example, choice between a desalination plant and a large dam is an issue of water governance. Good water governance is significantly linked to strong policy, legal, and regulatory frameworks; effective implementing organizations; community involvement to improve water governance; and appropriate investments. Good governance ensures appropriate linkages and processes between and within organizations and social groups involved in decision-making, both horizontally across sectors and between urban and rural areas, and vertically from local to international (Rogers and Hall, 2003). Water governance includes private sector and civil society in addition to the government. The differences between conventional and emerging concept of water governance is given in Table 1. Many decision-makers and water managers are currently not prepared enough to deal with new forms of governance issues. The new concept of water governance involves conflict mediation, mobilization of communities, partnership formation, managing processes of stakeholder dialogue and participation. Generation of knowledge and capacity building in water governance is therefore necessary.

There are mainly three forms of governance arrangements. These are public, public-private and private governance. In **public governance** system, government takes on all of the responsibilities and challenges of water and wastewater services. Throughout the world, about 85% of drinking water supply provision lies under the public governance. Municipal authority or City Corporation is generally responsible for water supply and wastewater schemes. Another form of public management involves cooperatives and user associations.

Conventional form of governance	Emerging form of governance
Emphasizes the government and bureaucracy	Emphasizes to civil society and markets. The government and bureaucracy are still important entities but with reduced authority
Political power monopoly	Co-steering
Steering	Steering diversity of actors and power diffusion
Hierarchical control	Horizontally shared control
Enforcement of rules and regulations	Inter-organizational relations and coordination
Control	Formal and informal institutions
Top-down management	Co-governing (distributed governance); Decentralization/bottom-up management
Formal institutions	Network governance
Inter-governmental relations	Process orientation; expansion of voluntary exchange; self-governance and market mechanisms; dialogue and partnership; participation and negotiation
Disciplinary knowledge based	Multidisciplinary knowledge based

Table 1. Differences between conventional and emerging forms of water Governance (based on Tropp, 2007)

In this system, customers have decision-making power through participation in elections for different water authorities. The system is externally audited annually.

A public-private governance mechanism in water sector involves transferring asset management or operations of a public water system into private sectors. Several public-private arrangements are service contracts, management contracts, leases, concessions and build own-transfer programs. In this governance system, ownership of water systems can be distributed between public and private shareholders in a corporate utility. Majority ownership is usually kept within the public sector. In service contract, a private company is responsible for a specific task, such as meter installation, distribution pipe maintenance or collection of bills. Under management contracts, government transfers certain operation and maintenance activities to a private company. Under the concession and Build Operate Transfer (BOT) models, capital investment, commercial risk, operations and management of the project are undertaken by the private sector. Concessions are usually long-term to allow the private company to recover its investments. At the end of the contract, assets are either transferred back to the public sector or another concession is granted. In the BOT model, the role of government is predominately regulatory. BOT models are usually used for water and wastewater treatment plants.

In the private governance system, government transfers the water business to the private sector through sale of shares or water rights of the public entity. In this system, infrastructure, capital investment, commercial risk, and operations and management become the responsibility of the private provider. This model is not generally adopted.

4. Water energy nexus

The close connection between energy and water is generally known as water energy nexus. Significant amount of water is required to create energy. For example, water is used for fuel extraction, refining and production; hydropower generation; and in cooling steam electric power plants fueled by coal, oil, natural gas and nuclear power. In the United States of America, besides agricultural consumption, energy production and power generation systems are major users of freshwater resources (Younos et al., 2009). On the other hand, huge amount of power (energy) is required for water supply schemes – from untreated water collection to treated water distribution. Power is used to operate pumps in water supply scheme and in operating water treatment plants. Irrigation of water requires power. Collection, treatment and disposal of wastewater require power. Uses of alternative water resources need power to operate dual reticulation system. For domestic water heating and cooling purposes, power is required. Shortage of any one or both of water and energy will reduce them. Because of climate change, it is anticipated that available water resources will be reduced in some parts of the world– which will then affect energy production. The impact of climate change on water energy nexus is sometimes called "climate, water and energy nexus". Interrelationships between water and energy are shown in Figure 2.

Table 3 shows average water consumption for different energy production and power generation techniques. The data presented in Table 3 are predominantly for the United States of America and were compiled by Younos et al. (2009). Natural gas production is found to be the most water efficient. Corn-based and soy-based biofuel production

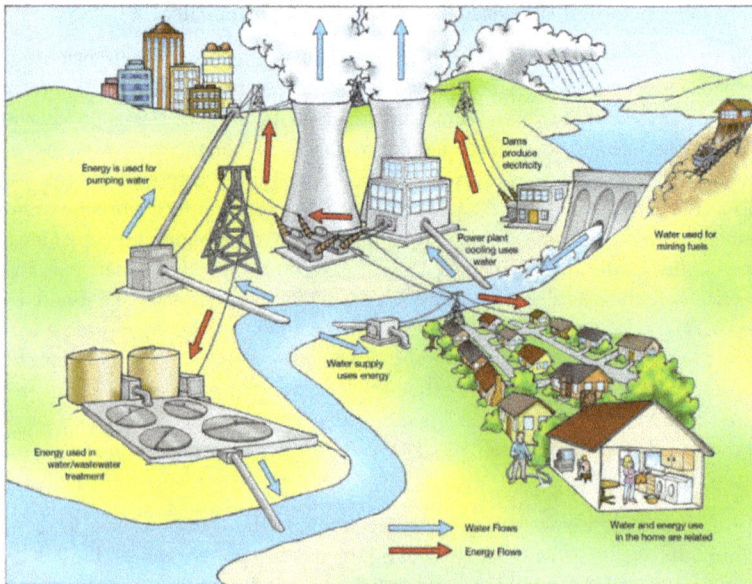

Figure 2. Interrelationships between water and energy (U.S. Department of Energy, 2006)

Fuel source	Water efficiency (Gallons/MBTU*)	Water consumptions
Coal	41 - 164	Coal mining operation, transport and storage, refining process; lubricate drilling equipment, post-mining activities such as land reclamation and revegetation
Natural gas	3	Drilling operation and gas purification, water exerted during drilling operation are reinjected to aquifer
Petroleum/Oil	1200 - 2420	Drilling operation, refining process, water exerted during drilling operation are reinjected to aquifer
Corn-Ethanol	2510 - 29100	Irrigation water demand for corn production
Soy-Biodiesel	14000 - 75000	Irrigation for Soya bean
Hydroelectric power generation	20	Evaporation water loss
Fossil fuel thermoelectric power generation	1100 - 2200	Steam turbine operation, cooling of turbine exhaust, condenser and reactor cooling
Nuclear power	2400 - 5800	Uranium mining and processing, nuclear reactor
Geothermal power	130	Steam turbine, evaporation loss
Solar thermoelectric power	230 - 270	Steam generation, coolant and cleaning purposes

(*MBTU is million British thermal unit)

Table 2. Water consumptions for energy production and power generation (based on Younos et al., 2009)

technologies are the most water intensive. Biofuels are also water intensive because significant amounts of water are required during crop growth. In terms of technology, hydroelectric power generation technique is the most water efficient technology. This is because used water is returned to the source.

Water supply and wastewater treatment systems are energy intensive. About 35% of total energy used by municipalities is used for operation of water and wastewater treatment plants. In the United States of America, about 1.4% of their total energy is consumed for water and wastewater treatment processes (Elliot, 2005). Water demand reduction can decrease energy uses in pumping and treating water. Reduction of volume of wastewater to wastewater treatment plant can also reduce the energy requirements. Adoption of energy efficient technologies can reduce water consumption, and ultimately energy use. Use of alternative sources of energy in water supply and wastewater schemes will reduce greenhouse gas emissions. For example, uses of solar and wind energy to pump water and to heat household hot water tanks can reduce energy consumption (Thrilwell et al., 2007).

5. Intelligent systems

Water treatment processes are governed by a set of complex non-linear relationships between physical, chemical, biological and operational parameters. Traditionally these process relationships are fitted with mathematical models using bench-scale data. These models often perform poorly when two or more key process parameters change simultaneously and in the application of real world treatment plants (Baxter et al., 2001). Plant operators require appropriate tools so that appropriate plant operation conditions are maintained in order to achieve desired effluent quality based on instantaneous monitoring of influent water quality. Coagulation, flocculation and sedimentation processes in water treatment are a prime example. The Jar test is conventionally applied for determination of optimal single coagulant dose. Optimization of multiple treatment chemicals for removal of both particulate matters and organics is a challenging task, particularly if influent water quality changes significantly. Plant operators are generally depending on deterioration of effluent water quality instead of pro-active optimization of treatment chemicals subject to influent water quality.

Artificial Intelligence (AI) is the technique that can control plant operations in a pro-active way. Recent significant development in computing methods opens the door for application of artificial intelligence techniques in water treatment process controls and optimization.

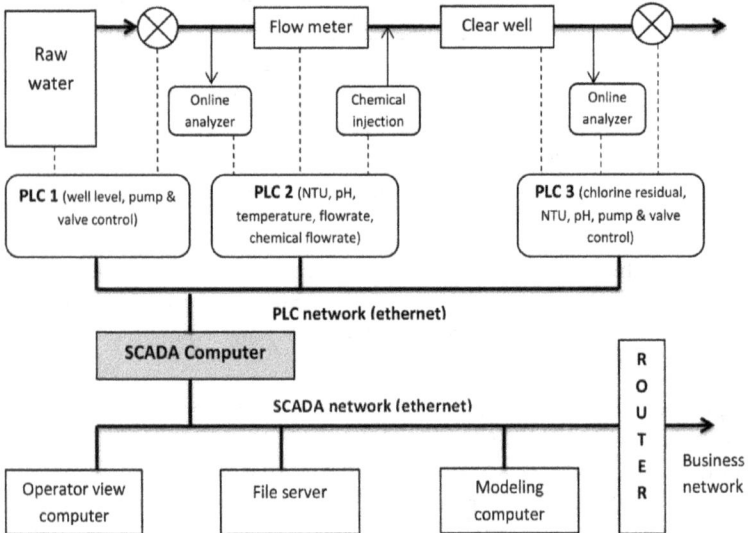

Figure 3. A schematic diagram of SCADA system in water treatment process (based on Baxter et al., 2001)

Application of Artificial Neural Network (ANN) in water treatment problems is an example of artificial intelligence.

The Supervisory Control and Data Acquisition (SCADA) system is the basic infrastructure for process control of water treatment plants (Grady et al., 1999). The SCADA system is capable of monitoring treatment process, real time collection of process data and works as a communication network between treatment plant and other utilities (Baxter et al., 2001). Typical components of a SCADA water treatment system are (1) communication networks (2) computers (3) Programmable Logic Controllers – PLCs and (4) online instruments. Conceptually the SCADA system has three layers namely hardware layer, network layers and application layer. Figure 3 shows a typical water treatment SCADA structure.

As an example, the EPCOR's E.L. Smith water treatment plant in Alberta, Canada (http://www. corp.epcor.com/watersolutions/operations/edmonton/Pages/el-smith-water-treatment-plant.aspx) applied neural network raw water quality classifier to monitor real time raw water quality. The online quality analyzer sends raw water quality data to the central process control computer through the SCADA system where the data are analyzed by raw water quality classifiers before entering to central database system. If any new data or error is detected, an alarm system is activated which requires a plant operator's investigation of sensors. Because of wireless SCADA system, plant operators can monitor the water treatment process operation from a remote computer.

6. Conclusion

Four management aspects were discussed in this chapter; these are (1) data management for water treatment and water supply scheme, (2) governance in water treatment and water supply scheme, (3) water energy nexus, and (4) intelligent systems in water treatment process.

For data management, the chapter discussed guidelines required in water treatment processes as a component of IUWM. The guidelines include [1] local guidelines originated from municipalities, water authorities, and environmental agencies, [2] monitoring database with different hydraulic, quality, and ecological records, [3] guidelines for water cycle component with relevant standards and operational and environmental data, and [4] data representation techniques. Monitoring programs and uncertainties involved in these programs were also discussed.

For water governance, it is found that lack of good governance in water management results in infrastructures not operated in an effective manner and hence benefits from appropriate technologies can't be fully utilized. The public governance, even though may not be the best choice in managing and operating some water sectors, it is found to be the most common system applied worldwide (about 85% applied in drinking water systems). Pure private governance system is not generally adopted.

Discussion of water energy nexus and data related to USA indicates that natural gas production is the most water efficient while corn-based and soy-based biofuel production technologies are the most water intensive. Hydroelectric power generation is also found the most water efficient technology from technology point of view where the used water is returned to the source.

As of Artificial Intelligence and its applications in managing water systems, an example of the EPCOR's E.L. Smith water treatment plant in Alberta, Canada, applied neural network and SCADA system to monitor and control real time raw water quality. Recent advancements in computing methods reflects the growing and significant contribution of applying artificial intelligence techniques in better management and control of water treatment process.

Author details

Rezaul K. Chowdhury and Walid El-Shorbagy
Department of Civil and Environmental Engineering, United Arab Emirates University, Al Ain, United Arab Emirates

7. References

Baxter, C. W., Tupas, R. T., Zhang, Q., Shariff, R., Stanley, S., Coffey, B and Graff, K. (2001) Artificial Intelligence Systems for Water Treatment Plant Optimization, American Water Works Association.

Elliot, R. N. (2005) Roadmap to Energy in the Water and Wastewater Industry, Report Number IE054, American Council for an Energy-Efficient Economy, United States of America.

Fletcher, T. D., Mitchell, V. G., Deletic, A. and Maksimovic, C. (2008) Chapter 1 - Introduction. In T. D. Fletcher & A. Deletic (Eds.), Data requirements for integrated urban water management. Paris: UNESCO Publishing and Taylor & Francis.

Global Water Partnership (2003) Effective Water Governance: Learning from the Dialogues, Global Water Partnership Secretariat, Stockholm, Sweden.

Grady, B., Farely, W. and Younos, T. (1999) Telemetry Options for Small Water Systems, Virginia Water Resources Research Center, Virginia.

Miranda, L, Hardijik, M, Malina, R. K. T (2011). Water Governance Key Approaches: An Analytical Framework, Literature Review.

Rogers, P. and Hall, A. (2003) Effective Water Governance. TEC Report No. 7, Global Water Partnership Secretariat, Stockholm, Sweden.

Thirlwell, G. M., Madramootoo, C. A. and Heathcote, I. W. (2007) Energy-water Nexus: Energy Use in the Municipal, Industrial, and Agricultural Water Sectors, Canada – US Water Conference, October 02, 2007, Washington D. C., United States of America.

Tropp, H. (2007) Water governance: trends and needs for new capacity development, Water Policy 9 Supplement 2: 19–30.

U.S. Department of Energy (2006) Energy Demands on Water Resources, A Report to Congress on the Interdependency of Energy and Water, U.S. Department of Energy, United Sates of America.

Younos, T., Hill, R. and Poole, H. (2009) Water Dependency of Energy Production and Power Generation Systems, VWRRC Special Report No. SR46-2009, Virginia Water Resources Research Center, Virginia.

Chapter 2

Some Details of Mathematical Modelling of Effluents in Rivers Downstream of a WWTP

Jiří Šajer

Additional information is available at the end of the chapter

1. Introduction

There are a number of directives to protect the environment, for example, in the EU see [1], [2]. Implementation of guidelines in accordance with these directives is based on environmental monitoring. Modelling based on existing data could look to reduce the necessary costs required for environmental monitoring in the future. This chapter was included in this book to reflect the adverse impacts of poor-treated and/or accidentally untreated water on nearby water bodies. Figure 1 shows an idealised sketch of mixing of pollutants in the river downstream of the outfall (0 m) of a wastewater treatment plant (WWTP).

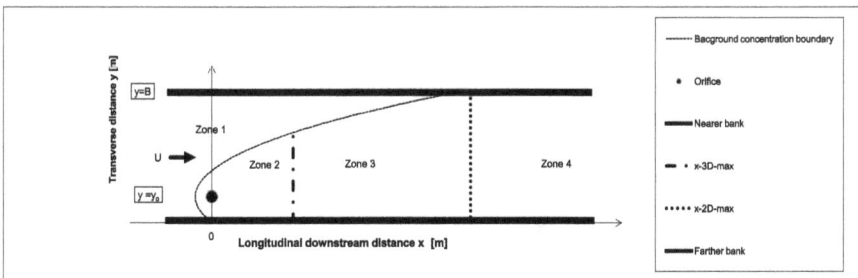

Figure 1. Sketch of mixing of pollutants in the river downstream of the WWTP outfall

In general four distinct zones exist downstream of the orifice. The first zone is a zone with background concentration. The second zone is a 3D mixing zone, the third zone is a 2D mixing zone and the fourth zone is a 1D mixing zone. In the initially three-dimensional mixing process (until the location where complete vertical mixing takes place) the drop-off of the maximum mass concentrationis relatively fast, $C_{max} \sim x^{-1}$, while it occurs more gradually in the vertically mixed, thus two-dimensional, mixing phase, $C_{max} \sim x^{-1/2}$ [3]. Mixing in rivers can be described in terms of the near-field and far-field mixing zones. The near-field is defined as the area between the discharge structure and the location where the effluent is vertically mixed in the river. It encompasses buoyant jet mixing and boundary interactions. Far-field mixing occurs after the effluent plume is vertically mixed, and encompasses transverse and longitudinal mixing in the river due to buoyant spreading and passive diffusion [4]. 2D mixing zone modelling will be explained using a simple model in the MS Excel file named Czech immission test (CIT), which is prepared for determination of regulatory mixing zones located downstream of WWTPs in the Czech Republic. CIT is an MS Excel spreadsheet model of the solution to the two-dimensional advection dispersion equation. Therefore, it may only be applied at a far-field distance where the plume is completely or nearly-completely vertically mixed. An environmental impact assessment (EIA) is an assessment of the possible positive or negative impact that a proposed project (in our case the example WWTP) may have on the environment. The only effective way of monitoring changes in biological quality is to perform continuous biological monitoring with instruments suitable for early warning purposes, for example see [5]. Due to the random occurrence of accidental spills, the only way to quickly detect hazardous situations is to perform continuous monitoring of surface water quality. That's the reason why the Daphnia Toximeter (a product of the German bbe Moldaenke company) has been installed at the monitoring station located on the borderline profile at the Odra River in the town of Bohumín. To monitor the biological impact of surface water quality on biota, this instrument uses young organisms of *Daphnia magna*. The organisms are exposed to the surface water in a flow-through chamber, to which the water from a monitored profile is pumped. New born daphnids aged less than 24 hours are placed into the chamber and stay there maximally for 7 days. They are then replaced by new ones. There are two reasons for this – older organisms are less sensitive than the young ones, and at the age of 9 to 11 days daphnids start to reproduce parthenogenetically, which leads to an undesirable change to the number of organisms in the chamber. This instrument can be checked through an Internet connection, thus making it possible to obtain online information about the biological impact of surface water quality in the monitored profile from anywhere with Internet access. Conversion of the daphnids' behaviour to numerical parameters is computed. If there is an accidental leakage of a toxic substance above the monitoring station, the output is a toxic index alarm curve. The second part of this chapter describes an example of toxic index alarm curve analysis.

2. Problem statement

The advective-dispersive equation for solute movement through a river forms the basis of the mathematical algorithm used by the riverine component. The surface-water flow is assumed to be steady and uniform; the algorithms are developed for the limiting case of unidirectional advective transport with three-dimensional (longitudinal, lateral, and vertical) dispersion. The advective-dispersive equation for solute movement in a river can be described by the following expression

$$\frac{\partial C}{\partial t} + U\frac{\partial C}{\partial x} = E_x\frac{\partial^2 C}{\partial x^2} + E_y\frac{\partial^2 C}{\partial y^2} + E_z\frac{\partial^2 C}{\partial z^2} + S + f_R(C,t) \qquad (1)$$

, where C is a dissolved instream contaminant concentration [kg m^{-3} or %]; U is the average instream flow velocity [m s^{-1}]; Ex, Ey, Ez are dispersion coefficients in the x-, y-, and z-directions, respectively [m^2 s^{-1}]. If source terms 'S' and 'f_R' are added as shown in the equation above, the so-called advection–diffusion reaction equation emerges. The additional terms represent [6]:

- Discharges or 'wasteloads' (S): these source terms are additional inflows of water or mass. As many source terms as required may be added to Equation (1). These could include small rivers, discharges of industries, sewage treatment plants, small wasteload outfalls and so on.
- Reaction terms or 'processes' (f_R). Processes can be split into physical and other processes.

Examples of physical processes are:

- settling of suspended particulate matter
- water movement not affecting substances, like evaporation
- volatilisation of the substance itself at the water surface.

Examples of other processes are:

- biochemical conversions like ammonia and oxygen forming nitrite
- growth of algae (primary production)
- predation by other animals
- chemical reactions on whether the flow is laminar orturbulent.

Dispersion is the scattering of particles or a cloud of contaminants by the combined effects of shear and transverse diffusion. Molecular diffusion is the scattering of particles by random molecular motions, which may be described by Fick's law and the classical diffusion equation. Turbulent diffusion is the random scattering of particles by turbulent motion, considered roughly analogous to molecular diffusion, but with eddy diffusion coefficients (which are much larger than molecular diffusion coefficients). The diffusion coefficients would either be molecular or turbulent, depending on whether the flow is laminar or turbulent [7]. In natural rivers, a host of processes lead to a non-uniform velocity field, which allows mixing to occur much faster than by molecular diffusion alone [7]. Under the assumptions of negligible momentum and buoyancy, and for a discharge near the stream's free surface or near the bottom, complete vertical mixing is expected to occur at distance

$$x_{3D-max} = 0,4\frac{Uh^2}{E_z} = 0,4\frac{Uh^2}{\beta h U_s} = 0,4\frac{hC_c}{\beta\sqrt{g}} \qquad (2)$$

, where x_{3D-max} is the distance to complete vertical mixing [m]; U is the mean velocity downstream of discharge outfall [m. s^{-1}]; h is the mean river depth downstream of discharge [m]; E_z is the vertical mixing coefficient [m^2.s^{-1}]; β is the vertical mixing coefficient constant [dimensionless]; U_s is the shear velocity [m.s^{-1}]; g is the gravity acceleration constant [m. s^{-2}] and C_c is the Chezy's coefficient [m^{0,5}.s^{-1}].

The constant β can be derived from the velocity profile (for example, see [8]), and its value is recommended at approximately 0.07 ± 50%. The distance to complete vertical mixing will be somewhat reduced if the finite dimensions of the discharge opening are considered or if the source location is varied within the water column, e.g. located at mid-depth. Thus, the discharge design can play a certain role in this initial region [3].

Complete transverse mixing is expected to occur at a distance

$$x_{2D-max} = \frac{0,1U\left(\left(2-2\left(\frac{y_0}{B}\right)\right)B\right)^2}{E_y} = \frac{0,1U\left(\left(2-2\left(\frac{y_0}{B}\right)\right)B\right)^2}{\alpha h U_s} \tag{3}$$

, where x_{2D-max} is the distance to complete transverse mixing [m]; U is the mean velocity downstream of discharge outfall [m. s^{-1}]; y_0 is the discharge distance from nearest shoreline [m]; B is the river mean width downstream of discharge [m]; E_y is the transversal mixing coefficient [m^2.s^{-1}]; α is the transversal mixing coefficient constant [dimensionless]; U_s is the shear velocity [m.s^{-1}] and h is the mean river depth downstream of discharge outfall [m].

The constant α is high in meandering and curved channels due to the presence of secondary currents. A value of 0.6 is recommended for most natural channels. Fischer [8] reports that this constant can range from 0.1 to 0.2 for straight artificial channels. Curves and sidewall irregularities increase the constant α such that in natural streams it is rarely less than 0.4. If the stream is slowly meandering and the sidewall irregularities are moderate, then the constant α is usually in the range of 0.4 to 0.8. Therefore, a value of 0.6 is usually recommended in natural channels. Uncertainty in this constant is usually at least ± 50 %.

In most practical problems we can start by assuming that the effluent is uniformly distributed over the vertical, or in the other words, we can analyse the two-dimensional spread from a uniform line source [8]. The longitudinal mixing term has very little influence on transverse mixing under the above condition and can be dropped. If the channel has the width B the effect of the boundaries, $\frac{\partial c}{\partial y}$ = 0 at y=0 and y=B, can be accounted by the method of superposition described in [8]. Volume friction of effluent in a sample at some sampling point in the effluent plume is independent on the ambient background concentration. The above facts were used to construct the Czech Immission Test (CIT) model. This CIT model is based on the following equation

$$VF_e = \frac{Q_e}{UBh\sqrt{4\pi x'}} \sum_{n=-\infty}^{\infty}\left\{e^{[-(y'-2n-y_0')^2/4x']} + e^{[-(y'-2n+y_0')^2/4x']}\right\} \tag{4}$$

, where VF_e is the volume friction of effluent in a sample at some sampling point in the effluent plume [dimensionless]; Q_e is the effluent discharge flow rate [m^3.s^{-1}]; U is the mean velocity downstream of discharge outfall [m. s^{-1}]; B is the river mean width downstream of discharge [m]; h is the mean river depth downstream of discharge outfall [m]; x' is defined by setting $x' = xE_y/UB^2$ [dimensionless]; y' is defined as y/B [dimensionless] and y_0' is defined as y_0/B [dimensionless].

Dilution factor DF is reciprocal of the VF_e.

The concentration of a pollutant of concern (or a dye tracer) at any sampling point is given by

$$C = \left(\frac{C_e - C_r}{DF} + C_r\right)e^{-kt} = (VF_e(C_e - C_r) + C_r)e^{-kt} \tag{5}$$

, where C_e is the solute concentration in the effluent [kg.m^{-3} or %]; C_r is the background concentration in the receiving water [kg.m^{-3} or %]; DF is the dilution factor [dimensionless]; VF_e is the volume friction of effluent in a sample at some sampling point in the effluent plume; k is the first order reaction coefficient [s^{-1}] and t is the time [s].

The closeness of the approach of model values to measured values is given by:

$$R_t^2 = 1 - \frac{\Sigma(C_f - C_m)^2}{\Sigma(C_f - C_{avgf})^2} \tag{6}$$

, where R_t^2 is the coefficient of determination; C_f is the field value; C_m is the model value and C_{avgf} is the mean of the field values. The perfect fit is indicated by $R_t^2 = 1$ and values decrease as the fit becomes poorer [9].

The unsteady solutions of the 1D advective dispersion equation (ADE) may be obtained using methods of Fourier transform, of Laplace transform and the Fourier method. Using the methods of calculus, analytical solutions are developed that provide the predicted solute concentration as a function of time and space. Analytical solutions are derived for conservative substance, constant velocity, constant discharge, constant cross-sectional area and constant dispersion coefficient. Point sources such as accidental spills may be viewed as instantaneous sources [10]. For a spill, the solution to ADE can be obtained using the method of Fourier transform. For a continuous rectangular input the solution can be obtained using the method of Laplace transform described by [8]. This solution is useable only for the duration of the continuous input. The principle of superposition may be used to develop the solution for all time periods after the termination of the continuous input [10]. Results of Fourier transform have greater error then results from Laplace transform [11]. The analytical solution of Laplace transform gives results comparable to the more time and calculation consuming Fourier method [11].

Sometimes the concentrations in the 1D zone are too low. The International System of Units (SI) doesn`t include ppb as a unit. Therefore it is convenient to compute with ratio R_C:

$$R_C = \frac{C}{C_{IN}} \tag{7}$$

, where C is the concentration at the station at time t [kg.m^{-3}] and C_{IN} is the input solute concentration after mixing over the cross-section of the stream computed specially for every monitoring station (effect of dead zones is included) [kg.m^{-3}]. Every sampling point has its own C_{IN} computed using Equation (8):

$$C_{IN} = \frac{m_s}{A_m U_r \tau} \tag{8}$$

where m_s is the mass of substance present in the accidental leakage cloud as it passed the sampling point [kg]; A_m is the model cross-section area [m^2]; U_r is the retarded velocity (dead zones included) [m.s^{-1}];τ is the duration of the continuous input [s]; x is the distance between source and monitoring station [m] and t_p is the time to pick value [s].

$$R_C(x,t) = 50\left[erfc\left(\frac{x-U_rt}{2\sqrt{E_xt}}\right) - Kerfc\left(\frac{x-U_r(t-\tau)}{2\sqrt{E_x(t-\tau)}}\right) + \right.$$
$$\left. +\exp\left(\frac{U_rx}{E_x}\right)\left\{erfc\left(\frac{x+U_rt}{2\sqrt{E_xt}}\right) - Kerfc\left(\frac{x+U_r(t-\tau)}{2\sqrt{E_x(t-\tau)}}\right)\right\}\right]$$

(9)

, where x is the distance between source and monitoring station [m]; t is the time [s]; U_r is the retarded velocity [m.s^{-1}]; τ is the duration of the continuous input [s]; E_x is the longitudinal dispersion coefficient [m^2.s^{-1}]; $erfc$ is the complimentary error function and K is coefficient [dimensionless]. It was convenient to define dimensionless coefficient K for computation as follows: if t< τ then K = 0; if t> τ then K = 1. Numerous studies on longitudinal dispersion have been conducted over the past few decades. Fick's second law predicts how diffusion causes the concentration to change with time. In actuality, immobile-flow zones (dead zones) may invalidate Fick's law. Chatwin [12] developed a method for determining longitudinal dispersion intended to address the problem of non-Fickian behaviour. Technically, the Chatwin's method is only really valid for impulse releases, but it does provide a reasonable approximation for longitudinal dispersion for pulse and continuous releases [13]. Chatwin's values b can be computed by equation

$$b = \sqrt{t \, \ln \frac{C_p\sqrt{t_p}}{C\sqrt{t}}}$$

(10)

, where b is the Chatwin's value [s$^{0.5}$]; t_p is the time to pick value [s]; C_p is the pick value [% or kg.m^{-3} or any other unit]; t is time [s] and C is value at value [% or kg.m^{-3} or any other unit].

Longitudinal dispersion is given by Equation (11)

$$E_x = \left(\frac{x}{2\,b^*}\right)^2$$

(11)

where b^* is the vertical axis (at $t = 0$ s) intercept of the straight-line fit to the early-time values b [s$^{0.5}$] and x is distance [m].

3. Some existing models used in water modelling and model choosing

There are some predictive models for examining the mixing from point sources and showing compliance with EQS-values:

- General water quality models may be required in more complex situations. Different methods for the far-field modeling exist, ranging from water quality models in estuary-type flows (e.g. model QUAL-2 of the U.S. EPA), to Eulerian coastal circulation and transport models (e.g. Delft3d of Delft Hydraulics) to Lagrangian particle tracking models. For rivers, the model AVG of the German ATV-DVWK, the model QUAL-2 of the U.S. EPA, and the model RWQM1 of the International Water Association are all examples of general water quality models. Such models also form the basis of management procedures for attaining a good quality status in the case of multiple sources, i.e. by following the principle of a distributed waste load allocation for individual water users [3]. Danish Hydraulic Institute MIKE Software is the result of years of experience and dedicated development. DHI Software models the world of water - from mountain streams to the ocean and from drinking water to sewage [14]. MIKE 11 is synonymous with top quality river modelling covering more application areas than any other river modelling package.

- Choosing of an initial dilution model - five models are described in [15]:

Three are theoretical (UM, UDKHDEN, and VSW), and two are empirical (RSB and CORMIX). UM is the current version of the earlier models UOUTPLM and UMERGE. It acts as a two-dimensional model for single ports, though a pseudo-three-dimensional version is employed when there is a multiport diffuser with potential merging. It uses the 3/2 power profile to calculate the ratio and determine the centerline concentration as a function of the top hat concentration that it predicts. The ratio changes continuously with each integration step along the trajectory. Merging is simulated with the reflection technique. The CORMIX model has three modules: CORMIX1 for submerged single-point discharges, CORMIX2 for submerged multi-port diffuser discharges, and CORMIX3 for buoyant surface discharges.

- Choosing a farfield model described in [15]:

There are two farfield models which are presently recommended for use. They are code named FARFIELD and RIVPLUM5. The appropriate farfield model to use in a particular mixing zone analysis depends on the combination of conditions involved:

1. The receiving water is sufficiently deep such that a plume will form and pass through the initial dilution phase without "Froude number less than 1", "overlap", or "boundary constraint" problems. Use FARFIELD as the algorithm (i.e., the version in 3PLUMES interface).
2. The receiving water is shallow and unidirectional; the effluent is thoroughly mixed surface to depth (i.e., no defined plume); and the discharge is a single port or short diffuser. Use RIVPLUM5.
3. There is/are bank constraint(s). Use RIVPLUM5, provided the conditions in 2. Above are also met.
4. Other shallow receiving waters (with no bank constraints) which occur with all other combinations of effluent plumes and discharger configurations. Use FARFIELD as a stand-alone model. A three-dimensional advective dispersion equation may also be appropriate.

- The QTRACER2 (program for tracer-breakthrough curve analysis for tracer tests in karstic aquifers and other hydrologic systems)[13] is fast and easy method for evaluating tracer-breakthrough curves (BTCs) generated from tracing studies conducted in hydrologic systems. It has been reviewed in accordance with U.S. Environmental Protection Agency policy and approved for publication. Results may then be applied in solute-transport modeling and risk assessment studies.

4. Some details and results of 2D modelling

First, the model must be calibrated; that is, its parameters must be adjusted to match the behaviour of the prototype. Second the model must be validated. This means that a calibrated model must be compared to data not used in the calibration to determine whether the model is applicable to cases outside the calibration data set [7]. The CIT model was tested against data from a dye tracer study of the City of Arlington WWTP discharge to the Stillaguamish River, described in reference [16]. This study was performed on 22 August, 2006 and documented in a Mixing Zone Study report (CEG, November 2006, revised May 2007). The field study included injection of Rhodamine WT dye into the WWTP effluent at a known concentration; collection of bottled fluorescence samples from within the effluent plume; and measurement of river bathymetry, width, and current velocity. At seasonal low flow conditions observed during the dye study, the river was approximately 121 feet (36.9 m) wide with an average depth of 4 feet (1.22 m). Average current speeds, measured with a Swoffer meter, were 1.5 feet per second (0.46 m.s^{-1}). The river channel is relatively straight and uniform downstream of the outfall, and river cross-section bathymetry is similar at other locations up to 500 feet (152.3 m) downstream of the outfall. The outfall consists of a single port discharge (12-inch-diameter) discharging horizontally at the river bottom. The outfall discharge is located approximately 52 feet (15.86 m) from the left (south) bank at an invert depth of 4.61 feet (1.406 m) during low flow conditions. Appendix A contains plan and profile record drawings of the outfall. Effluent discharge flow through the outfall was 2.2 million gallons per day (8,318.4 m^3.d^{-1}) during the study. Manning's roughness coefficient value (0.025) in the study report was based upon the average rock diameter observed at the site. Water column average dilution factors at plume centerline are summarized in Table 1. Except for the 50-foot distance, centerline profiles were measured over two time periods to better represent the time varying nature of the plume. The plume was observed to rise from the river bottom immediately following discharge to approximately river mid-depth, and was relatively unsteady with a billowing nature (wandering back and forth across river within a prescribed area). Between 100 and 304 feet from the outfall, complete vertical mixing of the plume was visually observed to occur, and the billowing nature of the effluent plume was less apparent. These observations are confirmed in the effluent volume fraction profiles at the mixing zone boundary (304 feet downstream), where both time period results are nearly indistinguishable, and effluent concentrations are nearly uniform from the top to bottom of the water column. Calibration of the RIVPLUM 5 model to the tracer study results produced a transverse mixing coefficient constant equal to 0.4.

Distance from outfall		Water column average dilution factor at plume centerline	Field concentration at point of interest	
x		DF_{field}	$C_{p\text{-}field} = 100/DF_{field} + C_{r\text{-}field}$	$C_{p\text{-}field} - C_{r\text{-}field}$
[ft]	[m]	[-]	[%]	[%]
0	0.00	1.0	100.00	100.00
30	9.15	12.8	7.81	7.81
50	15.25	13.4	7.46	7.46
100	30.50	27.0	3.70	3.70
304	92.72	41.1	2.43	2.43

Table 1. Field values

Calibration of the CIT model to the tracer study results produced a transverse mixing coefficient constant equal to 0.408 (see Figure 2). The closeness of approach of model CIT values to field values was computed using Equation (6). Resulting value of CIT model was about the same as resulting value of RIVPLUM 5 model and within the range of experiments reported by [8], from which CIT and RIVPLUM5 were developed.

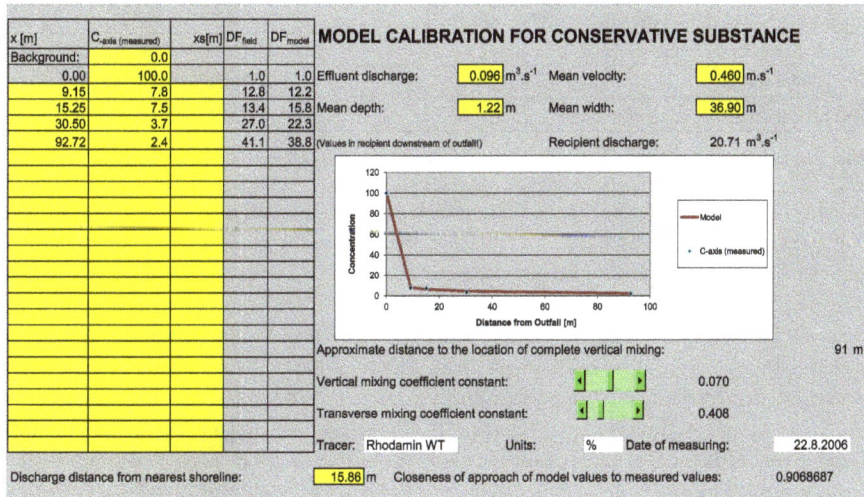

Figure 2. Water quality model downstream of Arlington WWTP – calibration procedure

Although physical processes play a large role in determining the fate of solutes, chemical and biological processes may be equally important. We would like to describe results of modelling for non-conservative substances now. The next fictive example shows possibility of investigating the impacts of different effluent quality using the CIT developer model to show its merits with different treatment efficiencies. We are interested about biological oxygen demand (BOD) in this case. We assume the calibrated and validated value of transverse mixing coefficient constant is equal to 0.408. Secondly we assume calibrated and validated value of the temperature dependent first order rate coefficient for BOD equal to $2.31.10^{-6}$ s^{-1} at a temperature 20°C. The CIT model computes concentration of BOD at point of interest using Equation (5). Resulting concentration of BOD at a downstream distance 200m can be found in Table 2.

Model variant No.	1	2	3
Influent BOD concentration [mg.l^{-1}]	120	120	120
Effluent BOD concentration [mg.l^{-1}]	7	30	120
Removal efficiency [%]	94.917	75	0
Effluent discharge rate [m^3.s^{-1}]	0.096	0.096	0.096
Background BOD concentration [mg.l^{-1}]	2	2	2
Mean stream depth [m]	1.22	1.22	1.22
Mean stream velocity [m.s^{-1}]	0.46	0.46	0.46
Mean stream width [m]	36.9	36.9	36.9
Manning's roughness coefficient [m$^{-1/3}$.s]	0.025	0.025	0.025
Discharge distance from nearest shoreline [m]	15.86	15.86	15.86
Distance downstream to point of interest [m]	200	200	200
Distance between point of interest and nearest shoreline [m]	15.86	15.86	15.86
Transverse mixing coefficient constant [dimensionless]	0.408	0.408	0.408
First order reaction coefficient [s^{-1}]	2.31E-06	2.31E-06	2.31E-06
VF_e [dimensionless]	0.00465	0.00465	0.00465
BOD concentration at point of interest [mg.l^{-1}]	2.02121	2.12805	2.54613

Table 2. Modelling BOD concentrations at a downstream distance of 200 m at a temperature 20°C.

Sudden changes usually occur in the regular channel of the natural river. These singularities can be natural, such as changes in roughness, meanders etc., or artificial, such as bridge piers, groynees etc. Burdych´s suggested method uses spare length x_s in these cases [17]. Spare length is a distance between the real location of discharge outfall and its fictive effective position. It is explained on the next fictive example. Effluent from the WWTP is discharged to the river at river kilometre 49.62. The river is approximately 50 m wide with an average depth of 3.32 m. The mean velocity of flow is 0.223 m.s^{-1}. The river is relatively straight and uniform downstream of the outfall. The outfall discharge is located at the bank of the river. Effluent discharge flow through the outfall is 0.455 m³.s^{-1}. Field values of conductivity are listed in Table 3 and can be seen in Figure 3.

For each distance listed in Table 3 the value F was computed. We can distinguish three characteristic reaches. The average transverse mixing coefficient E_y for each reach can be obtained by fitting a straight line through values F plotted against the distance. E_y is the factor which determines the slope of regression line (see Figure 4). The intercept of this regression line can be expressed as E_y*x_s [18].

x	C_r	C_p	$C_p - C_r$	$C_p - C_r$	DF	VF_e	$F = E_y(x+x_s) = (Q_e/VF_e)^2/(pUh^2)$
[m]	[mS.cm^{-1}]	[mS.cm^{-1}]	[mS.cm^{-1}]	[%]	[-]	[-]	[m^3.s^{-1}]
0	237	1027.00	790.00	100.00	1.00	1.000000	0.03
10		661.74	424.74	53.77	1.86	0.537650	0.09
80		387.17	150.17	19.01	5.26	0.190088	0.74
170		340.02	103.02	13.04	7.67	0.130399	1.58
330		310.94	73.94	9.36	10.68	0.093593	3.06
340		295.904	58.90	7.46	13.41	0.074559	4.82
390		293.26	56.26	7.12	14.04	0.071216	5.28
400		285.09	48.09	6.09	16.43	0.060877	7.23
450		283.62	46.62	5.90	16.94	0.059015	7.69
600		279.91	42.91	5.43	18.41	0.054311	9.08
760		276.78	39.78	5.04	19.86	0.050356	10.56
950		273.83	36.83	4.66	21.45	0.046620	12.32
1110		271.68	34.68	4.39	22.78	0.043899	13.90

xs – the difference between effective and real distance [m].

Table 3. Fictive field values and computing DF, VF_e and F values

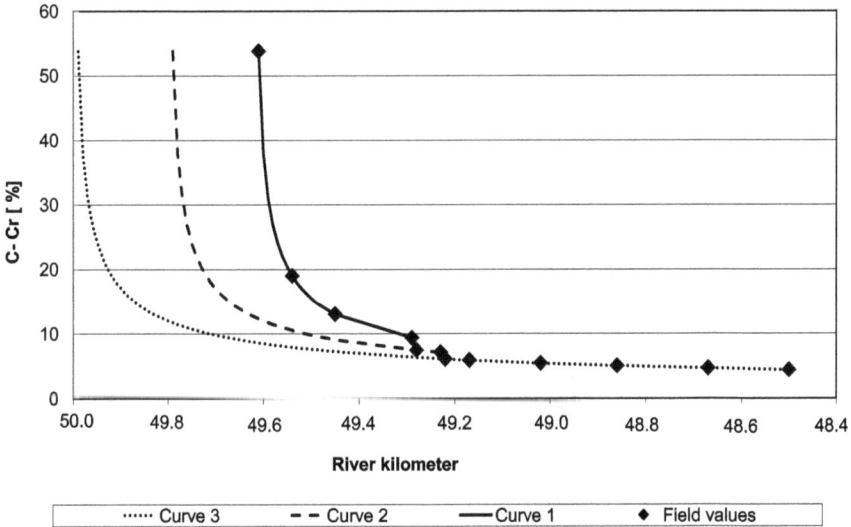

Figure 3. Centreline excess concentration above ambient background concentration

The transverse mixing coefficient constant α can be computed using Equation (12)

$$\alpha = \frac{E_y C_c}{hU\sqrt{g}} = \frac{E_y R^p}{nhU\sqrt{g}} \tag{12}$$

, where α is the transversal mixing coefficient constant [dimensionless]; E_y is the transversal mixing coefficient [m^2.s^{-1}]; C_c is the Chezy's coefficient [m$^{0.5}$.s^{-1}]; h is the mean river depth downstream discharge [m]; U is the mean velocity downstream of discharge outfall [m. s^{-1}]; g is the gravity acceleration constant [m. s^{-2}]; R is the hydraulic radius [m]; p is the Pavlovskij's or Manning's exponent and n is the roughness coefficient.

Figure 4 (graph):

$y = 0.0093x + 3.5208$

$y = 0.0093x + 1.6677$

$y = 0.0093x + 5E\text{-}16$

Y-axis: F [m³.s⁻¹] — labeled $F\ [m^3.s^{-1}]$, values 0, 2, 4, 6, 8, 10, 12, 14, 16

X-axis: x [m], values 0, 200, 400, 600, 800, 1000, 1200

Legend: ▲ Reach 1   ◆ Reach 2   ● Reach 3   —— Fit 1   ---- Fit 2   ——— Fit 3

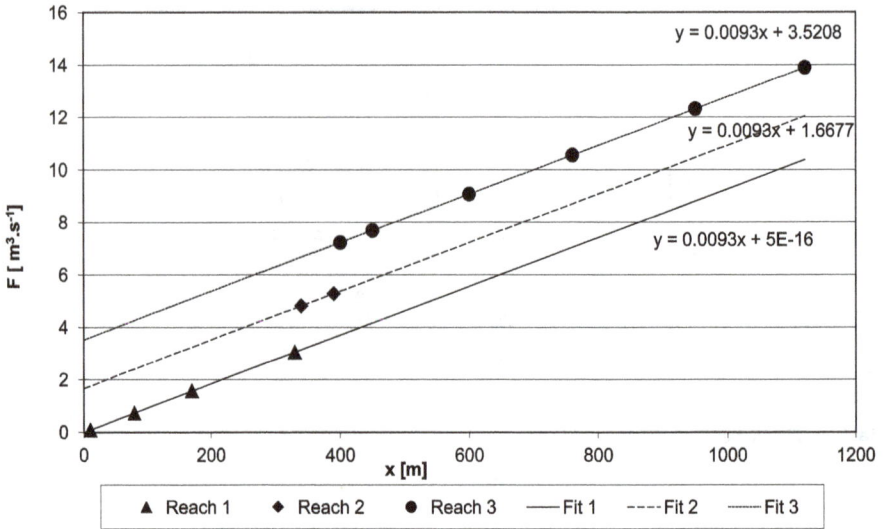

Figure 4. Regression analysis

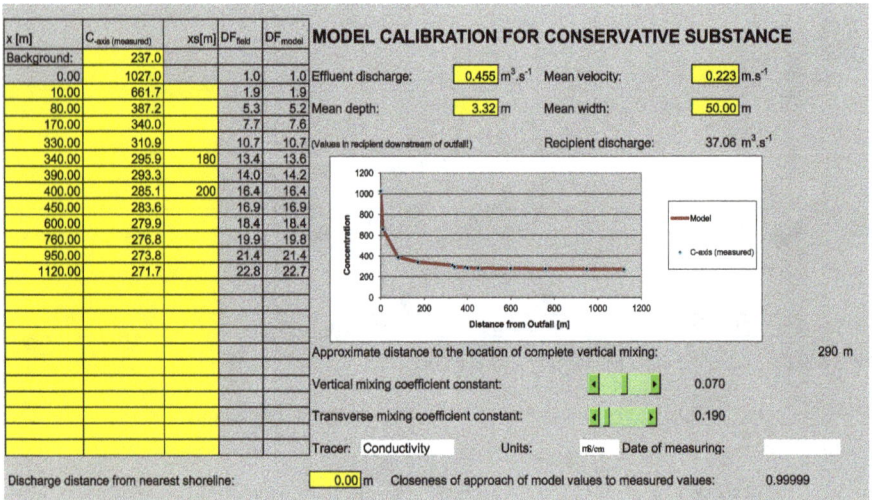

MODEL CALIBRATION FOR CONSERVATIVE SUBSTANCE

x [m]	$C_{\text{-axis (measured)}}$	xs[m]	DF_{field}	DF_{model}
Background:	237.0			
0.00	1027.0		1.0	1.0
10.00	661.7		1.9	1.9
80.00	387.2		5.3	5.2
170.00	340.0		7.7	7.6
330.00	310.9		10.7	10.7
340.00	295.9	180	13.4	13.6
390.00	293.3		14.0	14.2
400.00	285.1	200	16.4	16.4
450.00	283.6		16.9	16.9
600.00	279.9		18.4	18.4
760.00	276.8		19.9	19.8
950.00	273.8		21.4	21.4
1120.00	271.7		22.8	22.7

Effluent discharge: 0.455 m³.s⁻¹   Mean velocity: 0.223 m.s⁻¹

Mean depth: 3.32 m   Mean width: 50.00 m

(Values in recipient downstream of outfall)   Recipient discharge: 37.06 m³.s⁻¹

Chart: Concentration vs Distance from Outfall [m] — Model, C-axis (measured)

Approximate distance to the location of complete vertical mixing: 290 m

Vertical mixing coefficient constant: 0.070

Transverse mixing coefficient constant: 0.190

Tracer: Conductivity   Units: mS/cm   Date of measuring:

Discharge distance from nearest shoreline: 0.00 m   Closeness of approach of model values to measured values: 0.99999

Figure 5. Water quality model downstream WWTP – calibration procedure including x_s

There are a few methods on, how to calculate Chezy's coefficient, such as Manning's method, Pavlivskij's method etc. If we use Pavlivskij's method α is equal to 0.199 in our case. If we use Manning's method α is equal to 0.194 in our case. Another way to determine the average mixing coefficient constant is calibration using the CIT model (see Figure 5). The CIT model uses Manning's method. Pavlivskij's method is preferable for shallow rivers with great average rock diameter. Reach 1 needs negligible spare length, reach 2 needs the spare length equal to approximately 180 m and reach 3 needs the spare length equal to approximately 380 m.

5. An example of 1D modelling of accidental leakage

Thanks to the device for continuous monitoring of the biological water quality, which was installed at the Bohumín station located on the borderline profile at the Odra River, accidental leakage on the 9th of January 2012 was registered [19]. The first non-zero toxic index (TOX) was registered at 21:28:05. The maximal value of the toxic index was registered at 22:26:30. The difference between these two times is 3,505 s. The measured water discharge at the sampling station was 30.5 $m^3.s^{-1}$. There is usually a constant ratio between the time to pick of the value of the effluent cloud t_p and the time to beginning of the effluent cloud t_b at the far-field located sampling stations. This ratio is approximately between 1.1:1 and 1.3:1. You can find it in many studies based on tracer experiments, for example, see [20], [21]. In our case we find an optimal ratio of 1.17:1. The lag time of travel t_L is the time to the centroid of the toxic index in our case. One of the possible ways to compute t_L is shown in Table 4. The lag time of travel is particularly important because the centroid times of pollution concentration-time profiles at sites are often used to evaluate reach travel times (or velocities) for use in pollution incident and water quality models. Yet if the profiles so used are incomplete, the evaluation of the centroid by method of moments (the usual approach) is subject to error. Water and toxic pollution do not flow at one single advective velocity but experience a wide range of velocities, from rapid flow in the centres of large conduits to slow flow in adjacent voids. This variance of velocities is referred to as dispersion and is reflected in the width of the breakthrough curve.

In the second part of Table 4 the results for other characteristic times of effluent cloud passage at the sampling station are listed. The resulting lag time of travel (t_L=25,057 s) gives us distance between the source of accidental leakage and the sampling station of approximately 11,600 m. It is based on previous measurements of lag time of travel in the Odra River. An example of alarm curve is shown in Figure 6. The toxic index was calculated from the values of the parameters measured. The increasing part of toxic alarm curve has four characteristic values. The toxic index alarm curve is compared with the shape of the model Rc curve at Bohumín station on 9 January 2012. These solutions may be applied to our problem to determine the effects of advection and dispersion on solute concentration. Note that these equations do not consider any reactions the substance may undergo after entering the stream. As such, we are considering the worst case scenario, in which the toxic substance is transported conservatively and not allowed to degrade. This solution therefore provides an upper bound on the solute concentrations that are likely to be realized within the stream.

Chatwin's values b were calculated using Equation (10). Results are listed in Table 5.

Sample No.	$t-t_b$	t	dt	TOX	Q	$t*TOX*Q*dt$	$TOX*Q*dt$
	[s]	[s]	[s]	[-]	[m³.s⁻¹]	[m³.s]	[m³]
1	41185	20568		0	30.5		
2	0	20618	50	6	30.5	188651471	9150
3	41285	20668	50	6	30.5	189108971	9150
...	...	...	...	...	...	...	...
...	...	...	...	...	...	...	...
...	...	...	...	...	...	...	...
69	44589	23972	50	13	30.5	475237903	19825
70	44639	24022	50	13	30.5	476229153	19825
71	44690	24073	51	11	30.5	411895028	17110.5
72	44740	24123	50	14	30.5	515018515	21350
73	44790	24173	50	14	30.5	516086015	21350
...	...	...	...	...	...	...	...
...	...	...	...	...	...	...	...
...	...	...	...	...	...	...	...
229	52602	31985	50	1	30.5	48776587	1525
230	52652	32035	50	0	30.5	0	0
S						64058851053	2556510

t_b	t_L	t_L-t_b	t_p	t_p-t_b	t_p	t_L-t_1	t_L-t_p
[s]	[s]	[s]	[s]	[s]	[s]	[s]	[s]
20618	25057	4440	24123	3505	24123	4490	935

Table 4. Characteristic times

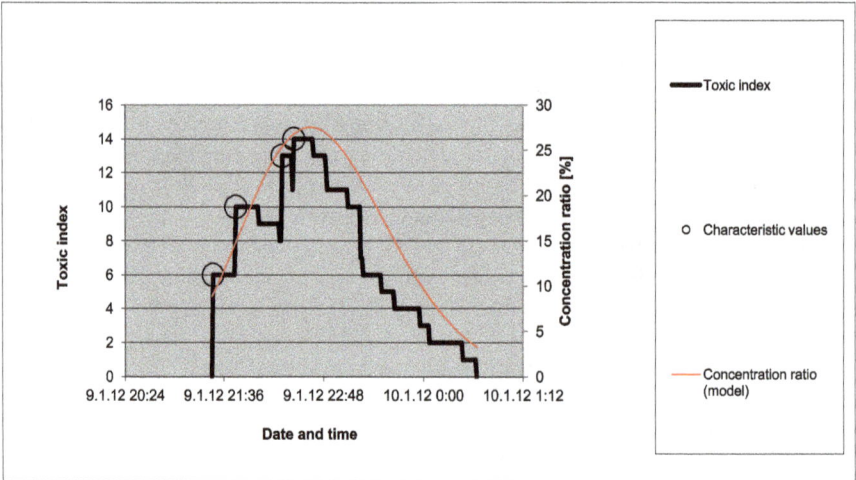

Figure 6. Comparison between the shape of toxic index curve and the shape of model concentration ratio curve at Bohumín station on 9 January 2012.

Sample No.	t [hod]	t [s]	TOX [-]	b [s^{0.5}]	Fit [s^{0.5}]
	0.00	0	0		862.23
1	5.71	20568	0		136.70
2	5.73	20618	6	138.16	134.94
22	6.01	21619	10	91.97	99.63
62	6.56	23622	13	44.70	28.97
72	6.70	24123	14	0.00	11.30

Table 5. Chatwin's values b

After fitting a straight line through the Chatwin's values we obtained value b^* equal to 862.23 $s^{0.5}$ (see Figure 7). The Equation (11) allows for solution of the longitudinal dispersion coefficient, provided that the plot against early-time data reasonably falls as a straight line. The late-time data will depart from the straight line due to non-Fickian dispersion characteristics (e.g., dead zones). For the distance of 11,600 m we calculated a longitudinal dispersion coefficient 45.25 $m^2.s^{-1}$ using Equation (11). Ratio Rc was computed using Equation (9). Duration of accidental leakage τ was estimated so, that the shape of the Rc curve in Figure 6 is about the same as the shape of the toxic index curve. The potential source of accidental leakage could be at the bank of the Odra River approximately 11,600 m upstream of the sampling station, or at the bank of any tributary within this distance.

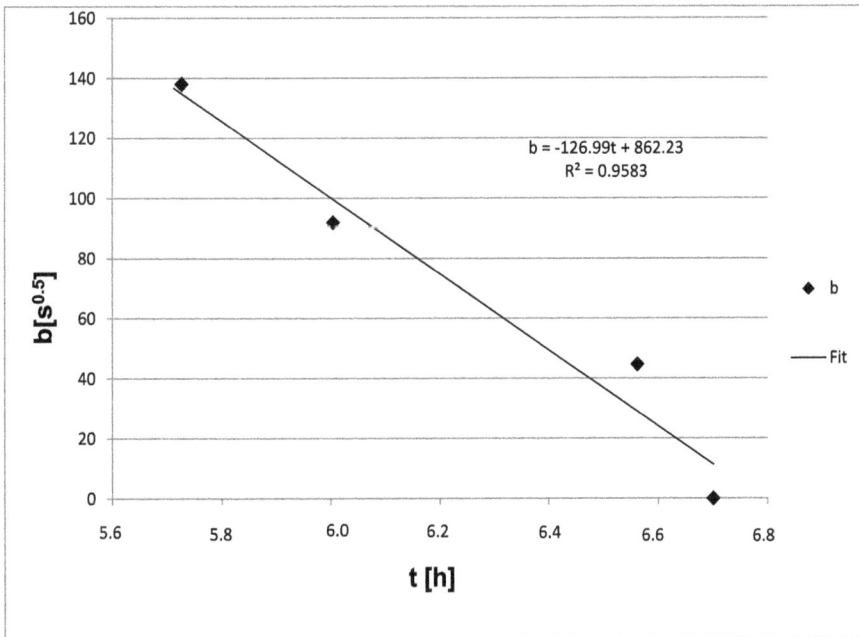

Figure 7. Chatwin's analysis

6. Conclusion

Water quality modelling can help us in environmental impact assessment. One of many 2D water quality models is the CIT model, which is described in this chapter. Every water quality model should be calibrated and validated for the conservative substance at first. Afterwards, this model may be calibrated and validated for non-conservative substance. The CIT model was successfully tested against data from the dye tracer study of the City Arlington WWTP discharge to the Stillaguamish River. CIT model may only be used at a far-field distance where the plume is completely or nearly-completely vertically mixed. Near-field mixing driven by jet velocity can be included to computing using spare length x_s. The buoyancy of discharge is neglected. Therefore the model conservatively underestimates mixing that occurs in the near-field if the plume is vertically mixed. The method described in the second part of this chapter could help us to find the source of accidental leakage, which is recorded at the sample station. The reason of this was to achieve about the same shape of the model R_C curve as the shape of toxic index alarm curve. Analytical solution of advection-dispersion equation was used. If numerical solution of advection-dispersion equation is used, then effect of numerical dispersion should be excluded. Described model might be used for early forecast of concentrations of conservative harmful substances in Odra River at Bohumín boundary profile.

Author details

Jiří Šajer

T. G. Masaryk Water Research Institute, p. r. i., Ostrava Branch Department Macharova 5, Ostrava, Czech Republic

Acknowledgement

For the first part of this chapter financial support for the research plan MŽP 0002071101 "VÝZKUM A OCHRANA HYDROSFÉRY" from the Ministry of the Environment of the Czech Republic is gratefully acknowledged. For the second part of this chapter financial support for the research project TA01020714 "VÝVOJ NÁSTROJŮ VČASNÉHO VAROVÁNÍ A REAKCE V OBLASTI OCHRANY POVRCHOVÝCH VOD" from the Technology Agency of the Czech Republic is gratefully acknowledged.

7. References

[1] Directive 2008/105/EC Of The European Parliament And Of The Council of 16 December 2008on environmental quality standards in the field of water policy, amending and subsequently repealing Council Directives 82/176/EEC, 83/513/EEC, 84/156/EEC, 84/491/EEC, 86/280/EEC and amending Directive 2000/60/EC of the European Parliament and of the Council, Available: http://eur-lex.europa.eu/LexUriServ/LexUriServ.do?uri=OJ:L:2008:348:0084:0097:EN:PDF. Accessed 2012 Feb 23.

[2] EIA Directive 85/337 EEC as amended by 97/11/EC and 2003/35/EC, Available: http://www.environmentcentre.com/rte.asp?id=85.Accessed 2012 Feb 23.

[3] Jirka G.H., Bleninger T., Burrows R. , Larsen T. (2004) Environmental Quality Standards in the EC-Water Framework Directive: Consequences for Water Pollution Control for Point Sources. Available:http://www.ewaonline.de/journal/2004_01h.pdf.Accessed 2012 Feb 23.

[4] Marks, B. J. (1996) Initial Dilution of a Horizontal Jet in a Strong Current, The University of British Columbia. Available: https://circle.ubc.ca/bitstream/handle/ 2429/4621/ubc_1996-0434.pdf.Accessed 2012 Feb 16.

[5] van der Schalie, W. H. (1986) Can biological monitoring early warning systems be useful in detecting toxic materials in water?, Aquatic toxicology and environmental fate: Ninth volume, ASTM STP 921. In: T. M. Poston and R. Purdy (Eds.), American Society for Testing and Materials (107-121). Philadelphia: American Society for Testing and Materials.

[6] Loucks, P.D. and van Beek, E. with contributions from Jery R. Stedinger, R.J. and Dijkman, P.M.J. (2005) Planning and Management An Introduction to Methods, Models and Applications ,Chapter Water Quality Modelling and Prediction, UNESCO and WL , Delft Hydraulics, ISBN 92-3-103998-9. Available: http://ecommons.library.cornell.edu/ bitstream/1813/2804/9/12_chapter12.pdf.Accessed 2012 Feb 23.

[7] Socolofsky, S. A., Jirka, G.H. (2005) Special Topics in Mixing and Transport Processes in the Environment, Engineering – Lectures, fifth ed. Coastal and Ocean Engineering Division, Texas A & M University.Available: https://ceprofs.civil.tamu.edu/ssocolofsky/ ocenx89/downloads/book/socolofsky_jirka.pdf.Accessed 2012 Feb 16.

[8] Fischer, H.B., E.J. List, R.C.Y. Koh, J. Imberger and N.H. Brooks(1979)Mixing in Inland and Coastal Waters, Academic Press, Inc., New York, xiv + 483 p.

[9] Manson J.R. and Wallis S.G. (1995). An accurate numerical algorithm for advective transport. Commun. Numer. Methods. Eng., 11, 1039–1045.

[10] Runkel, R. L., Bencala, K. E. (1995) Transport of reacting solutes in rivers and streams.Environmental Hydrology. Netherlands. pp. 137-164.

[11] Jandora J., Daněček J. (2002) Příspěvek k použití analytických metod řešení transportně disperzní rovnice, (Contribution to application of analytical methods solving advective-dispersion equation) J.Hydrol. Hydromech., Vol. 50, No. 2, pp. 139-156

[12] Chatwin, P.C. (1971) On the interpretation of some longitudinal dispersion experiments. J. Fluid Mech. 48(4);689–702.

[13] Field, M. (2002) The QTRACER2 program for Tracer Breakthrough Curve Analysis for Tracer Tests in Karstic Aquifers and Other hydrologic Systems. – U.S. Environmental protection agency. Available: http://cfpub.epa.gov/ncea/cfm/recordisplay.cfm?deid=54930. Accessed 2012 Feb 16

[14] Modelling the world of water MIKE by DHI software, Available: http://www.mikebydhi.com/ Accessed 2012 Jun 04

[15] Guidance for conducting mixing zone analyses, Available: http://water.epa.gov/ scitech/swguidance/standards/mixingzones/upload/2006_07_19_standards_mixingzone_ WA_MZ_Guide.pdf Accessed 2012 Jun 04

[16] Fox, B.(2009) RIVPLUM5 Validation Study, Cosmopolitan Engineering Group, PROJECT #: HER001, Available: http://www.wsdot.wa.gov/NR/rdonlyres/ 0B027B4A-F9FF-4C88-8DE0-39B165E4CD94/61143/BA_RIVPLUMvalidationStudy.pdf Accessed 2010 Jun 29

[17] Hyánek, L., Rešetka, D., Koller, J. , Nesměrák, I. (1991) Čistota vod (Water Purity), Alfa Bratislava 264 pp. ISBN 80-05-00700-0

[18] Šajer, J. (2010) Modelová interpretace výsledků měření mísící zóny v Labi pod vypouštěním z ČOV Hradec Králové. (Model interpretation of results of measurement mixing zone in the Elbe River downstream from the WWTP Hradec Králové.) Journal of Hydrology and Hydromechanics, Vol. 58, No. 2, pp. 126—134. ISSN 0042-790X

[19] Šajer, J. (2012) Analýza záznamu úniku toxických látek do řeky Odry.(Analysis of the record of leakage of toxic substances into Odra River) VTEI, Vol. 54, No. 2, pp. 12–15, ISSN 0322-8916, příloha Vodního hospodářství č. 4/2012.

[20] Atkinson, T.C., Davis, P.M. (2000) Longitudinal dispersion in natural channels: 1. Experimental results from the River Severn, U.K. Hydrol. and Earth Sys. Sci. 4(3);345– 353. Available: http://www.hydrol-earth-syst-sci.net/4/345/2000/hess-4-345-2000.pdf. Accessed 2012 Feb 16

[21] Lippert, D., Mai, S., Barjenbruch, U. (2004) Traceruntersuchungen in der Elbe zur Validierung des operationellen Schadstofftransportmodells Alamo. Available: http://www.bafg.de/nn_222616/M1/DE/04__Aktuelles/Archiv/tracer__elbe__2004,templ ateId=raw,property=publicationFile.pdf/tracer_elbe_2004.pdf. Accessed 2012 Feb 16

Chapter 3

Optimal Design of the Water Treatment Plants

Edward Ming-Yang Wu

Additional information is available at the end of the chapter

1. Introduction

The design of a water treatment system represents a decision about how limited resources should be used to achieve specific objective, and the final design is selected from various proposals that would accomplish the same objectives [1].

A design must satisfy a number of technical considerations thus of a good design for a water requires technical competence in the related areas. While engineers may take this fact to be self-evident, it often needs to be stressed to industrial or political leaders motivated by their hopes for what a proposed water treatment system might accomplish, rather than what is possible with the resources available [2]. Moreover economics and other values must also be taken into account in the choice of a design,which cannot be determined by technical considerations alone. Moreover, these non-technical issues tend to dominate the final choice of a design for a water treatment system.

The traditional approach to designing water treatment systems uses the average (mode or median) of water quality data.

However the operation of such water treatment plants may lead to a number of significant dangers [3], as the input water quality is usually not at a constant level. This leads to uncertainties that can only be addressed using a stochastic optimal design model. In the stochastic model [4], a small probability (α) leads to lower risk and higher reliability, with the α-value being chose by a decision maker.

The objective of this paper is to demonstrate how to apply systems analysis to the design of a water treatment system based on the concept and practice of optimization theory.

Models for solving environmental system problems are generally nonlinear, including objective functions and constraints. Such the models should thus be solved using nonlinear methods, although NLP problems are more difficult to solve, as they have been studied by researchers for more than 20 years, no ideal solutions have yet been found.

This is mainly because many certain factors cause the solution to stop at a non-optimal point.

The rest of this paper is organized as follows. Section 2 examines the mathematical theory underlying the concept of flexible tolerance, which forms the basis for all that follows. Section 3 applies the model a case study of an existing water treatment plant, while Section 4 describes the system optimization procedure. Section 5 gives the results of a sensitivity analysis, an essential part of any practical optimization approach, since mathematical descriptions of reality are inherently in exact. Finally, Section 6 presents the suggestions for academics and practitioners, and the conclusions of this work.

2. Flexible tolerance concept

2.1. Concept of tolerance

While using the concept of tolerance in a flexible simplex method to slove NLP problems is theoretically feasible, when the number of variables exceeds seven or eight, the simplex deteriorates and becomes much less efficient [5, 6]. Therefore, methods based on the concept of flexible tolerance have not been proposed in the literature on NLP.

2.2. The concept of flexible tolerance

Kao et al. proposed the concept of flexible tolerance, in which the tolerance is gradually reduced in the process of calculations, and approaches zero when the optimal solution is reached. Using this method, many pull-in operations are not needed [7], and thus intuitively this is a feasible approach.

2.3. The multiplier method with flexible tolerance

This study uses the following four methods: the feasible directions method, flexible simplex method, quadratic approximation method, and multipliers method, which are all implemented to cope with the concept of flexible tolerance. A computer program is developed in this work that makes use of these approaches based on factors such as convergence, rate of convergence, accuracy, core memory needed, and the ease of use. Of all four methods, the multipliers approach has been shown to have the best performance with regard to all these factors [7].

Due to space limitations, this paper only presents the basic theory of the multipliers method and the procedure used to apply it, based on concept of flexible tolerance, in order to solve NLP problems.

There are two types of difficulty that arise when solving NLP problems. First, in the process of calculation, the constraints are often difficult to satisfy in order to reach the optimal solution. Next, even if the optimal solution is obtained, many pull-in operations are required to meet all the constraints at all times. Therefore, the concept of flexible tolerance is proposed in this work to allow a tolerance for each constraint. In the process of calculation, the tolerance is gradually reached. And then approaches zero when the optimal solution is obtained. The problem is as follows:

$$\text{min.} \quad -x_1 - x_2$$
$$\text{s.t.} \quad -x_1^2 - x_2^2 + 4 = 0, \tag{1}$$
$$x_1 \geq 0$$
$$x_2 \geq 0$$

As shown in Fig. 1, the feasible region is an arc. If the initial point is X^0=(2, 0), then it needs to move through the arc $-X_1^2 - X_2^2 - X_2^2 + 4 = 0$, and finally converges at the optimal point $X^* = (2/\sqrt{2}, 2/\sqrt{2})$. The point moves along the line and travels a very short way at first and must enter the feasible region. In the next step, the point moves forward in a straight line direction and repeats the same pull-in operations. However, this process wastes much computational time, because the convergence rate is too slow. Therefore, a new concept of tolerance based on the work of Paviani and Himmeblau is proposed in this study [7,8].

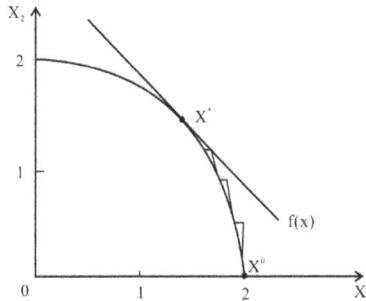

Figure 1. Model solution when the constraint is curved.

At the beginning of the solving procedure, a tolerance range is introduced for every constraint, and the constraint is assumed to be satisfied within this. There are three cases to be considered:

1. $h(x)$=0, then it is **feasible**.
2. $-\epsilon \leq h(x) \leq \epsilon, \epsilon > 0$ is tolerance, it is **near feasible**.
3. $h(x) > \epsilon$ or $h(x) < -\epsilon$, it is **infeasible**.

Where $h(x)$ is the constraint.

In each step, the tolerance is gradually reduced so that it can finally approach zero when the optimal solution is reached. Figure 2 shows the geometry of the concept of tolerance. When the problem has more than one constraint, as shown by Equation (2), all the constraints can be combined to consider their overall tolerance, which is presented as follows:

$$\text{min.} f(x)$$
$$\text{s.t.} \, g_i(x) \geq 0, i = 1,2,....I \tag{2}$$
$$h_j(x) = 0, j = 1,2,....J$$

Suppose that Ui is a **Heaviside operator,**

$$U_i = \begin{cases} 0, & if \ \ g_i(x) \geq 0 \\ 1, & if \ \ g_i(x) < 0 \end{cases} \qquad (3)$$

$T(X)$, which is always positive, is defined as follows:

$$T(X) = \sum_{i=1}^{I} U_i g_i^2(x) + \sum_{j=1}^{J} h_j^2(x) \qquad (4)$$

which means that all constraints should be satisfied. When $T(X) = 0$, all constraints are satisfied, and X is a feasible solution. When $0 \leq T(X) \leq \epsilon$ X is almost feasible; when $T(X) > \epsilon$, X is infeasible so it should move toward the feasible region.

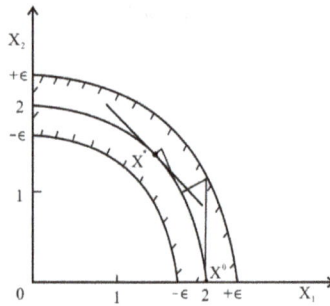

Figure 2. Geometry of the tolerance concept.

When the problem is presented in the form of equation (1) with $\epsilon = 1/3$, Fig. 3 represents the near feasible region. The two semicircles are the tolerance for $X_1 \geq 0$ and $X_2 \geq 0$.

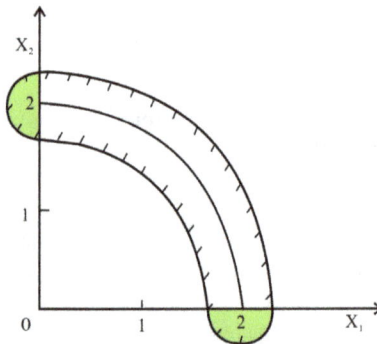

Figure 3. The quasi-feasible region obtained by combining all the constraints.

2.4. Hestenes' multiplier method

One of the ways to solve an NLP problems is the use of a combined technique, which the objective function and constraints are combined by using a special algorithm to become an unconstrained NLP. While the solution remains the same as for the original problem, the unconstrained NLP is much easier to solve than the constrained one. Therefore, this method of functional conversion is favored by most researchers, and many methods have been proposed to convert functions, using the interior penalty value, exterior penalty value, and multipliers, with this last one being considered the best [9,10].

For the NLP problem, as in Eq. (2), the objective function and constraints are combined into Eq. (5):

$$\min.p(x) = f(x) + R\left[\sum_{i=1}^{I} U_i g_i^2(x) + \sum_{j=1}^{J} h_j^2(x)\right] \tag{5}$$

where

U_i: Heaviside operator

$$U_i = \begin{cases} 0, & if \ g_i(x) \geq 0 \\ 1, & if \ g_i(x) < 0 \end{cases} \tag{6}$$

R: penalty value

When R approaches?, then the optimal solution X^* for Equation (5) is the X^* for Equation (2).

With regard to the concept of flexible tolerance, the tolerance is defined as follows:

$$T(X) = \sum_{i=1}^{I} U_i g_i^2(x) + \sum_{j=1}^{J} h_j^2(x) \tag{7}$$

The minimum value of $F(X, l, m, R)$ is calculated for each stage.

Assuming

$$F(X, \lambda, \mu, R) = f(x) + \sum_{j=1}^{J} \lambda_j h_j(x) + R\sum_{j=1}^{J} h_j^2(x)$$
$$- \sum_{i=1}^{I} \mu_i \overline{g_i}(x) + R\sum_{i=1}^{I} \overline{g_i}^2(x) \tag{8}$$

where

λ; μ: Lagrange multiplier

R: penalty value, a constant during the solution process

$$\begin{cases} \lambda_{n+1,j} = \lambda_{n,j} + 2Rh_i(x) \\ \mu_{n+1,i} = \mu_{n,1} - 2R\overline{g_j}(x) \end{cases} \tag{9}$$

The difference between two functions (F) in two consecutive stages should be equal to or greater than zero, or

$$F(X, \lambda_{n+1}, \mu_{n+1}, R) - F(X, \lambda_n, \mu_n, R) \geq 0 \tag{10}$$

This is the basis of the convergence for the multipliers method. Based on the research conducted by Equation (2) can be transformed as follows:

$$F(X, \sigma, \tau, R) = f(x) + R\sum_{i=1}^{I}[< g_i(x) + \sigma_i >^2 - \sigma_i^2] + R\sum_{j=1}^{J}[(h_j(x) + \tau_j)^2 - \tau_j^2] \tag{11}$$

Where

< >: means that if the value inside the < > is greater than zero, then < >=0; if it is smaller than zero (e.g., a), then < >=a (a<0).

And

$$\begin{cases} \sigma_{n+1,i} = < g_i(x) + \sigma_{n,i} > & ; \quad i = 1,2,....,I \\ \tau_{n+1,j} = h_j(x) + \tau_{n,j} & ; \quad j = 1,2,...,J \end{cases} \tag{12}$$

The minimum value for Equation (11) can be found in every step. Meanwhile, the values of σ and τ can be modified based on Equation (12).

When $\sigma_{n+1} = \sigma_n, \tau_{n+1} = \tau_n$, then the convergence is in the optimal solution.

If $\nabla F(X^n, \sigma_n, \tau_n, R) = 0$, and $\sigma_{n+1} = \sigma_n, \tau_{n+1} = \tau_n$, then Xn is the Kuhn-Tucker stationary point for Equation (2).

The Hessian matrix included in Equation (11) is expressed as follows:

$$\begin{aligned} \nabla^2 F(X) = \nabla^2 f(x) + 2R\sum[< g_i(x) + \sigma_i > \nabla^2 g_i(x) + \nabla^2 g_i^2(x)] \\ + 2R\sum[(h_j(x) + \tau_j)\nabla^2 h_j(x) + \nabla h_j^2(x)] \end{aligned} \tag{13}$$

If gi (x) and hj(x) are linear constraints, then Equation (13) can be simplified to

$$\nabla^2 F(X) = \nabla^2 f(x) + 2R\sum_{i=1}^{I}\nabla g_i^2(x) + 2R\sum_{j=1}^{J}\nabla h_j^2(x) \tag{14}$$

$\nabla^2 F(x)$ is independent of either σ or τ, and thus the convergence will not be influenced by the shape of function F. Figure 4 shows a flow chart of the multipliers method using the concept of tolerance.

3. Case study

Figure 5 is a Flowchart of the existing water treatment facilities in Taiwan.

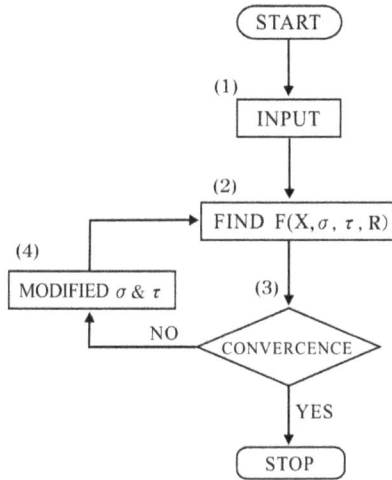

Figure 4. Flowchart of the multiplier method with the concept of tolerance.

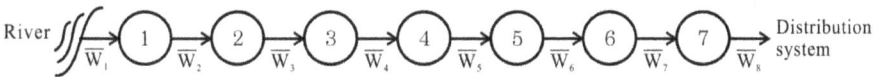

Figure 5. Flowchart of the existing water treatment facilities in Taiwan.

Where;

W_j: Vector of water quality parameters into unit "J"; 1:

Prechlorination; 2: Alum feeders; 3: Rapid mixing basin; 4: Flocculation basin; 5: Tube-settler sedimentation; 6: Modified greenleaf type filter; 7: Postchlorination.

Figure 5 shows a flow chart of the existing water treatment facilities in Taiwan.

The variables for various unit operations or processes are as follows:

X_1: Feedrate of prechlorination (kgs/h)
X_2: Feedrate of alum feeder (kgs/h)
X_3: Volume of the rapid mixing basin (M^3)
X_4: Volume of the flocculation basin (M^3)
X_5: Surface area of the tube-settler sedimentation (M^2)
X_6: Surface area of the modified-greenleaf filter (M^2)
X_7: Feedrate of postchlorination (kgs/h)

3.1. The objective function

Cost functions for water treatment plants have been proposed by Clark [11] and Wiesner [12]. Most of the cost functions used in the United States and Taiwan are given as power functions. The costs associated with water treatment plants include: (1) initial capital (construction) cost; (2) maintenance and operation cost; and (3) inflation adjustment factor (real discount rate). In Taiwan, the cost function of a typical treatment plant is normally given as a quadratic function, as in Lee and Wu [13].

The actual objective function for an existing plant is as follows:

$$
\begin{aligned}
\text{min.}\ \ Z = {} & 0.572\ X_1 - 0.00289\ X_1^2 + 0.0422\ X_2 - 0.0000449\ X_2^{\,2} + 0.00386\ X_3 \\
& + 0.000136\ X_3^{\,2} + 0.00775\ X_4 - 0.00000052\ X_4^2 + 0.001449\ X_5 \\
& - 0.000000063\ X_5^{\,2} + 0.0148\ X_6 - 0.000025\ X_6^{\,2} \\
& + 0.844\ X_7 - 0.00406\ X_7^{\,2}
\end{aligned}
\tag{15}
$$

The coefficients in Equation (15) can be obtained, given a specific feed rate and plant capacity (parameter of units design criteria), from various regression curves available in the government water supply design manuals produced by the Taiwan Water Supply Company [14].

The structural constraints are the parameters representing the input water quality, output water quality, treatment efficiency, detention time, operating limit, and treatment characteristics.

3.2. Stochastic constraints

The input water quality is not always at a constant level, and thus the model will generate a probability problem. Due to space limitation, this work will only consider the chance-constrained stochastic model.

Corrosion control relationship

$$
-\frac{32314}{Q}X_1 - \frac{10669}{Q}X_2 - \frac{32314}{Q}X_7 \geq 5.5 - 0.133\ F_{A1}^{-1}(\alpha) + F_{C_1}^{-1}(1-\alpha)
\tag{16}
$$

$$
-\frac{32314}{Q}X_1 - \frac{10669}{Q}X_2 - \frac{32314}{Q}X_7 \geq 5.5 - 0.133\min(A_1) + \max(C_1)
\tag{17}
$$

where

A_1: Total alkalinity of input water (mg/l)

C_1: Free carbon dioxide of input water (mg/l)

$\min(A_1)$: The minimum value of raw water total alkalinity obtained from past records.

$\min(C_1)$: The minimum value of raw water free carbon dioxide obtained from past records.

Q: Design flow rate (CMD)

$F_{A_1}^{-1}(\alpha) = \max\{A_1 | F_{A_1}(\alpha) \leq \alpha\}$, the solution for α in the equation

$F_{A_1}(a) = \alpha$, or the inverse of the marginal cumulative distribution function of input total alkalinity (mg/l)

$F_{c1}^{-1}(\alpha)$ =the solution for C in the equation $F_{c1}(C) = 1 - \alpha$

α: Probability that can take value between zero and one.

$1-\alpha$: complementary probability of α.

Equation (16) is formally identical to Equation (17) (i.e., they are deterministically equivalent) except that the former includes the random nature of total alkalinity and the free carbon dioxide.

The alum for coagulation can be obtained from:

$$\frac{x_2}{0.00043Q} \geq \log[F_{T1}^{-1}(1-\alpha)] + 0.281 \tag{18}$$

where:

T_1: Raw water turbidity (T.U)

$F_{T1}^{-1}(1 - \alpha)$ =the solution for t in the equation $F_{T1}(t) = 1 - \alpha$

Output turbidity

$$K_T \frac{84}{Q} X_5 + \frac{28}{Q} X_4 + \frac{5.97}{Q} X_3 + 1.4\ X_6 \geq \log[F_{T1}^{-1}(1-\alpha)] + 364 \tag{19}$$

where

K_T: Turbidity removal rate in sedimentation basin (1/h)

Effluent coliform bacteria

$$0.055\frac{84}{Q} X_3 + 0.25\frac{28}{Q} X_4 + K_B \frac{3.5}{Q} - 0.5\ X_6 \geq \log[F_{B1}^{-1}(1-\alpha)] + 433.25 \tag{20}$$

where

B_1: Coliform bacteria of input water (MPN #/100 ml)

K_B: Removal rate for coliform bacteria in sedimentation basin (1/day)

$F_{B1}^{-1}(1 - \alpha)$ =the solution for b in the equation $F_{B1}(b) = 1 - \alpha$

Total alkalinity for coagulation

$$-\frac{19200}{Q} X_1 - \frac{9840}{Q} X_2 \geq 35 - F_{A1}^{-1}(\alpha) \tag{21}$$

Detention times

$$X_3 \geq 0.0007Q$$
$$X_4 \geq 0.021Q \tag{22}$$
$$X_5 \geq 0.012Q$$

Hydraulic filter breakthrough

$$X_6 \geq 0.00563Q \tag{23}$$

Chlorine disinfection

$$X_1 \geq F_{D1}^{-1}(1-\alpha) \cdot Q \cdot \frac{10^3}{24}$$
$$\tag{24}$$
$$X_7 \geq F_{D7}^{-1}(1-\alpha) \cdot Q \cdot \frac{10^3}{24}$$

where:

D_1: Design dosage for prechlorination
D_7: Design dosage for postchlorination

3.3. Stochastic constraints transformation

Equations (16) to (24) show that the stochastic characteristic has been included in each constraint so that the inverse of the marginal cumulative distribution function of the input water quality parameters needs to be solved. In this situation, the stochastic constraints can be transferred to certain ones. The relationship between probability and b_i^α is shown in Fig. 6. If $\Pr[g_i(x) \leq B_i] \geq \alpha$, and $\alpha = 0.15$, $b_i^\alpha = 100$, $b_i^{1-x} = 180$ diagram (b) can be used to convert the constrain, which contains probability the constraint, into a deterministic equivalent. Since pr[.] is greater than 0.15, the answer is in the curve interval (mn), and the corresponding b_i^α is equal to 100; the solution is $g_i(x) \geq 100$, which is inconsistent with the original problem statement that $g_i(x) \leq B_i$. Diagram (c) thus explains the conversion procedure. Since pr[.]$\geq$0.15, the answer is on the curve (m₀n₀); the corresponding b_i^{1-x} equals 180; so the solution is $g_i(x) \leq 180$. This coincides with the original problem statement $g_i(x) \leq B_i$.

Equation (21) is expressed as

$$-\frac{19200}{Q}X_1 - \frac{9840}{Q}X_2 \geq 35 - F_{A1}^{-1}(1-\alpha) \tag{25}$$

which is the same as Equation (24)

$$pr[A_3 \leq 35 + \frac{19200}{Q}X_1 + \frac{9480}{Q}X_2] \leq \alpha \tag{26}$$

Using the above conversion algorithm, we can obtain $K = b_i^{1-x}$, where K is constant, and K>0, and hence

$$K = \begin{bmatrix} b_i^{\alpha} \\ b_i^{1-\alpha} \end{bmatrix} \text{ when } \begin{array}{l} \text{using diagram}(b) \\ \text{using diagram}(c) \end{array} \quad (27)$$

and $1 - \alpha$ is the complementary probability of α.

$$A_3 = A_1 - \frac{19200}{Q}x_1 - \frac{9480}{Q}x_2 \geq 35 \quad (28)$$

$$A_1 = K \quad (29)$$

$$K \geq 35 + \frac{19200}{Q}x_1 + \frac{9480}{Q}x_2 \quad (30)$$

Rearranging the terms in Equation (29), one obtains:

$$\frac{-19200}{Q}x_1 - \frac{9480}{Q}x_2 \geq 35 - K \quad (31)$$

3.4. Deterministic equivalent for the stochastic model

Since the objective function excluded the probability item, it is not convertible.

The following Constraints are used.

The total alkalinity for coagulation is presented as follows:

$$\frac{-19200}{Q}x_1 - \frac{9840}{Q}x_2 \geq 35 - F_{A_1^{-1}}(\alpha) \quad (32)$$

that is,

$$F_{A1}(35 + \frac{19200}{Q}x_1 + \frac{9480}{Q}x_2) \leq \alpha \quad (33)$$

or

$$pr[A_1 \leq 35 + \frac{19200}{Q}x_1 + \frac{9480}{Q}x_2] \leq \alpha \quad (34)$$

Using diagram (b) in Figure 6, if the decision maker lets $\alpha = 0.25$ or $1 - \alpha = 75\%$, and $Q = 75000^{CMD}$, then Equation (35) can be easily obtained:

$$35 + \frac{19200}{75000}x_1 + \frac{9480}{75000}x_2 \leq 48.5 \quad (35)$$

The corrosion control relationship is presented as follows when $\alpha < 0.6$:

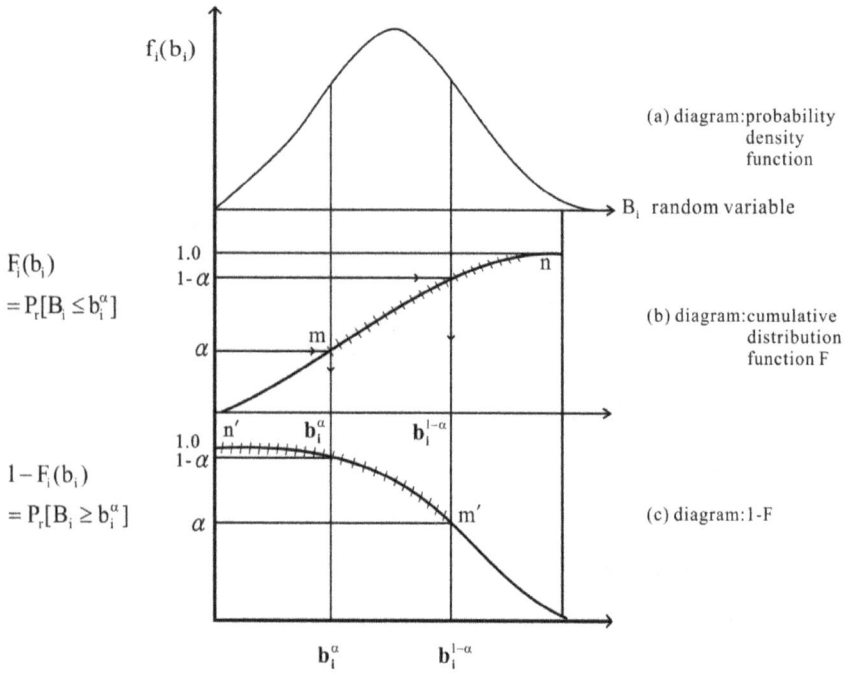

Figure 6. CDF transformation concept.

$$0.215 \ x_1 + 0.071 \ x_2 + 0.215 \ x_7 \leq -0.54 \ \text{(inconsistent)} \tag{36}$$

If the value of α is increased to 1.0 ($\alpha = 1.0$), Equation (37) can be obtained:

$$0.215 \ x_1 + 0.071 \ x_2 + 0.215 \ x_7 \leq 4.14 \ \left(\text{consistent but unreasonable}\right) \tag{37}$$

The effluent water quality of the selected existing plant should be corrosive in some respects. If this constraint is not neglected in the structural constraints, a feasible solution for this model is not available, and thus the sensitivity analysis is necessary to make the conclusion significant and meaningful. The following equations are established using $\alpha = 0.25$: For the alum dosage for coagulation:

$$X_2 > 109.4 \tag{38}$$

For the effluent turbidity:

$$0.213 \ x_3 + x_4 + 0.126 x_5 + 7500 \ x_6 \geq 1847491 \tag{39}$$

For the effluent coliform bacteria:

$$-x_3 - 4.55 \ x_4 - 0.057 x_5 + 16200 \ x_6 \leq 13982985 \tag{40}$$

3.5. Computer solution

3.5.1. Input data

The software developed by Kao et al. [7] was tested and modified to ensure that it was appropriate for use in this research. The data input was carried out using the following two methods:

1. Data such as N1(number of variables), NEQ (number of constraints containing "="), NGE (number of constraints containing "<" or ">"), R(penalty value), and X_0 (coordinates of the beginning and ending points) are input using the READ command.
2. The objective function and constraints are input using the "SUB ROUTINE" function; the objective function is designated as SUBROUTINE FUN, whereas the constraints are designated as SUBROUTINE CON.

Besides;

Initial values are "0" for σ_0 and τ_0 so that $F(X, \sigma_0, \tau_0, R)$ is made a standard penalty function. And the value of penalty value R is input intuitively using values of 1, 10, 10^2, 10^{-1}, 10^{-2}, ..., and so on. It is kept a constant during the problem solving process. In the actual scope of solution, the probability value a varies from 0.1 to 0.4 with intervals of 0.05.

3.5.2. Outputs for the converged solution

1. Number of iterations: 23
2. Tolerance $T(x^*) = 0$
3. Running time: 330.448secs.
4. Computing facility: CDC Cyber 830(in National Cheng Kung University, Taiwan)

3.6. Analysis of the results

Two methods based on the concept of tolerance, the method of feasible directions and the multipliers method, have been developed in this research, and solved using computer software. Due to the limited length of this article, only the multipliers method that leads to better results is presented in this work to demonstrate the theory and application of flexible tolerance.

The quadratic regression cost function is used to obtain the nonlinear objective function. The computer program outlined above is used to execute the calculation based on a 10% interest rate, 0.11017 of capital recovery factor, and 25 years design life for the water treatment plant. The results show that when α is less than 0.25 (α<0.25), the mode has no solution, and the optimal solution can only be obtained when α equals 0.25. This indicates that the actual operation of the water treatment plant is subject to 25% risk or 75% reliability. Compared with the 50% reliability for a water treatment plant designed using the traditional approach, the results obtained using the method proposed in this research are much better.

Table 1 lists the original design values and those obtained in this research. The solution of the proposed model is considered "an exact" one. The observation that the proposed model has significant solutions only when the α value exceeds 0.25 is significant, as it indicates that the lack of appropriate equipment for adding alkaline chemicals in the existing water treatment plant causes corrosive, treated effluent. In the flowchart shown in Figure 5, if a new unit operation, that is, Unit #2', is added between Unit #2 and Unit #3, a new variable, that is, X_2', will be included in the model to indicate the lime dosage in kgs/h. As shown in Fig. 7, these modifications lead to the optimal solution. As shown in Table 2, the modified model can be solved even if the αvalue equals 0.01, thus indicating that the reliability is raised to 99.9%.

The sensitivity analysis is usually conducted by varying the price coefficient (c_i), constants (b_i), and coefficient constant (a_{ij}) used in a model and changing the constraints and decision variables. [15] Although not discussed in this article due to length limitations, the results of the sensitivity analysis are more significant with regard to effluent turbidity, corrosion control, and fecal coliform removal.

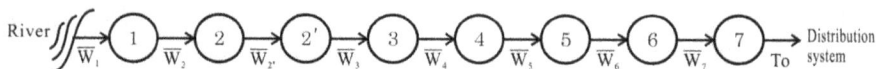

Figure 7. Adding Unit#2' to the original treatment flowchart.

Objective Function	min $z - 0.572x_1 - 0.00289x_1^2 + 0.0422x_2 - 0.0000449x_2^2 + 0.00386x_3 + 0.000136x_3^2 + 0.00775x_4 -$ $0.00000052x_4^2 + 0.001449x_5 - 0.000000063x_2^2 + 0.0148x_6 - 0.000025x_6^2 + 0.844x_7 - 0.00406x_7^2$									
Design period	Comparison	Objective Value (NT$10⁶)	The optimal solution of decision variables							α -value
			X_1	X_2	X_3	X_4	X_5	X_6	X_7	
25yrs	Original design alues	—	45.25	175.4	34.8	3093.2	1596	819	9	—
	Stochastic model	55.60036	31.25	109.4	104	3925	1800	844.5	5	α =0.25

Table 1. Solution for a case study

Decision Variables Probability (α)	X_1 Pre-chlo rination dosage (kg/hr)	X_2 Alum dosage (kg/hr)	X_2' Lime Dosage (kg/hr)	X_3 Rapid mixing tank voluem (M³)	X_4 Flocculati on Tank Volume (M³)	X_5 Clairifer surface area (M²)	X_6 Rapid filter area (M²)	X_7 Post chlorination dosage (kg/hr)
0.01	31.25	172.8	91.26	104	3125	1800	844.5	5
0.02	31.25	169.18	89.50	104	3125	1800	844.5	5
0.05	31.25	160.3	84.73	104	3125	1800	844.5	5
0.10	31.25	147.12	78.07	104	3125	1800	844.5	5

Table 2. Final solution obtained after the original water treatment is modified

4. Systems optimization procedure

4.1. The problem

No single procedure can deal completely with all aspects of a system, and systems analyst, with the responsibility to carefully investigate the entire situation, we must incorporate all the important elements. The question thus arises, how can this be done efficiently? This section presents a procedure for using all the optimization elements to achieve the best design.

4.2. Design procedure

There are four main steps for systems, as follows.

4.2.1. Screening

Screening of the feasible solutions to obtain a small set of non-inferior ones, using a screening models [16]. The screening process in effect defines regions of optimality, and the results are best interpreted as first-order estimates or the nature of the actual best designs for a system.

4.2.2. Sensitivity analysis

Sensitivity analysis of these best solutions is then carried out, to determine their performance in realistic situations.

In the formal process, a specific sensitivity analysis should be conducted to determine how the optimum design would change if the problem were formulated differently. Similarity, the opportunity costs should be examined to see if the optimum design is likely to change, given the known or anticipated changes in the parameters of the objective function.

Overall, the sensitivity analysis general reveals many ways in which the "optimum" solutions derived in the screening process can be improved, demonstrating that some designs perform better over a wide range of likely conditions. The analysis may also indicate the importance of certain factors that are otherwise assumed away.

4.2.3. Dynamic analysis

Dynamic analysis is used to establish the optimal pattern of development over time, and can be done reasonably easily after the screening and sensitivity analysis.

Dynamic programming is typically best suited for this analysis, as it deals effectively with nonconvex feasible regions such as those generated by exponential growth and economies of scale.

4.2.4. Presentation

Presentation is the organization of the final results in a way that makes sense to the client, as the client needs to see why the proposed plan is preferable to alternative, to the appreciates, and that the trade-offs between objectives are reasonable.

5. Sensitivity analysis

5.1. Concept

Sensitivity analysis is the process of investigating the dependence of an optimal solution to changes in the way a problem is formulated. Doing a sensitivity analysis is a key part of the design process, equal in importance to the optimization of the process itself.

The significance of sensitivity analysis stems from the fact that the mathematical problem solved in any optimization is only an approximation of the real problem, and no mathematical models will ever represent systems exactly, each differing from reality in any or all of the following ways:

- Structurally, because the overall nature of the equations does not correspond precisely to the actual situation.
- Parametrically, as we are not able to determine all coefficients precisely.
- Probabilistically, in that we typically assume that the situation is deterministic when it is generally variable. In this work, the author has used a stochastic model to solve the problem of uncertainty.

This section presents the sensitivity analysis principally in the context of linear programming. This is because the solutions to linear programming problems automatically include most of the sensitivity information a designer needs, and thus linear programming is the main basis for sensitivity analysis. In addition the linearity of linear programming makes it easier to explain key concepts, which the reader can then extend to other forms of optimization.

Most of this section is devoted to the two most important aspects of sensitivity analysis, the concept and use of:

- shadow prices.
- opportunity costs.

5.2. Shadow prices

A shadow price is the rate of change of the objective function with respect to a particular constraint, an essentially equivalent to the Lagrangian multiplier. The shadow price has no necessary connection with money, despite its name, and its units are those of the objective function divided by the constraint. The shadow price is expressed in dollars only when the objective function is also expressed in dollars or profit.

5.2.1. Use of shadow price

Shadow prices enable the designer to:

- *identify* which constraints might most beneficially be changed, and to *initiate* these changes as fundamental ways to improve the design.
- *react* appropriately when external circumstances create opportunities or threats to change the constraints.

5.2.2. Sign of shadow prices

A key practical question with regard to shadow prices is: what is the sign of the shadow price, and in which direction can one change a constraint to improve the design? The relationship between the nature of the shadow prices and the changes in constraints is that:

- *Relaxing* the constraints leads to improvements in the optimum design, either increasing a maximum or decreasing a minimum.
- *Changes* in constraints that "raise the roof" or "lower the floor" will tend to improve the optimum design. A constraints is relaxed if it is changed so as increase the size of the feasible region, that is, if an upper bound is increased or a lower bound is decreased.

It is important to note that there is no simple relationship between the sign of the change in constraint and the sign of the shadow price. This is because an increase in the constraint can either relax or tighten a constraint, depending on whether it is an upper or lower bound and if the constraint is a maximizing or minimizing one.

5.2.3. Range of shadow prices

In general, the shadow price is the instantaneous change in the objective function with respect to a specific constraint, $\partial Y / \partial b_j$ ($Y = g(x)$, b_j is the r.h.s of the j-th constraint).

This rate can vary with the decision variables and normally will when the constraints are nonlinear.

The peculiarity of linear programming in this regard is that the shadow prices are constant over a range, rather than varying continuously. If we really describe the problem accurately with the appropriate nonlinear equations, the shadow prices will usually vary instantaneously. Even though the range of constancy of the shadow prices is thus an artificial result, the concept is very useful in practice, because it indicates how sensitive the optimum solution is to the constraint. Indeed, if the range is narrow, this means that even small changes in the constraint could lead to quite different shadow prices, thus that the shadow prices may change rapidly.

The range of the shadow price is defined by the intersections of the constraints adjacent to the one that defines the optimum solution of the linear program. In general there can be a limit to the range of a shadow price for both increases or decreases.

5.3. Opportunity costs

5.3.1. Definition of opportunity cost

Opportunity costs, in the context of sensitivity analysis, are related to the coefficients of the decision variables in the objective function. In general terms they define the "cost" of using decision variables that are not part of the optimal design.

In general, the set of optimal decision variables, X^*, can be divided into two categories. The two categories for the optimum set of the decision variables, X^* are thus the

- *optimal variables* those with nonzero values at the optimum($X^* \neq 0$). These are said to be "in the solution".
- *non-optimal variables*, those equal to zero at the optimum($X_i^* = 0$). These are said to be "not in the solution".

With this distinction in mind, we can now formally define opportunity costs in the sensitivity analysis: The *opportunity cost* is the rate of the degradation of the optimum per unit use of a non-optimal variable in the design. The notion of degradation here is important, as it refers to the worsening of an optimum solution. This may either be a decrease, if we are trying to maximize, or an increase, if we are trying to minimize.

5.3.2. Use of opportunity costs

Opportunity costs are thus used to define the coefficient of the decision variables which would lead to a change in design. The designer, having defined the optimum design, then continuously monitors the situation to determine when it has changed enough so that a new design ought to be used.

6. Conclusion

This study was presented under the assumptions that readers have some knowledge of and experience with the following: (1) mathematical models for systems optimization, (2) engineering economics, (3) cost-benefit analyses, and (4) water supply engineering, especially the functional design of water treatment systems.

This work proposes a new method based on the concept of flexible tolerance to solve problems involving nonlinear conditions that unavoidably arise when mathematical models for optimizing water treatment plant design are implemented. The significant contribution of this paper is that the proposed method can be used to obtain optimal solutions rapidly and accurately by allowing approximate solutions to approach exact ones. Additionally, this work also proposed proactive and improved concepts for the sensitivity analysis and systems optimization procedures, which can help that enable readers to implement the method presented in this work and thus optimize water treatment design by drawing inferences about their use from the examples given in earlier sections.

Author details

Edward Ming-Yang Wu
I-Shou University, Kaohsiung, Taiwan

Acknowledgement

The author would like to express his gratitude to Dr. Chiang Kao, who offered assistance with the computer programming. The author is also very grateful to the. President of Carnegie Mellon University, Dr. Jared Cohon for his helpful comments and suggestions on an earlier draft of this work.

7. References

[1] Li, K.C. and Wu, E. M., (1988). Application of stochastic NLP model on the optimal design of the water treatment facilities. J. Water Supply 6, 137.

[2] Dykstra, D.P. (1984). Mathematical Programming for Natural Resource Management. New York: McGraw-Hill.

[3] Wu, E.M. and Chu, W.S. (1991). System analysis of water treatment plant in Taiwan. J. Water Resour. Plann. Manag. 117, 536.

[4] Fletcher, R. (1981). Practical Methods of Optimization, Vol. 2, Constrained Optimization. New York: John Willey & Sons.

[5] Paviani, D. and Himmelban, D.M. (1969). Constrained nonlinear optimization by heuristic programming. J. Oper. Res. 17, 872.

[6] Himmeblau, D.M. (1972). Applied Nonlinear Programming. New York: McGraw-Hill.

[7] Kao, C., Chiou, S.T. and Wu, E.M. (2005). Some flexible tolerance methods for solving NLP; Research Report of National Cheng-Kung University, Taiwan, R.O.C.

[8] Bett, J.T. (1975). An improved penalty function method for solving constrained parameter optimization problems. JOTA 16, 1.

[9] Hwang, F.A. (1993). Using the penalty function method to solve the NLP; master thesis of National Cheng-Kung University. Department of Mathematics Science, Taiwan, R.O.C.

[10] Martinez, J.M. and Svaiter, B.F. (2003). A practical optimality condition without constraint qualifications for nonlinear programming. J. Optimization Theory Appl. 118, 117.

[11] Clark, R. M. (1982), Cost estimating for conventional water treatment. J. Environ. Engrg. Div., ASCE, 108(5), 819.

[12] Wiesner, M. R., O'Melia, C. R., and Cohon, J. L. (1987), Optimal water treatment plant design. J. Environ. Engrg. Div., ASCE, 113(3), 567.

[13] Lee, K. C., Wu, E. M., (1980). Research on the construction cost for sanitary system in Taiwan, Report No. 16, Grduate School of Environmental Engineering, National Taiwan University, Taiwan.

[14] Taiwan Water Supply Company, (1985). Water supply engineering construction and operation cost estimates.

[15] Xu, Y., Huang, G.H., and Qin, X. (2009). Inexact two-stage stochastic robust optimization model for water resources management under uncertainty. Environ. Eng. Sci. 26, 1765.

[16] Neufville, D.R. (1990). Applied systems Analysis-Engineering Planning and Technology Management, Singarpore, Mc-Graw-Hill.

Advanced Treatment Processes

Chapter 4

Natural Zeolites in Water Treatment – How Effective is Their Use

Karmen Margeta, Nataša Zabukovec Logar, Mario Šiljeg and Anamarija Farkaš

Additional information is available at the end of the chapter

1. Introduction

Natural zeolites are environmentally and economically acceptable hydrated aluminosilicate materials with exceptional ion-exchange and sorption properties. Their effectiveness in different technological processes depends on their physical-chemical properties that are tightly connected to their geological deposits. The unique tree-dimensional porous structure gives natural zeolites various application possibilities. Because of the excess of the negative charge on the surface of zeolite, which results from isomorphic replacement of silicon by aluminum in the primary structural units, natural zeolites belong to the group of cationic exchangers. Numerous studies so far have confirmed their excellent performance on the removal of metal cations from wastewaters. However, zeolites can be chemically modified by inorganic salts or organic surfactants, which are adsorbed on the surface and lead to the generation of positively charged oxi-hydroxides or surfactant micelles, and which enables the zeolite to bind also anions, like arsenates or chromates, in stable or less stable complexes. Natural zeolites have advantages over other cation exchange materials such as commonly used organic resins, because they are cheap, they exhibit excellent selectivity for different cations at low temperatures, which is accompanied with a release of non-toxic exchangeable cations (K^+, Na^+, Ca^{2+} and Mg^{2+}) to the environment, they are compact in size and they allow simple and cheap maintenance in the full-scale applications. The efficiency of water treatment by using natural and modified zeolites depends on the type and quantity of the used zeolite, the size distribution of zeolite particles, the initial concentration of contaminants (cation/anion), pH value of solution, ionic strength of solution, temperature, pressure, contact time of system zeolite/solution and the presence of other organic compounds and anions. For water treatment with natural zeolites, standard procedures are used, usually a procedure in column or batch process. Ion exchange and adsorption properties of natural zeolites in comparison with other chemical and biological processes have the advantage of removing impurities also at relatively low concentrations and allows conservation of water chemistry, if the treatment is carried out in the column process [1]. Subject of further academic and industrial research should be to improve the chemical and physical stability of modified zeolites and to explore their catalytic properties, which would allow their use in catalytic degradation of organic pollutants. More careful consideration of their superb metal removal properties and awareness of possible regeneration or further use of contaminant/metal-loaded forms can considerably increase their environmental application possibilities, with a focus the reduction of high concentrations of cations and anions in drinking water and wastewater, for surface, underground and public municipal water treatment independently or in combination with others physical - chemical methods [2].

1.1. Water treatment using natural zeolites

1.1.1. Wastewater treatment

The use of natural zeolites in wastewater treatment is one of the oldest and the most perspective areas of their application. The presence of heavy metals (Zn, Cr, Pb, Cd, Cu, Mn, Fe, etc.) in wastewater is a serious environmental problem and their removal by natural zeolites have been extensively studied along with other technologies, including chemical precipitation, ion exchange, adsorption, membrane filtration, coagulation flocculation, flotation and electrochemical methods [3]. Recent investigations of natural zeolites as adsorbents in water and wastewater treatment, their properties and possible modification of natural zeolites have been a subject of many studies. Various natural zeolites around the world have shown good ion-exchange capacities for cations, such as ammonium and heavy metal ions. Modification of natural zeolites can be performed by several methods, such as acid treatment, ion exchange, and surfactant functionalization. The modified zeolites can show high adsorption capacity also for organic matter and anions [4].

1.1.2. Surface waters, ground and underground water treatment

The applicability of natural zeolites for the simultaneous removal of ammonia and humic acid, two of the most encountered current contaminants, from the surface waters was also investigated. Their removal depends on pH value, initial concentrations of humic acid and ammonia, temperature and contact time. The obtained results indicated that zeolite showed best performance for simultaneous removal of ammonia and humic acid at the pH close to that of natural waters [5]. The use of natural and modified zeolites has been further investigated for the simultaneous removal of Fe and Mn ions from underground water samples. In particular, Fe and Mn removal levels are between 22-90% and 61-100% for natural zeolite - clinoptilolite [6]. The development of new and cost effective methods to remove As from ground water and drinking water also becomes one of the research priorities. The occurrence of arsenic in natural ground waters is due to geological composition of soil.

1.1.3. Drinking and grey water treatment

Several conventional methods are used for the removal of pollutants from drinking water, such as coagulation followed by filtration, membrane processes and ion exchange. Adsorption methods proved to be effective, economically efficient, easy to perform and construct. Some experiments were conducted to study the efficiency of natural zeolite clinoptilolite and of the clinoptilolite-Fe system in removal of Cu, Mn, Zn, which are simultaneously found in water samples. A very unique property of natural zeolites is their selectivity towards cationic. The excellent results of adsorption experiments, especially for the modified forms along with the fact that the clinoptilolite–Fe system is inexpensive, easily synthesized and regenerated, harmless for human beings, as well as for the environment, we can consider it as a very promising selective metal adsorbent [7]. Using iron/aluminum hydroxide to remove arsenic from water is a proven technology. An alternative method to enhance the performance is to use coarse-grained sorbents to increase the flow rate and throughput of the process. The removal of arsenic from drinking water was studied by using modified adsorbents (natural zeolite) prepared by the use of different iron solutions. The arsenic sorption on the Fe-exchanged zeolite could reach up to 100 mg/kg [8,9]. The high surface area of modified natural zeolite (clinoptilolite)-iron oxide system in strongly basic conditions, can also enhance the removal of cations, like Cu from drinking water. The specific surface area of modified clinoptilolite increased up to 5-times (from 30 to 151m^2 /g) and the maximum amount of adsorbed Cu ions was 13.6 mg/g zeolite for natural clinoptilolite and 37.5 mg/g for modified clinoptilolite [10]. In spite of many scientific evidences of the effectiveness of zeolites in anion removal, not many of them are used on larger scales up to date. High concentrations of fluoride ions in groundwater up to more than 30 mg/L, occur widely, notably in the United States of America, Africa and Asia. More than 260 million people worldwide consume drinking water with a fluoride content of >1.0 mg/L. The available techniques for the removal of F$^-$ -anions from drinking water are membrane techniques, dialysis, electro-dialysis and finally adsorption techniques. Clinoptilolite-type natural zeolite was pre-conditioned with nitric acid solution before loading with Al^{3+}, La^{3+} or ZrO^{2+}. Aluminium-loaded low-silica zeolites as adsorbents for fluorides showed that modified zeolites were able to defluoridate water to below WHO's maximum allowable concentration (MAC) of 1.5 mg/L. The maximum fluoride adsorption was in the pH range of 4–8 [11]. High nitrate concentrations in drinking water sources can lead to a potential risk to environment and public health. Removal efficiency of NO$_3^-$ ions can be increased by treatment of the clinoptilolite samples with HDTM$^+$ (hexadecyltrimethylammonium cation) or cetylpyridinium bromide (CPB) [12]. Grey water is wastewater originated from bathroom and laundry in households. Ammonium is one of the most significant grey water contaminants. Natural and modified zeolites are used for their purification and they shows good performance with up to 97% of ammonium removal depending on contact time, zeolite loading, initial ammonium concentration and pH value. The desorption–regeneration studies demonstrated that the desorption of ammonium on the zeolite is sufficiently high [13,14].

1.1.4. The technological application of natural zeolites in water treatment

Numerous and excellent research results in the last 10 years have shown that natural zeolites have practical use, which is confirmed by a large number of patents, especially for the two naturally occurring zeolite minerals: clinoptilolite and modernite shown in Figure 1. The number of patents is substantial for both zeolite types, which gives a clear notice that the interest of researcher in natural zeolites is strongly encouraged by the commercial sector covering the use in households or in industrial/large-scale processes and treatments.

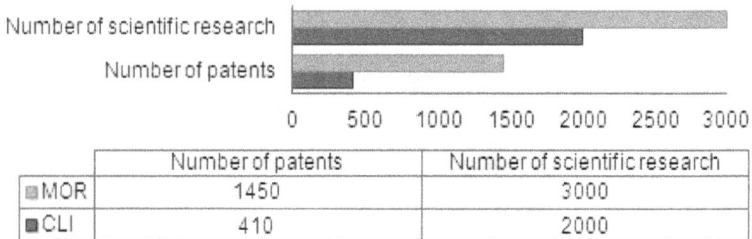

	Number of patents	Number of scientific research
▦MOR	1450	3000
▪CLI	410	2000

Figure 1. Number of patents for clinoptilolite (CLI) and mordenite (MOR), (The last ten years) [15].

2. Properties of natural zeolites

The structure of natural zeolite is very interesting and complex. The *primary building units* (PBU) of zeolites are the SiO_4 and AlO_4 tetrahedra. They connect via oxygen ions into *secondary building units* (SBU), which are then linked into a three-dimensional crystalline structure of zeolite. Substitution of Si by Al defines the negative charge of the zeolite framework, which is compensated by alkaline and earth alkaline metal cations. Therefore natural zeolites appear as cation exchangers because they have negative charge on the surface. In the zeolite lattice, substitution is not limited to Si-Al substitution. Atoms of iron, boron, chromium, germanium, and titanium may also substitute silicon. Water molecules can be present in voids of large cavities and bonded to framework ions and exchangeable ions via aqueous bridges. One of the most investigated zeolite in basic and applied research is clinoptilolite. The characteristic way of linking of PBUs and the formation of unique structural units ultimately results in the fact that these materials are highly porous with channels and cavities in the structure that have characteristic pore sizes and shapes. In the structure of clinoptilolite, there are three types of channels, of which two are parallel, and made of ten and eight-membered rings of Si/AlO_4, while one, defined by eight-membered rings, is vertical. In these channels the hydrated cations can occupy the following places: **I** - cation (Na- and Ca-ions) is located in the 10-member ring channels (free diameters 0.44 x 0.72 nm); **II** - cation (Na- and Ca-ions) is located in the 8-member ring channels (free diameters 0.41 x 0.47 nm); **III** - cation (K- ion) is located in the 8-member ring vertical channels (free diameters 0.40 x 0.55 nm); **IV** - cation (Mg-ion) is located in the channel of 10-member rings and it is located in the center of the channel (Figure 2).

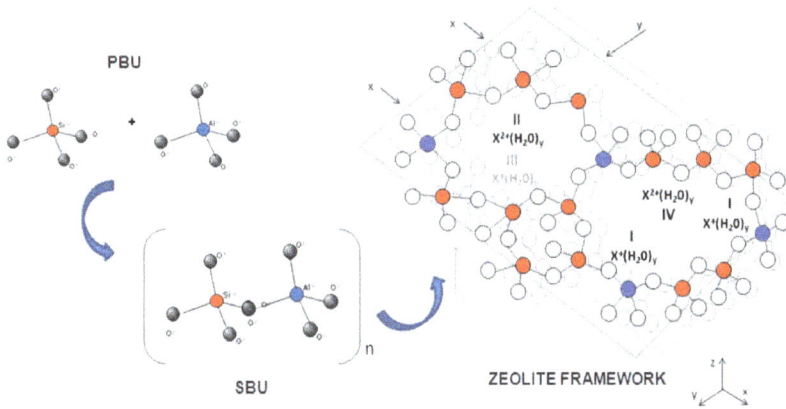

Figure 2. Binding of building units (PBU and SBU) in three-dimensional zeolite- clinoptilolite structure.

Usually the number of water molecules in the zeolite structure does not exceed the number of oxygen atoms. Ratio (Si + Al): O is 1:2, and the number of aluminum atoms in the tetrahedrons is equal to the sum of the positive charges (x + 2y) of exchangeable cations. Replacement of silicon and aluminum atoms in zeolites ranges from a minimum ratio of 1:5 (mordenit), up to a maximum 1:1 (erionit) [16].

2.1. Categorization and characterization of natural zeolites

Natural zeolites are divided into seven main groups (Table 1) according to their crystal structure, based on morphology, their physically properties, different ways of binding secondary units in the three-dimensional framework, the free pore volume and types of exchangeable cations in zeolite structure. These diverse types of zeolite are a reflection of the fascinating structures of these microporous materials. Each time a new zeolite framework structure is reported, it is examined by the Structure Commission of the International Zeolite Association (IZA-SC), and if it is found to be unique, it is assigned a 3-letter framework type code, like CLI, MOR, ANA etc. This code is part of the official IUPAC nomenclature for microporous materials [17]. Characterizations of natural zeolites include chemical and instrumental analyses of the samples and are crucial for their further application in water treatment. The *chemical composition,* usually determined by several different methods: classical chemical analysis – gravimetric method, Atomic Absorption Spectrometry or X-ray Fluorescence Spectrometry, is very important for the efficiency of the water treatment processes and provides insight into the main amount of basic oxide components (SiO_2 and Al_2O_3), exchangeable cations (Na^+, K^+, Ca^{2+}, Mg^{2+}, Ba^{2+}, Sr^{2+}) and other elements present in smaller concentrations (like Ti atoms). According to the proportion of exchangeable cations, we can sometimes already determine the type of zeolite. The proportion of the oxide components in natural zeolite materials depends on the geological deposits. Table 2 gives the basic information of the chemical composition of natural zeolites, from different countries expressed as mass fraction of oxide components.

Zeolite[1]	Primary cell formula, Structure[2], Crystal system	Channel dimensions[3]	Free Volume[4]	Exchangeable cations
GROUP 1				
Analcime (ANA)	$Na_{16}(Al_{16}Si_{32}O_{96})\cdot16H_2O$ Cubic	0,16 x 0,42	0,18	Na, K, Ca, Rb, Cs
Laumontite (LAU)	$Ca_4(Al_8Si_{36}O_{48})\cdot16H_2O$ Monoclinic	0,40 x 0,53	-	K, Na, Ca
Phillipsite (PHI)	$K_2(Ca_{0,5},Na)_4\cdot$ $(Al_6Si_{10}O_{32})\cdot12H_2O$ Monoclinic	0,38 x 0,38	0,31	Na, K, Ca
GROUP 2				
Erionite (ERI)	$NaK_2MgCa_{1,5}(Al_8Si_{28}O_{72})$ $\cdot28H_2O$ Hexagonal	0,36 x 0,52	0,35	K, Na, Ca, Mg
GROUP 3				
Zeolite A	$Na_{12}[(AlO_2)_{12}(SiO_2)_{12}]\cdot27$ H_2O	-	0,47	-
GROUP 4				
Chabazite (CHA)	$Ca_2(Al_4Si_8O_{24})\cdot12H_2O$ Hexagonal	0,38 x 0,38	0,47	Na, Ca, K
GROUP 5				
Natrolite (NAT)	$Na_{16}[(AlO_2)_{16}(SiO_2)_{24}]$ $\cdot16H_2O$ Ortho-rhombic	0,25 x 0,41	0,23	Na, K, Ca
GROUP 6				
Mordenite (MOR)	$Na_3KCa_2(Al_8Si_{40}O_{96})$ $\cdot28H_2O$ Ortho-rhombic	0,65 x 0,70	0,28	Na, Ca, K
GROUP 7				
Heulandite (HEU)	$(Na,K)Ca_4(Al_9Si_{27}O_{72})$ $\cdot24H_2O$ Monoclinic	0,44 x 0,72	0,39	Na, K, Ca, Sr, Ba
Clinoptilolite (CLI)	$(Na,K)_6(Al_6Si_{30}$ $O_{72})\cdot20H_2O$ Monoclinic	0,44 x 0,72	0,34	Na, K, Ca, Sr, Ba

[1]Web pictures (http://www.galleries.com/Zeolite_Group), [2]Atlas of Zeolite Framwork [16]., [3]Dimension (nm) is the dimension of the largest channel, [4]mL H_2O / mL zeolite

Table 1. Categorization and structural properties of seven main groups of natural zeolites.

Zeolite	Components, wt (%)								Ref.
	SiO₂	Al₂O₃	Fe₂O₃	Na₂O	K₂O	CaO	MgO	LOI*	
Croatia (CLI)	64,93	13,39	2,07	2,40	1,30	2,00	1,08	9,63	[18]
Serbia (CLI)	65,63	12,97	1,48	1,20	1,33	3,21	1,41	12,96	[19]
Australia (CLI)	68,26	12,99	1,37	0,64	4,11	2,09	0,83	8,87	[20]
Mexican (CLI)	70,17	11,07	1,12	0,83	4,90	1,73	0,35	-	[21]
China (CLI)	65,72	13,50	1,30	1,16	3,14	3,10	0,63	11,12	[22]
Turkey (CLI)	64,28	12,07	0,84	5,62	0,83	2,47	2,07	-	[23]
Cuba (CLI)	64,30	11,00	1,4	1,40	1,20	3,70	0,50	-	[24]
Bulgaria (CLI)	64,20	11,67	1,03	2,36	3,84	7,42	0,35	8,66	[25]
Greece (HEU)	68,62	11,80	0,07	1,13	2,92	2,14	0,75	12,34	[26]
Ukraine (CLI)	68,64	11,50	1,57	0,29	3,12	2,38	0,89	-	[27]
Equador (CLI)	65,80	11,32	3,42	4,10	0,45	1,23	0,96	12,29	[28]
Brazil (STI)	68,79	11,71	5,25	2,75	0,62	3,34	1,31	5,84	[28]
Italia (PHI)	52,47	17,57	3,70	0,92	7,47	4,99	1,50	9,48	[29]
Argentina (CLI)	64,51	11,25	0,97	3,60	1,21	4,38	0,60	13.14	[29]
USA (CLI)	66.61	12,91	1,7	0,39	2,44	3,18	1,54	10,72	[30]

CLI – clinoptilolite; STI – stilbite; PHI – philipsite; HEU – heulandite
*LOI – Loss of ignition

Table 2. Chemical composition of natural zeolites from different deposits.

Because of the complex mineralogical composition of natural zeolites and consequently uneven distribution of different phases and elements in the zeolite tuff, a combination of microscopic, spectroscopic, and other *instrumental techniques* must be used to fully characterize the material. The basic information about the accessibility of the pores for different ions and molecules can be obtained from BET Analysis based on nitrogen physisorption, which gives information about the BET surface and accessibility of the pores for different ions and molecules. The BET values usually range from 15 to 40 m²/g [19]. X-ray powder diffraction analysis allows quantitative determination of mineralogical composition of zeolitic tuffs, including the type of zeolite, and is crucial for any further application of natural zeolites. A typical X-ray powder diffractogram is shown on Figure 3. Detailed studies of some zeolite tuffs from Croatia and Serbia by powder X-ray diffraction and subsequent quantitative Rietveld refinement phase analyses are shown in Figure 4.

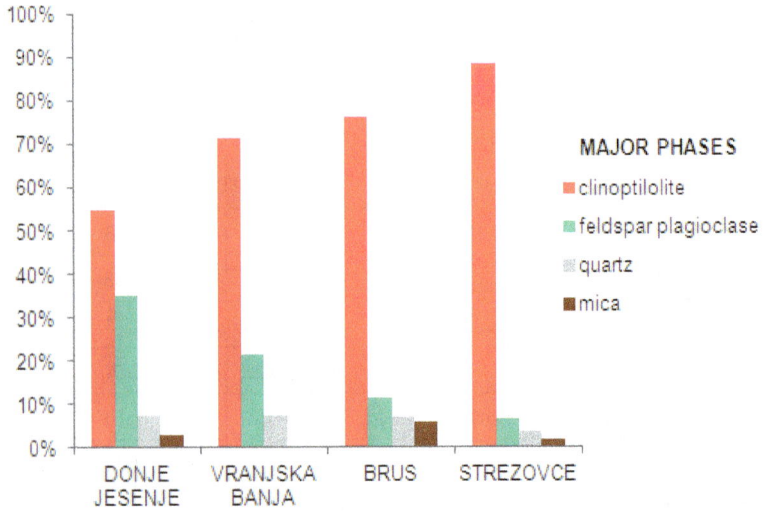

Figure 3. The phase composition of zeolite tuffs from different deposits

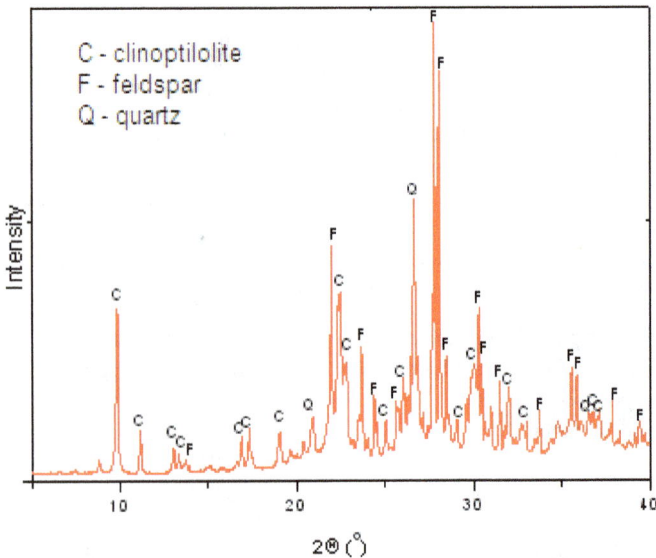

Figure 4. XRPD of natural zeolite [18].

The size and the morphology of the crystals in the zeolite samples is usually studied by scanning electron microscopy and accompanied Energy Dispersive X-ray Spectroscopy

(EDXS) analysis system, that is attached to the scanning electron microscope. The obvious advantage of EDXS elemental analysis over conventional chemical analysis is that we can obtain elemental composition of selected phase in the tuff, not only in the bulk sample. An average elemental composition of the sample using EDXS is usually obtained by a data collection at three or more different mm²- sized windows on the sample surface. Typical SEM photographs of zeolite tuffs are shown in the figure below (Figure 5).

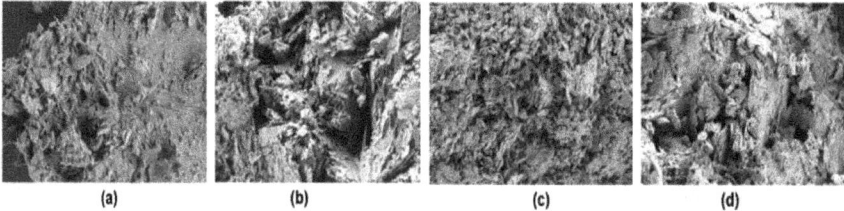

(a)   (b)   (c)   (d)

Figure 5. SEM of zeolite tuffs from Donje Jesenje (a), Vranjska Banja (b), Brus (c) and Strezovce (d). Magnification of app. 20.000x was used in the measurements. Beside the typical plate-like morphology of clinoptilolite crystals, some fiber-like particles of mordenite could be also observed in all four samples.

The X-ray Photoelectron Spectroscopy (XPS or ESCA) analysis is used to determine the type and oxidation state of elements on the surface of natural zeolite by X-ray radiation from anode, e.g. Al anode, mostly by putting the sample in the form of 1 mm thick pressed pallet. XPS depth profiling can be performed by alternating cycles of ion sputtering to remove surface layers of zeolite and acquisition of photoelectron spectra (ion sputtering can be performed with 1 keV Ar^+ beam rastering over 3 x 3 mm² area). In this way a depth distribution of elements can be obtained. The relative error for calculated concentrations of metals is estimated to be about 20%.

The X-ray absorption spectroscopy (XAS) is a particularly useful method for the characterization of metal-loaded zeolite samples. Information on the oxidation state of metals, like copper, zinc, chromium, arsenic and many others, in natural zeolites can be obtained by XANES (X-ray Absorption Near Edge Structure). The energy position of the metal absorption edge is shifted to higher energies with increasing oxidation state. The relation can be calibrated with XANES spectra of reference compounds with the same type of metal ligands as in the investigated sample. More information on the local environment of metal atoms in the zeolite samples can be obtained by EXAFS (Extended X-ray Absorption Fine Structure) analysis. The EXAFS spectra can be quantitatively analyzed for the coordination number, distance, and Debye- Waller factor of the nearest coordination shells of the metal.

3. Modifications of natural zeolite

Natural zeolite can be modified by single or combined treatment such as heating and chemical modification (acids, bases and inorganic salts). Chemical and thermal treatment of zeolite may result in cation migration and thus affect the cation location and pore opening. "Pore engineering" is a popular term for methods used in zeolite modification in which some of its sorbent properties are manipulated. The processes of ion exchange and adsorption in zeolite/solution contact occur concurrently.

3.1. Modification with solution of inorganic salts

Chemical modification with inorganic salts (NaCl, CaCl$_2$, BaCl$_2$, NH$_4$Cl, FeCl$_3$) or a cationic surfactant (hexsadecyltri-methylammonium (HDTMA) - bromide) give to improve zeolite properties and increase its efficiency in water treatment [19, 31-38]. For successful modification high-concentration solutions of inorganic salts on the surface of zeolite is significant as shown in Figure 6. Under normal conditions, large cavities and entries to the channels inside the zeolite framework are filled with water molecules forming hydration spheres around exchangeable cationic (Figure 6-A). After the contact of zeolite with an inorganic salt solution such as NaCl, exchange of cations (H$^+$ or Na$^+$) from solution with exchangeable cations (Na$^+$, K$^+$, Ca^{2+}, Mg^{2+}) from the zeolite framework occurs (Figure 6-B). To remove anions from the water, zeolite surface has to be modified with a solution of inorganic salts (for example FeCl$_3$) whose adsorption on the zeolite surface leads to the formation of oxi-hydroxides, which then form stable complexes with anions in solution. This modification can result in – to a smaller or greater extent – the creation of an adsorption layer on zeolite surface and modification of surface charge on zeolite surface (from negative to positive) (Figure 6-C) [39,40]. Specific surface area (BET) of natural and modified zeolite (deposits from Croatia and Serbia) after pre-treatment with inorganic salts is shown in Table 3.

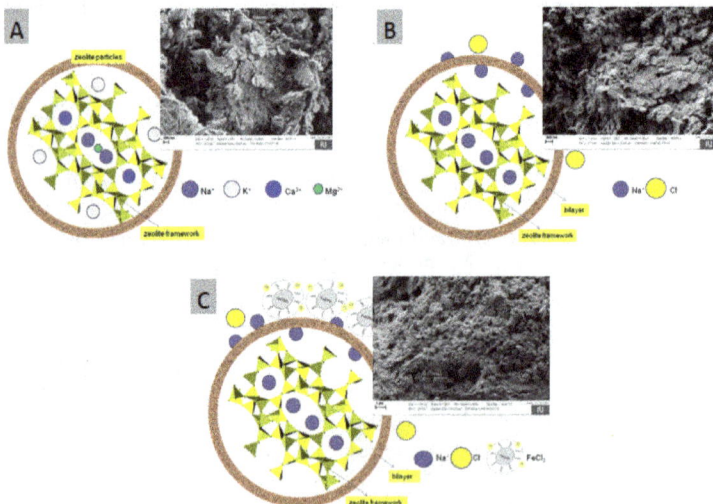

Figure 6. Zeolite particles in natural and modified zeolites (Na and Fe forms of clinoptilolite from the deposit V. Banja, Serbia) and SEM image of zeolite surface after the implementation of chemical modification.

Zeolite	BET Specific surface area, m²/g	Ref.
Natural CLI (VB)	24	[8]
Natural CLI (DJ)	16	[8]
Ca-CLI (VB)	20	[19]
Na-CLI (VB)	23	[19]
H-CLI (0,5 M) (DJ)	40	[18]
H-CLI (1,0 M) (DJ)	42	[18]
NaOH-CLI (DJ)	19	[18]
NaFe-CLI (VB)	91	[8]
NaFe-CLI (DJ)	51	[8]
Natural MOR	42	[41]
HDTMA (0,25 mmol/g)-MOR	6	[41]
HDTMA (0,85 mmol/g)-MOR	5	[41]

Table 3. Specific surface area of natural and modified zeolites (clinoptilolite and mordenite) from Croatia (DJ) and Serbia (VB).

This confirms the theoretical study of the suitable position of Na^+ ions in zeolite structure and possibilities of zeolite exchange with metal ions from solutions. Na-modifications have shown the highest selectivity for zinc ions when Zn^{2+} ions are mixed together with Fe^{3+} ions, which is highly dependent on the acidity of the solution and cation hydratation enthalpy. Chemical modification of zeolite with $FeCl_3$ solutions is defined by the parameter system: pH solution, ionic strength of the solution, oxidation-reduction conditions, iron concentration and the type of salts used (chlorides, sulphates, nitrates, perchlorates, etc.). Fe^{2+} and Fe^{3+} ratio results in sorption of iron ions and iron oxi-hydroxides on the surface and in the pores of clinoptilolite. However, regardless of the portion of different iron forms present in zeolite structure, the common property of all iron modified zeolites is a high increase in sorption capacity for arsenic oxi-anions present in water solutions. According to literature data, two mechanisms of arsenic binding on the surface of iron oxi- hydroxide are: the mechanism of surface precipitation and the mechanism of surface complexation. Surface complexation mechanism can be monodentate, dominant at a low surface coverage of modified zeolite, or bidendate at higher surface coverage when iron forms complexes with arsenic [42-44].

3.2. Modification with acid/basic treatment

Zeolite structure and its chemical and physical properties can be modified with either inorganic basic (NaOH, Ca(OH)₂) or acid solutions (HCl, HNO₃). Acid treatment is among the most common and simple methods for zeolite modification. The effectiveness of acid treatment depends on the chemical composition, structure, mineral purity, and the working conditions. Dissolution of some amorphous materials that block the pores of natural zeolites is another consequence of acid modification. According to the Brønsted and Lewis theory, dissolution of natural zeolites in acid solution occurs as a result of the acidic/basic behaviour of the aluminosilicate structure in the presence of H+ or OH- ions in the solution. Brønsted and Lewis acidic/basic sites present in the zeolite framework are also responsible for their chemical behaviour in aqueous solutions. Interactions of natural high-silica zeolites (e.g. clinoptilolite, heulandite, mordenite, erionite and ferrierite) with acidic and basic aqueous media generally occurring at low dissolution rates and are adequately acid-resistant [26]. By the dealumination process, Al^{3+}-ions can be progressively removed from the aluminosilicate structure. The reactions favour lower pH values and the $AlOH_2^+$ species formed are detached because of their high degree of surface protonation. Decationization (exchange of zeolite cations with H+ ions) is minimal in solutions with high cation concentrations but is also significantly dependent on the solution acidity and cation hydratation enthalpy [45]. Hydrochloric acid solution treatment leads to decationization (obtaining the so-called "hydrogen form" zeolites), dealuminization and sometimes destruction of the crystal lattice. The effect of hydrochloric acid on various zeolites varies. As an example, the HCl acid modification of the natural materials mordenite and erionite conducted under similar conditions led to weak decationization and almost no dealuminization of mordenite, while the extent of alkaline and alkaline earth metals uptake, as well as of aluminium uptake from erionite was more than 90 % [25]. Changes in the chemical composition and structure of zeolites as a result of the decationization and dealumination steps lead to changes of minerals properties. The nature of the mineral and the exchangeable cation content, as well as impurities, have a significant influence on acid modification of structurally identical natural zeolites. Zeolite/solution contact time, heating before and after modification, pre-treatment with water or other solutions, such as NH_4Cl, influence the efficiency of modification as well. Low siliceous zeolites are unstable in the acid and their decationization is conducted by other methods. Ion exchange with more soluble ammonium salts (usually NH_4Cl) is the initial stage, followed by heating of the samples rich in ammonium ions to eliminate ammonia and hydrogen. This method of decationization is also applied to highly siliceous zeolites. Influence of the concentration of modification solution and the reaction time are also taken into consideration.

4. Hydrothermal treatment of natural zeolite

Thermal treatment at high temperature, depending on the solid sample and temperature used can enhance pore volume by removing water molecules and organics from pore channels. Water present in cages and channels of the zeolite framework contributes 10 – 25 % to the total mass of zeolites. To enable efficient use of zeolites in water treatment, it is important to know the properties of dehydration and structural stability of particular zeolitic materials. Stability in the structure of some natural zeolites is given in Table 4.

To acquire information about the mass loss change and adsorption or crystallization, thermal analysis methods are used: thermogravimetric /differential thermogravimetric analysis (TG/DTG) and differential thermal analysis (DTA). A stable zeolite structure, such as clinoptilolite, results in continuing, but reversible loss of water. The dehydration process of natural and modified zeolites and zeolite mass loss rate by increasing temperature are shown in Figure 7.

Natural zeolite	Structural stability
Analcime	up to 700 ºC
Laumontite	up to 500 ºC
Erionite	up to 750 ºC
Mordenite	up to 800 ºC
Heulandite	up to 300 ºC
Clinoptilolite	up to 750 ºC

Table 4. Structural stability of some natural zeolites [16].

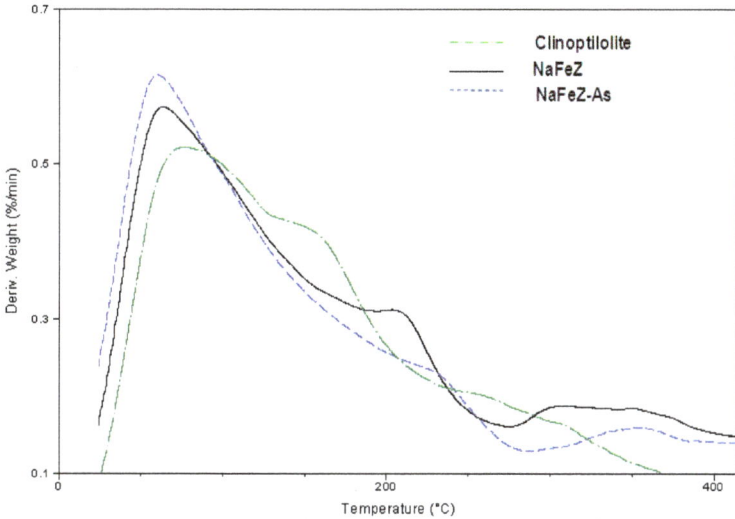

Figure 7. Loss of mass for natural and Fe-modified and Na-modified zeolite by the DTG method.

Zeolite water can be removed by heating to approximately 400 °C. The hydration of iron modified zeolite was 17 % higher than that of Na-modified zeolite and natural zeolite.

5. Adsorption and ion exchange processes in water treatment

A large number of parameters can influence the process of ion sorption/removal from water treatment. Conductivity, pH, temperature of treated water, ionic strength, initial concentration of cations and anions in solution, zeolite mass, zeolite particle size are all important parameters.

5.1. Processes of ion exchange, adsorption and hydrolysis

Ion exchange process is characterized by the capacity, selectivity and kinetics of the exchange. Capacity is needed in several determinations, such as the normalization of equilibrium isotherms and the application of kinetic models in order to determine the ion-exchange diffusion coefficients (Table 5). Differences in the exchange isotherms derived by different researchers for the same ion-exchange systems are mainly due to the assumption of different cation-exchange capacities of zeolitic materials.

IUPAC recommendations for ion-exchange nomenclature define the following capacity types: "Theoretical (specific) capacity, apparent capacity (effective capacity)", "Practical (specific) capacity", "Useful capacity" and "Breakthrough capacity" [46].

Zeolite	Si/Al ratio	*CEC, meq/g zeolite
Analcime	1,4 – 3,0	4,5
Chabasite	1,4 – 4,3	3,7
Clinoptilolite	2,7 – 5,7	2,2
Erionite	2,3 – 3,4	3,1
Heulandite	4,0 – 6,1	3,2
Laumontite	1,3 – 3,3	4,3
Mordenite	4,17 - 5,0	2,3
Phillipsite	1,7 -3,3	3,9
Natrolite	1,5	5,2

Table 5. Cation-exchange capacities of different zeolitic materials (*CEC is operationally defined – determine the amount of a cation that can be removed by a specific substance once the material and solution have come to equal) [16].

Zeolites selectivity related to cations and anions is an important property in water treatment procedure. Selectivity is a property of the exchanger to show different preferences for particular ions and it depends of field strength in zeolite pore. Zeolites with low field strength and with higher Si content, such as clinoptilolite, are more selective for cations with lower charge density (K^+, NH_4^+, Ag^+, Cs^+). Zeolites with high field strength, i.e. higher Al content, are more selective toward the high charge density cations (Na^+, Li^+). At room temperature and low concentration of the solution ions are exchanged, the advantages of the amendment have ions with higher charge. Increasing the concentration of the solution, the difference in ion exchange affinities of different charges is reduced. If the solution contains different ions of the same charge, the selectivity increases with increasing atomic number (Li^+, Na^+, NH_4^+, K^+). The selectivity of clinoptilolite towards alkali metals exist in the sequence: $Cs^+> K^+> Rb^+> Na^+> Li^+$, and the alkaline earth metals: $Ba^{2+}> Sr^{2+}> Ca^{2+}> Mg^{2+}$. The selectivity of clinoptilolite towards heavy metal ions (cations) exist in the series: $Pb^{2+}> Cd^{2+}> Cu^{2+}> Co^{2+}> Cr^{2+}> Zn^{2+}> Mn^{2+}> Hg^{2+}$ and selectivity by anions exists in the series: $SO_4^{2-}> I^-> NO_3^-> HCrO_4^-> Br^-> Cl^-> OH^-$ [47,48]

The processes of ion exchange and adsorption on natural zeolite occur concurrently with the process of hydrolysis in aqueous solutions. Determination of hydrolytic activity and stability of zeolites according to previous studies showed a very important aspect of technological applications, and hydrolytic activity indicates the chemical stability.

The hydrolysis process (1) is a reaction following the process of ion exchange. Understanding and studying zeolite hydrolysis is of great importance to understanding the properties of zeolite. The hydrolysis process of zeolite is usually observed by monitoring the pH levels and electric conductivity during which a sudden increase in the pH value can be seen at beginning of the hydrolysis process after which the zeolite-water system tends to stabilize the pH value.

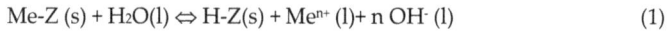

$$Me\text{-}Z\,(s) + H_2O(l) \Leftrightarrow H\text{-}Z(s) + Me^{n+}\,(l) + n\,OH^-\,(l) \qquad (1)$$

n – cation charge, Me – exchangeable cations (Na^+, K^+, Ca^{2+}, Mg^{2+}), Z – zeolite

The created OH^- ions cause an increase in the pH value of the system. A reaction of the metallic ions occurs at the same time (2).

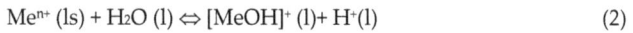

$$Me^{n+}\,(ls) + H_2O\,(l) \Leftrightarrow [MeOH]^+\,(l) + H^+(l) \qquad (2)$$

Increase in the concentration of OH^- ions at the beginning of hydrolysis causes thus created OH^- ions to adsorb onto the surface of zeolitic particles, which in turn causes melting of the surface layer of zeolitic particles. Anions on the zeolite surface form with exchangeable cations more or less stable complexes, depending on the stability constant (3).

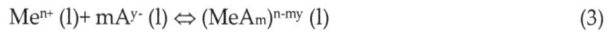

$$Me^{n+}\,(l) + mA^{y-}\,(l) \Leftrightarrow (MeA_m)^{n\text{-}my}\,(l) \qquad (3)$$

Hydrolysed cations in the channels have good mobility and ability to exchange with the cations from the solution because they are connected by weak electrostatic bonds to the basic aluminium-silicate structure. The rate of exchange and the quantity of exchangeable ions depend on the size of cations and their charge, cation concentration in the solution, ionic strength of the solution, anions present in the solution, temperature, pH value, physical and chemical properties of the solvent, structural characteristics of zeolite and exchange kinetics. Exchange of ions is preceded by the adsorption of ions from the solution to the surface of zeolite particles. Ion exchange (4) is a reversible process where a balance between the solid and fluid phases is achieved. During the process of achieving balance within the zeolite–solution system, a reversible process occurs.

$$Me_1Z\,(s) + Me_2^{n+}\,(l) \Leftrightarrow Me_2Z\,(s) + Me_1^{n+}\,(l) \qquad (4)$$

Me_1Z and Me_2Z - concentrations of exchangeable cations Me_1 and Me_2 in zeolite Z

Me_2^{n+} and Me_1^{n+} - concentrations of exchangeable cations in the solution

n - the charge number of exchangeable cations

The process of diffusion in the system of zeolite/ aqueous solution can be divided into several phases (Figure 8): (I) Diffusion in solution, (II) Diffusion through the film, (III) Diffusion in pores, (IV) Ion exchange.

It is known that the rate of a multistep chemical reaction is defined by the slowest stage. The rate of these processes can be changed by altering the physical characteristics of the heat exchanger or the solution containing the ions. Thus the rate of diffusion of ions in the film

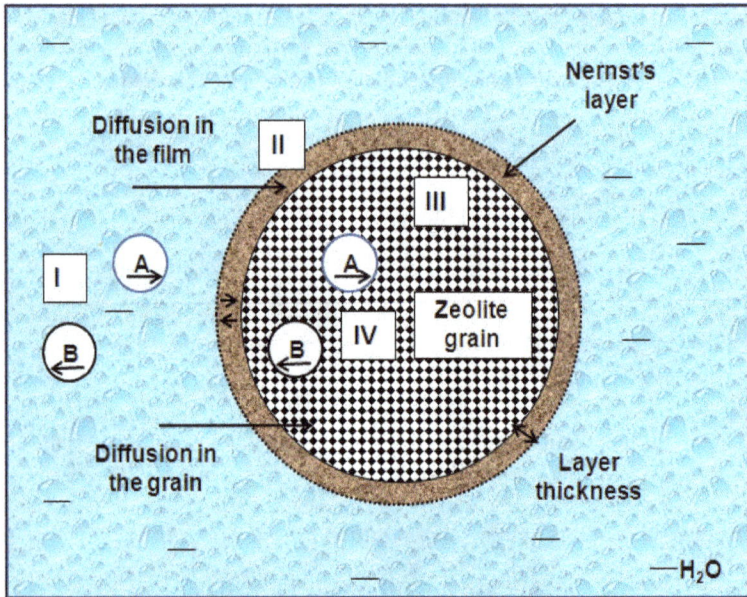

Figure 8. Diffusion processes in the system of zeolite/aqueous solution.

can be increased by intensive mixing of the solution, with increasing solution temperature and concentration. Rate of diffusion through macro and micro pores can be increased by reducing the grain size and concentration of the solution. It is therefore necessary to assess the degree of the ion-exchanging process, which is important for the overall process of ion exchange and hence affects its rate [50].

5.2. Adsorption isotherms

The equilibrium distribution of metal ions between the natural or modified zeolites and the solution is important in determining the maximum sorption capacity. Several isotherm models (Table 6) are available to describe the equilibrium sorption distribution. In most studies the following two models, Langmuir and Freundlich models, were used [51].

The Langmuir isotherm:

$$q_c = \frac{K_L C_e}{1 + Q_{max} C_e} \qquad (5)$$

Linear form		Transformed		Slope	Intercept
		X-values	Y-values		
$\dfrac{C_e}{q_e} = \dfrac{1}{Q_{max} K_L} + \dfrac{C_e}{Q_{max}}$		C_e	C_e/q_e	Q_{max}/K_L	$1/K_L$

where q_e is metal concentration on the zeolite at equilibrium (mg of metal ion/g of zeolite); Q_{max} (mg/g) and K_L (1/mg) are Langmuir constants related to the maximum adsorption capacity corresponding to complete coverage of available adsorption sites and a measure of adsorption energy (equilibrium adsorption constant) respectively. These constants are found from the slope and intercept of c_e/q_e vs. c_e linear plot so that Q_{max} = 1/slope and K_L = slope/intercept.

The Freundlich isotherm:

$$q_e = K_F C_e^{1/n} \tag{6}$$

Linear form	Transformed		Slope	Intercept
	X-values	Y-values		
$ln q_e = ln K_F + \left(\dfrac{1}{n}\right) ln C_e$	$\ln(C_e)$	$\ln(q_e)$	$1/n$	$\ln(K_F)$

where K_F and n are Freundlich constants determined from the slope and intercept of plotting $\ln q_e$ vs. $\ln C_e$.

Isotherm models	Equilibrium	Ref.
Sips	$q_e = \dfrac{q_m (a_s C_e)^{n_s}}{1 + (a_s C_e)^{n_s}}$	[52]
Langmuir-Freundlich	$q_e = \dfrac{K_{LF} C_e^{n_{LF}}}{1 + (a_{LF} C_e)^{n_{LF}}}$	[53]
Temkin	$q_e = \dfrac{RT}{b_T} \ln(A_T C_e)$	[54]
Toth	$q_e = \dfrac{K_t C_e}{(a_t + C_e)^{1/t}}$	[55]
Redlich-Peterson	$q_e = \dfrac{K_R C_e}{1 + a_R C_e^{b_R}}$	[56]
Dubinin-Radushkevich	$q_e = q_D \exp\left(-B_D \left[RT \ln\left(1 + \dfrac{1}{C_e}\right)\right]^2\right)$	[57]

Table 6. Other important isotherms models are used to describe the equilibrium sorption distribution.

Gibbs free energy of sorption

The apparent Gibbs free energy of sorption (ΔG^0) is the fundamental criterion of spontaneity. Reaction occurs spontaneously at a given temperature if ΔG^0 is negative. The standard Gibbs free energy change (ΔG^0) for the adsorption of ions by natural or modified zeolite can be calculated using the following thermodynamic equation (7):

$$\Delta G^0 = - RT \ln K \tag{7}$$

where T is the absolute temperature and R is the gas constant (8.314 J/molK) [44].

Table 7 shows the calculated values of Gibbs free energy for several different ions adsorbed to natural and modified zeolites

Natural zeolite	Ions	ΔG^0	Ref.
CLI	Fe^{3+}	-16.98	[51]
Modified CLI	As (V)	-18,14	[58]
CLI	NH_4^+	-19,52	[59]
Modified CLI	Cu^{2+}	-29,3	[60]
CLI	Pb^{2+}	-6,86	[61]
Modified CLI	Mn^{2+}	-9,8	[62]

Table 7. Gibbs free energy of sorption (ΔG^0) different metal ions.

The negative sign for ΔG^0 is indicative of the spontaneous nature of metal ions adsorption on the natural zeolite at certain temperatures.

Kinetics of the adsorption process

The kinetic results obtained from experiments can be analyzed using different kinetic models such as *Lagergen pseudo first-order* (8) and pseudo second-order models (9). Lagergen pseudo first-order model is given by:

$$\ln(q_e-q_t) = \ln q_e - k_{1,ads} t \qquad (8)$$

where q_t is metal concentration adsorbed on natural zeolite at any time (mg of metal ions/g of zeolite) and k_1 is the adsorption rate constant (min^{-1}). A linear plot of $\ln (q_e - q_t)$ against t gives the slope= k_1 and intercept = $\ln q_e$.

The equation that describes the *pseudo-second order model* is given in the following linear form:

$$t/q_1 = 1/k_{2,ads}q_e^2 + t/q_e \qquad (9)$$

where k_2 is the adsorption rate constant (g/mgmin). K_2 and q_e are found from the intercept and slop of t/q_t vs. t linear plot so that q_e = 1/slope and k_2 = slope2/intercept.

The degree of goodness of the linear plot of these kinetic models can be judged from the value of the coefficient of determination of the plot, which can also be regarded as the criterion in the determination of adequacy of a kinetic model [51,63-64].

6. Application efficiency of natural zeolite in water treatment

Wastewaters obtained from processes of many industries contain pollutants (inorganic cations, anions, oils, organic matter, etc.) have a toxic effect on the ecosystem. It is necessary to treat the metal contaminated wastewater prior to discharge into the environment and the removal of these pollutants requires economically justifiable and efficient technologies and techniques [4].

Natural zeolites in wastewater treatment are quite effective in comparison with other methods. The conditions of purification process are very important for wastewater treatment. Thus possible regeneration of the zeolite, re-use of the zeolite and re-use of the obtained concentrate metal ions after regeneration of the zeolite, very positively impact on the environment without creating new waste. Table 8 shows the efficiency of the removal of metal cations from wastewater using natural zeolite which is supplemented with other physical and chemical methods. The efficiency of removing metal ions from wastewaters depends on many factors such as initial concentration of metal ions in wastewater, the pH value of the system, the possibility of formation of metal hydroxyl anion, previous chemical and thermal modification of zeolite and the amount of water that should be purified. Besides the already mentioned properties (selectivity, capacity, etc.) natural zeolites have excellent resistance to chemical, biological, mechanical or thermal changes.

Methods		Removal efficiency, %									Ref.
		Cd(II)	Cr(III)	Cu(II)	Ni(II)	Zn(II)	Fe(III)	Mn(II)	Pb (II)	As(III)	
Physical / chemical methods	Precipitation	96-99	99	80	71-85	99	-	99	92	-	[65-67]
	Ion exchange	100	100	100	100	100	100	100	-	-	
	Membrane filtration*	93-99	86-95	98-100	60-100	95-99	90	85	99	20-55	
	Coagulation/ Flocculation	99	-	99	-	99	-	99	-	-	
	Flotation	-	95-98	85-99	70-98	99-100	-		-	-	
	Electrochem. methods	13	77-99	98-99	69-90	96	-	78	-	99	
Natural zeolite	CLI	90	90	90	75	85	70	-	95	-	[68-75]
	M-CLI	90-99	88	80	37	92	90	70	90-99	90	
	CHA	-	-	98	98	98	-	-	-	-	
	SCO	59	96	-	40	-	-	75	-	-	
	PHI	-	-	-	-	74	88	-	-	-	

* Membrane processes: Ultrafiltration, Nanofiltration, Reverse osmosis
CLI – clinoptilolite; M-CLI modified clinoptilolite; CHA – chabazite; SCO – scolecite; PHI - philipsite

Table 8. Comparison of the efficiency removal metal ions from waste water using standard techniques and natural zeolites.

6.1. Procedures for removing contaminants from water using natural zeolite

Two standard procedures are usually applied to remove contaminants from water using zeolites: *batch* and *column* procedures. Before any practical ion exchange application, some specific studies of representative zeolite samples from the deposit for its exploitation potential should be carried out. The cation exchange capacity of the zeolite is distinctly dependent on its original cationic composition, because not all the cationic sites in the zeolite structure are available for cation exchange [48]. Removal efficiency of ions from water solution increases with higher portion of the main component of natural zeolite.

6.1.1. "Batch" process

In the *batch process* a certain amount of natural or modified zeolites is used, which are placed in contact with a solution of synthetic or real samples of water at certain times. Bach procedure is usually carried out at constant temperature and hydrodynamic conditions. The amount of immobilized ions on the zeolite is presented as equilibrium distribution of ions between the natural or modified zeolites and the solution.

The removal efficiency of metal ions by natural zeolite depends on the quantity and exchange capacity as well as the presence of other cations and anions in the treated water. With increasing concentration of metal ions in water the removal efficiency increases of these ions using natural zeolite (from 500 mg/L to 1 g/L). Natural and modified zeolites are efficient at low concentrations of different metal ions like Zn^{2+} (concentration from 1 to 5 mg/L) and Fe^{3+} (concentrations from 0.2 to 2.0 mg/L) about 50-60%. In many studies revealed the values of the maximum capacity adsorption, Q_{max} for metal ions removal from synthetic and real water. On Figure 9 Q_{max} values on cations uptake from water by natural zeolites from different deposits are summarized [76].

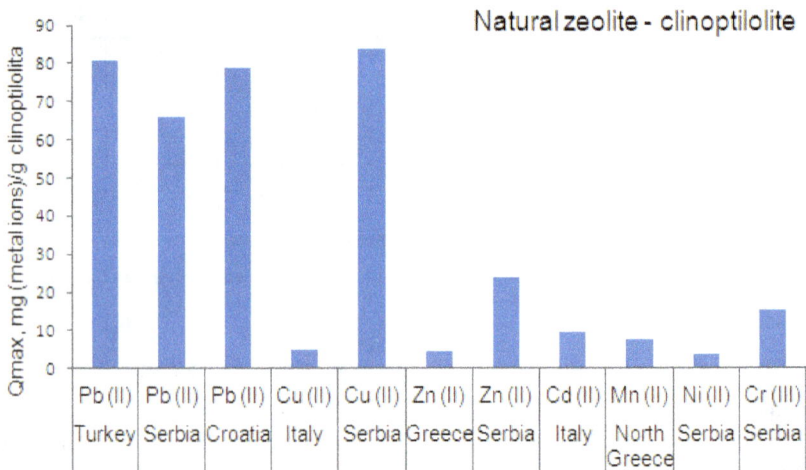

Figure 9. The uptake of some cations from water solution by natural zeolites from different deposits

Chemical pretreatment of natural zeolite (clinoptilolite) increases the removal efficiency of cations from water. The increase (in %) in the removal efficiency for different cations by modified zeolite compared to the natural zeolite is shown in Table 9 [76].

Zeolite	Treatment	Ion	Increasing of removal efficiency on cation uptake in relation to the natural zeolite - clinoptilolite
Clinoptilolite	NaCl	Pb^{2+}	34 %
	$NaNO_3$		10 %
	$FeCl_3$		50 %
	NaCl	Cu^{2+}	60 %
	NaCl	Zn^{2+}	44 %
	$NaNO_3$	Ni^{2+}	64 %
	NaCl	Cd^{2+}	33 %
	$NaNO_3$		34 %
	NaCl and NaOH	NH_4^+	45 %
	NaCl		33 %
	KOH and $Fe(NO_3)_3$	Mn^{2+}	71 %

Table 9. Effect of chemical pretreatment on the removal efficiency of cations by natural zeolite – clinoptilolite.

For efficient water treatment an important property of natural zeolite is regeneration through cyclic processes and their re-use. Recent studies of desorption efficiency of metal ions and regeneration natural zeolite - clinoptilolite are presented in Table 10 and indicate that the process is reversible in most cases [77].

Metal ion	Desorbing solution	Desorption efficiency , %
Pb^{2+}	3 M KCl	> 99,5
	0,5 M NaCl	95
Zn^{2+}	0,34 NaCl	24
	5 g/L EDTA	29
Cd^{2+}	1 M KNO_3	92
	1 M NaCl	97
Cu^{2+}	0,34 NaCl	20
	5 g/L EDTA	60
Ni^{2+}	0,1 M HCl	93

Table 10. Desorption efficiency of metal ions and regeneration natural zeolite-clinoptilolite.

Sorption of ions on to iron modified natural zeolite depends on the success of chemical modification and creation of iron hydroxide on its surface. Effect of solution pH and initial concentration of arsenic ions in water are very important for removal efficiency of arsenic ions [7, 44]. The sorption efficiency of arsenic ions on to iron modified zeolite (Serbia) decreases in the order: Pb (II)> As (V)> Mn (II)> As (III) ≥ Cr (VI) (Figure 10 (left). With increasing initial concentration of As (V) ions removal efficiency of As (V) ions decreases continuously with the use of natural zeolites from deposit D.Jesenje (Croatia) and V. Banja (Serbia). Also, at higher initial concentration of arsenic ions (300 mg/L) removal efficiency of

As (V) ions is 30% lower for natural zeolite (Croatia) in relation to the natural zeolite (Serbia) (Figure 10 (right)).

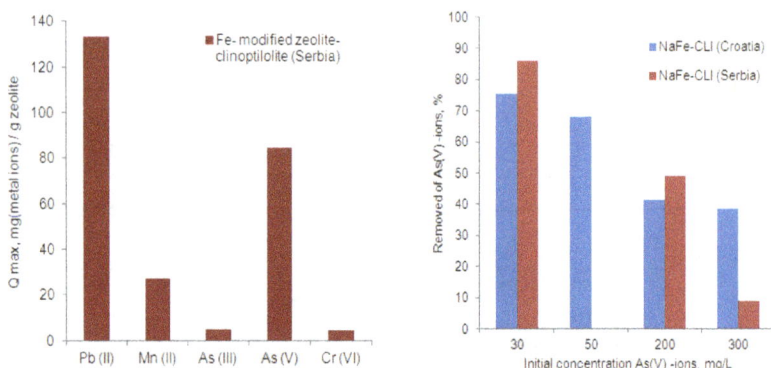

Figure 10. Sorption efficienty of ions on to iron modified zeolite – clinoptilolite

On the other hand, several anions (e.g. acetate, chloride, nitrate, phosphate or sulphate) have been witnessed as interfering with arsenate sorption. For arsenic ions binding on iron modified zeolite, the process was irreversible, i.e. stable complexes of arsenic and iron occurred, which is very important for permanent removal of arsenic ions from water solutions.

Ammonium is one of the main pollutants contained in municipal sewage, fertiliser factory wastewater and agricultural wastes. Although non-toxic, it is dangerous for the environment being one of the main causes of eutrophication. Cation-exchange has been identified as the NH_4^+ sorption mechanism with operating CEC values comprised from 0.1 to 2.3 meq/g (from 1.8 to 41.4 mg/g). Specifically, the highest values have been registered for sodium-exchanged clinoptilolite. Some studies have shown that ammonium ions bind on the zeolite framework through the combination of two parallel processes: ion exchange and adsorption. Zeolite modification by NH_4^+ ions is established by increasing ammonium ion concentrations and is determined by cation exchange capacity and concentrations of metal ions released from zeolite structure. Na^+ ion is most rapidly exchanged with ammonium ions from water solutions. Total CEC of zeolite was found to be up to 13 mg NH_4^+/g zeolite (deposit D. Jesenje, Croatia). Chemical treatment of natural zeolite with different concentrations of HCl has a significant influence on ammonium ion removal from solutions. Namely, the higher the acid concentration with which zeolite is treated, the weaker is the exchange capacity for NH_4^+ ions (Figure 11) [78].

Anions as well as in-dissociated or poorly dissociated compounds, both inorganic and organic (e.g. cyanide, flouride, borate, hydrocarbon derivatives) are noxious contaminants present in water mainly as a consequence of chemical-intensive industry or are of natural origin. Two main modifications have been noticed to make zeolites effective also for sorption of these species: i) metal doping and ii) treatment with cationic organic surfactants.

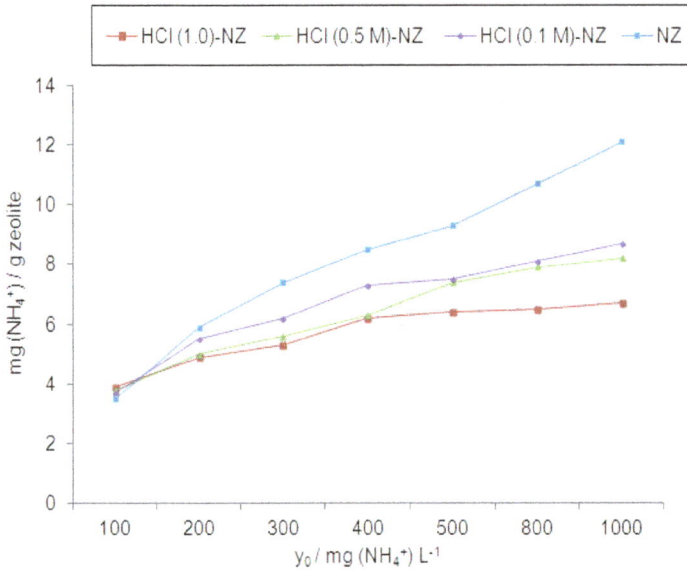

Figure 11. Removal of NH_4^+ -ion from solution using chemicaly treated natural zeolite with different concentrations of HCl.

In spite of that, clinoptilolite mineral as such has been surprisingly reported as sorbent for the removal of naturally occurring fluoride from well water Fluoride removal efficiency higher than 50 % (from 2 to 6 mg/L initial fluoride concentration) has been claimed. Metal doping (*e.g.* addition of oxides of lanthanides, aluminium or zirconium) has been cited as a possible way to improve natural zeolite removal efficiency. Two mechanisms have been inferred for fluoride sorption: hydroxide/fluoride exchange (due to steric and electrostatic similarities between these anions) [61] and fluoride occlusion [79].

6.1.2. Column process

In the *column process* there are several aspects that affect the dynamics of the cation uptake by zeolite, such factors are mainly solution (temperature, pH, the flow rate, the initial cation concentration being removed by zeolite, the pretreatment solution, the presence of other competing ions in the solution, characteristics of the heavy metal being removed by the zeolite) and solid specific factors (particle size, surface dust, impurities found in the zeolite sample, the pretreatment procedure type applied to the zeolite). Results of examinations in the column procedure are presented by breakthrough curves [80,81]. In Table 11 shown the removal efficiency of cations from different polluted water in column process.

Zeolite / Water	Process conditions	Ion	Removal efficiency	Regeneration	Ref.
Clinoptilolite (USA) wastewater	pH 6,5 two zeolite columns in series	NH_4^+	99%	NaCl, $CaCl_2$	[82]
Clinoptilolite (USA) municipal wastewaters	pH 4 – 8 Different particle size from 0.3 to 0.84 mm Flow rate 15 BV/h.	NH_4^+	95.7%	NaCl, NaOH $CaCl_2$	[83]
Clinoptilolite (Turkey)	Different particle size from 0,3 to 3,35 mm Flow rates 22-78 BV/h	NH_4^+	*QB = 0,085 - 0,891 meq NH_4^+ / g zeolite	-	[84]
Clinoptilolite (Turkey)	Different particle size from 0.3 to 0.6 mm	NH_4^+, Cs^+	*QB = 0.036 - 1.08 meq NH_4^+/g zeolite	-	[85]
Clinoptilolite (Cuba) piggery manure wastewater	Particle size - 5 mm Modified zeolite with NaCl, KCl, $CaCl_2$, $MgCl_2$	NH_4^+	90%	-	[86]
Clinoptilolite, Mordenite (New Zealand) wastewater	Different particle size from 0,25 to 2,83 mm	NH_4^+	95 % (Clinoptilolite) 55 % (Mordenite)	-	[87]
Phillipsite, Chabazite (Spain, Israel) wastewater	Flow rate 0,35 mL/min Flow rate 27 BV/h	NH_4^+	82 % (Phillipsite/ Chabazite) 95 % (Chabazite)	-	[88,89]
Clinoptilolite (USA) battery manufacturing wastewater	pH (5,5-6) Different particle size from 0.2 to 0.5mm Modified zeolite with NaCl Flow rate 10 BV/h	Pb^{2+}	≈90%	NaCl	[90]
Clinoptilolite (Slovakia) drinking water	Different particle size from 3 to 5 mm Thermaly treated clinoptilolite	NH_4^+	20%-40%	NaCl	[91]
Clinoptilolite (Turkey)	Different particle size from 0,875 to 2,0 mm Flow rates 10, 25, 50, 75 BV / h	NH_4^+,	*QB = 0,57 meq / g zeolita (smaller particle size) *QB = 0,38 meq / g zeolita (larger particle size)	-	[92]

Zeolite /Water	Process conditions	Ion	Removal efficiency	Regeneration	Ref.
Clinoptilolite (Greece)	Different particle size from 1,14 to 1,18 mm Flow rates 5, 10, 15 BV / h	Pb^{2+}, Cr^{3+}, Fe^{3+}, Cu^{2+}	*QB = 0.433 meq Pb^{2+} / g zeolite	-	[93]
Clinoptilolite (Greece) groundwater	Different particle size from 0,15 to 1mm	NH_4^+, Cu^{2+},Pb^2 +	82% (except Cu^{2+} ion)	-	[94]
Clinoptilolite, (New Zealand) wastewater	-	NH_4^+	*QB = 18.7 - 20.1 g NH_4^+/kg zeolite	HCl	[95]
Clinoptilolite (Turkey) drinking water	Different particle size from 1 to 2,8 mm	NH_4^+	-	NaCl, NaOH	[96]
Clinoptilolite (Iran) wastewater	Different particle size from 0,42 to 0,84 mm	NH_4^+	*QB = 16.31 - 19,5 mg NH_4^+/ g zeolita	NaCl	[97]
Clinoptilolite (Turkey) municipal wastewater	Different particle size from 1 to 2 mm Flow rate 0.5 mL/min	NH_4^+	*QB = 1,646 meq NH_4^+/ g zeolita	-	[98]
Clinoptilolite (Serbia)	Different particle size from 0,6 to 0,8 mm Flow rates 1, 2, 3 mL / min	Pb^{2+}	96 %	-	[99,100]
Clinoptilolite, mordenit (New Zealand) wastewater	Different particle size from 0,5 to 1,4 mm Modified zeolite with NaCl	NH_4^+	50 % (Clinoptilolite) 36 % (Mordenite)	-	[101]
Clinoptilolite (Croatia and Serbia) ground water	Particle size < 63 μm Flow rate 420 L/h Modified zeolite with NaCl	NH_4^+	99 -100%	NaCl	[102]
Clinoptilolite (China)	Different particle size from 52 to 61 μm	Pb^{2+}	*QB = mg Pb2+/g zeolite	-	[103]

Table 11. The removal efficiency of cations in column process (*QB - breakthrough capacity)

Two examples from the Eastern Croatia show a successful application of research results at a pilot plant for potable water. The ground water from the Valpovo region was previously treated by aeration and sand filtration for the removal of iron, manganese, ammonium and other present pollutants. The physical and chemical parameters of the ground water and ammonium concentration were determined in the ground water after the pre-treatment. The ammonium sorption efficiency of 100% was obtained after 10 to 90 h with the use of MNZ–SRB, but it was sharply decreased after 70 h with use of NZ–CRO or MNZ–CRO (Figure 12).

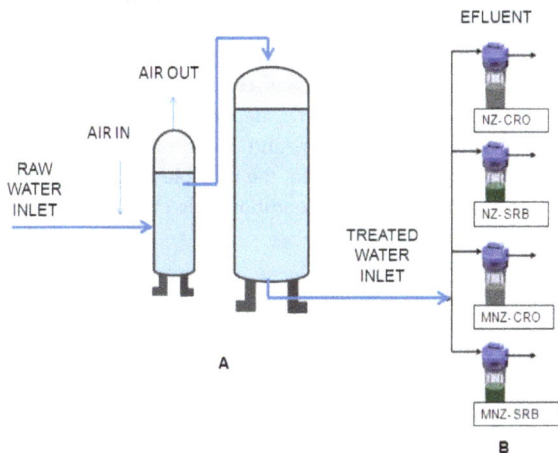

Figure 12. Scheme of drinking water production plant (A) and pilot plant "filter guards" (B) (NZ-, MNZ-CRO natural and modified zeolite from Croatia; NZ-, MNZ-SRB natural and modified zeolite from Serbia) [102].

Natural zeolite modification was carried out by applying a manganese dioxide layer, which is a technologically protected/patented procedure. Obtained results have shown that the removal efficiency of iron ions from drinking water can be increased by about 90 % by using modified zeolite.

The MnO_2-modified clinoptilolite (New Mexico, USA) appears to be a promising adsorbent for removing trace arsenic amounts from water. The removal efficiency obtained with the modified zeolite was doubled as compared to that obtained with the unmodified zeolite [30].

7. Conclusions

The unique ion exchange and adsorption properties, high porosity and excellent thermal stability of zeolites make them very suitable for many applications, also in water treatment processes. Many different studies have demonstrated their effectiveness in reducing the concentrations of contaminants (heavy metals, anions and organic matter) in water. The complexity of aquatic systems demands special attention in the selection and preparation of materials for water purification. The chemical behavior of natural zeolites in different aqueous environments, which was also a subject of recent geochemical and technological studies, additionally proved their applicability, although monitoring of pH and its changes, remains very important for their use of real environments. Namely, zeolites can interact with hydrogen or hydroxyl ions present in solutions and, as a consequence, certain physicochemical phenomena such as hydrolysis of solids, degradation, dissolution and even phase transformations can occur. All these phenomena again depend on the structural characteristics and the chemical composition of the used zeolite. Nowadays, modified natural zeolites are increasingly used also for biological treatment of water, precisely for surface binding of biological agents from water. Further research should be focused on the optimization of the surface modification procedures to raise their efficiency and to enhance the capability of regeneration. Furthermore, detailed characterization of natural and modified zeolites is needed to better understand the structure-property relationship. To open up new possibilities for their application, the possible further uses of used zeolites as well as the behavior of zeolites at extreme conditions, also at low temperatures, should be examined.

Author details

Karmen Margeta
University of Zagreb, Faculty of Chemical Engineering and Technology, Croatia

Nataša Zabukovec Logar
National Institute of Chemistry, Ljubljana, Slovenia

Mario Šiljeg
Croatian Environment Agency, Zagreb, Croatia

Anamarija Farkaš
Institute for International Relations, Zagreb, Croatia

Acknowledgment

The work described in the chapter was partly supported by the Ministry of Science, Education and Sports of the Republic of Croatia through the research project 125-1253092-3004, by the Ministry of Science, Education and Sports of the Republic of Croatia through the bilateral project (Croatia-Slovenia) "Natural zeolite in water nanotechnology", by the Slovenian Research Agency research program P1-0021, and by the EUREKA E!4208 project "PUREWATER". The financial support from the founding agencies as well as the scientific contribution of researchers involved in the projects is greatly acknowledged.

8. References

[1] Caputo D, Pepe F (2007) Experiments and data processing of ion exchange equilibria involving Italian natural zeolites: a review. Micropor.Mesopor.Mater. 105:222-231.

[2] Misaelides P (2011) Application of natural zeolites in environmental remediation: A short review. Micropor.Mesopor.Mater. 144:15-18.

[3] Fu F, Wang Q (2011) Removal of heavy metal ions from wastewaters: A review. J.Environ.Manage.92:407-418.

[4] Wang S, Peng Y (2010) Natural zeolites as effective adsorbents in water and wastewater treatment. Chem. Engin.J. 156:11-24.

[5] Moussavia G, Talebi S, Farrokhi M, Sabouti RM (2011) The investigation of mechanism, kinetic and isotherm of ammonia and humic acid co-adsorption onto natural zeolite. Chem. Engin.J. 171:1159-1169.

[6] Inglezakis VJ, Doula MK, Aggelatou V, Zorpas AA (2010) Removal of iron and manganese from underground water by use of natural minerals in batch mode treatment. Desalin. Water Treat. 18:341-346.

[7] Doula MK (2009) Simultaneous removal of Cu, Mn and Zn from drinking water with the use of clinoptilolite and its Fe-modified form. Water Res. 43:15:3659-3672.

[8] Li Z, Jean JS, Jiang WT, Chang PH, Chen CJ, Liao L (2011) Removal of arsenic from water using Fe-exchanged natural zeolite. J.Hazard. Mater. 187:318-323.

[9] Habuda-Stanić M, Kalajdžić B, Kuleš M, Velić N (2008) Arsenite and arsenate sorption by hydrous ferric oxide/polymeric material. Desalination 229:1-9.

[10] Doula MK, Dimirkou A (2008) Use of an iron-overexchanged clinoptilolite for the removal of Cu2+ ions from heavily contaminated drinking water samples. J.Hazard. Mater. 151:2-3:738-745.

[11] Mohapatra M, Anand S, Mishra BK, Giles DE, Singh P (2009) Review of fluoride removal from drinking water. J.Environ.Manage. 91:67-77.

[12] Bhatnagar A, Sillanpaa M (2011) A review of emerging adsorbents for nitrate removal from water. Chem. Engin. J. 168:493-504.

[13] Widiastuti N, Wu H, Ming Ang H, Zhang D (2011) Removal of ammonium from grey water using natural zeolite, Desalination 277:15-23.

[14] Widiastuti N, Wu H, Ming Ang H, Zhang D (2008) The potential application of natural zeolite for greywater treatment, Desalination 218:271-280.

[15] Margeta K, Vojnović B, Zabukovec Logar N (2011) Development of natural zeolites for their use in water-treatment systems. Recent patents of nanotech. 5:2:89-99.

[16] Breck DW (1974) Zeolite molekular sieves, John.Wiley&Sons, New York

[17] Baerlocher Ch, McCusker LB, Olson DH (2007) Atlas of Zeolite Framwork, Sixth Revised Edition, Elsavier, Netherlands.

[18] Farkaš A, Rožić M, Barbarić-Mikočević Ž (2005) Ammonium exchange in leakage waters of waste dumps using natural zeolite from the Krapina region, Croatia. J.Hazard.Mater. 117:1:25-33.

[19] Š. Cerjan-Stefanović, N. Zabukovec Logar, K. Margeta, N. Novak Tušar, I. Arčon; K. Maver, J. Kovač, V. Kaučič (2007) Structural investigation of Zn2+ sorption on clinoptilolite tuff from the Vranjska Banja deposit in Serbia. Micropor. Mesopor.Mater. 105:3:251-259.

[20] Wang S, Zhub ZH (2006) Characterisation and environmental application of an Australian natural zeolite for basic dye removal from aqueous solution. J.Hazard.Mater. 136:946-952.

[21] Davila-Jimenez MM, Elizalde-Gonzalez MP, Mattusch J, Morgenstern P, Perez-Cruz MA, Reyes-Ortega Y., Wennrich R, Yee-Madeira H (2008) In situ and ex situ study of the enhanced modification with iro of clinoptilolite-rich zeolitic tuff for arsenic sorption from aqueous solutions. J.Colloid.Interf.Sci. 322:527-536.

[22] Liang Z, Ni J (2009) Improving the ammonium ion uptake onto natural zeolite by using as integrated modification process. J.Hazard.Mater. 166:52-60.

[23] Coruh S, Senel G, Nuri Ergun O (2010) A comparison of the properties of natural clinoptilolites and their ion-exchange capacities for silver removal. J.Hazard.Mater. 180:486-492.

[24] Farías T, Ruiz-Salvador AR, Velazco L, Charles de Ménorval L, Rivera A (2009) Preparation of natural zeolitic supports for potential biomedical applications. Mater.Chem.Phy.118:322-328.

[25] Allen SJ, Ivanova E, Koumanova B (2009) Adsorption of sulfur dioxide on chemically modified natural clinoptilolite. Acid modification, Chem.Engin.J. 152:389-395.

[26] Filippidis A, Kantiranis N (2007) Experimental neutralization of lake and stream waters from N. Greece using domestic HEU-type rich natural zeolitic material. Desalination, 213:47–55.

[27] Sprynsky M, Golembiewski R, Trykowski G, Buszewski B (2010) Heterogeneity and hierarchy of clinoptilolite porosity. J.Phy.Chem.Solids, 71:1269-1277.

[28] Calvo B, Canoira L, Morante F, Martínez-Bedia JM, Vinagre C, García-González JE, Elsen J, Alcantara R (2009) Continuous elimination of Pb2+, Cu2+, Zn2+, H+ and NH4+ from acidic waters by ionic exchange on natural zeolite. J.Hazard.Mater. 166:2-3:619-627.

[29] Ruggieri F, Marín V, Gimeno D, Fernandez-Turiel JL, García-Valles M, Gutierrez L (2008) Application of zeolitic volcanic rocks for arsenic removal from water. Engineering Geology. 101:3-4: 245-250.

[30] Camachoa LM, Parrab RR, Denga Sh (2011) Arsenic removal from groundwater by MnO2-modified natural clinoptilolite zeolite: Effects of pH and initial feed concentration. J.Hazard.Mater. 189:286–293.

[31] Kumar Jha V, Hayashi Sh (2009) Modification on natural clinoptilolite zeolite for its NH4+ retention capacity. J.Hazard.Mater. 169:1-3:29-35.

[32] Oliveira CR, Rubio J (2007) New basis for adsorption of ionic pollutants onto modified zeolites. Mineral. Engin. 20:6:552-558.

[33] Ortega EA, Cheeseman Ch, Knight J, Loizidou M (2000) Properties of alkali-activated clinoptilolite. Cement.Concrete.Res. 30:10:1641-1646.

[34] Dávila-Jiménez MM, Elizalde-González MP, Mattusch J, Morgenstern P, Pérez-Cruz MA, Reyes-Ortega Y, Wennrich R, Yee-Madeira H (2008) In situ and ex situ study of the enhanced modification with iron of clinoptilolite-rich zeolitic tuff for arsenic sorption from aqueous solutions. J.Colloid.Interf. Sci. 322:2:527-536.

[35] Šiljeg M, Cerjan Stefanović Š, Mazaj M, Novak Tušar N, Arčon I, Kovač J, Margeta K, Kaučič V, Zabukovec Logar N (2009) Structure investigation of As(III)- and As(V)-species bound to Fe-modified clinoptilolite tuffs, Micropor. Mesopor.Mater. 118:1-3:408-415.

[36] Jeon CS, Baek K, Park JK, Oh YK, Lee SD (2009) Adsorption characteristics of As(V) on iron-coated zeolite. J.Hazard.Mater. 163:2-3:804-808.

[37] Chutia P, Kato S, Kojima T, Satokawa S (2009) Adsorption of As(V) on surfactant-modified natural Zeolites. J.Hazard.Mater. 162:1:204-211.

[38] Li Z, Kirk Jones H, Zhang P, Bowman RS (2007) Chromate transport through columns packed with surfactant-modified zeolite/zero valent iron pellets. Chemosphere 68:1861–1866.

[39] Stefanakis AI, Akratos CS, Gikas GD, Tsihrintzis VA (2009) Effluent quality improvement of two pilot-scale, horizontal subsurface flow constructed wetlands using natural zeolite. Micropor.Mesopor.Mater. 124:1-3:131-143.

[40] Bosinceanu R, Sulitanu N. (2008) Synthesis and characterization of FeO(OH)/Fe3O4 nanoparticles encapsulated in zeolite matrix, J. Optoelectron. Advanc. Mater. 10:3482-3486.

[41] Rožić M., Miljanic S (2011) Sorption of HDTMA cations on Croatian natural mordenite tuff. J.Hazard.Mater. 185:1:423-429.

[42] Grossal PR, Eick MJ, Sparks DL, Goldberg S, Ainsworth CC (1997) Environ.Sci.Techol. 31:321-326.

[43] Fenforf S, Eick MJ, Grossal PR, Sparks DL (1997) Environ. Sci. Technol, 31: 315-320.

[44] Kragović M, Daković A, Sekulić Ž, Trgo M, Ugrina M, Perić J, Gatta GD (2012) Removal of lead from aqueous solution by using the natural and Fe(III) modified zeolite. Appl.Surf.Sci. 258:8:3667-3673.

[45] Rozić M, Cerjan-Stefanović Š, Kurajica S, Mačefat M, Margeta K, Farkaš A (2005) Decationization and dealumination of clinoptilolite tuff andammonium exchange on acid-modified tuff. J. Colloid Interf. Sci. 284:1:48-56.

[46] Inglezakis VJ (2005) The concept of "capacity" in zeolite ion-exchange systems. J.Colloid Interf. Sci. 281:68-79.

[47] Armbruster T (2001) Clinoptilolite-heulandite:applications and basic research. Stud.Surf.Sci.Cataly. 135:13-27.

[48] Langella A, Pansini M, Cappelletti P, de Gennaro B, de Gennaro M, C. Colella (2000) NH4+, Cu2+, Zn2+, Cd2+ and Pb2+ exchange for Na+ in sedimentary clinoptilolite,North Sardinia, Italy. Micropor. Mesopor.Mater. 37:337-343.

[49] Collela C (1996) Ion Exchange Equilibria in Zeolite Materials. Mineral. Deposita 31:554.

[50] Hafez MB, Nazmy AF, Salem F, Eldesoki M (1978) Fixation mechanism between zeolite and some radioactive elements. J.Radioanal.Chem. 47:115-122.

[51] Al-Anbera M, Al-Anberb ZA (2008) Utilization of natural zeolite as ion-exchange and sorbent material in the removal of iron. Desalination 225:70-81.

[52] Lina J, Zhan Y, Zhu Z, Xing Y (2011) Adsorption of tannic acid from aqueous solution onto surfactant-modified zeolite. J.Hazard.Mater. 193:102-111.

[53] Jiménez-Cedillo MJ, Olguín MT, Fall Ch, Colín A (2011) Adsorption capacity of iron- or iron–manganese-modified zeolite-rich tuffs for As(III) and As(V) water pollutants. Appl. Clay Sci. 54:206-216.

[54] Perić J, Trgo M, Vukojević Medvidović N (2004) Removal of zinc, copper and lead bynatural zeolite-a comparison of adsorption isotherms. Water Res. 38:1893-1899.

[55] Yousefa RI, El-Eswed B, Al-Muhtasebc AH (2011) Adsorption characteristics of natural zeolites as solid adsorbents for phenol removal from aqueous solutions: Kinetics, mechanism, and thermodynamics studies. Chem. Eng.J. 171:1143-1149.

[56] Noroozifar M, Khorasani-Motlagh M, Fard PA (2009) Cyanide uptake from wastewater by modified natrolite zeolite–iron oxyhydroxide system: Application of isotherm and kinetic models. J.Hazard.Mater. 166:1060-1066.

[57] Misaelides P, Zamboulis D, Sarridis Pr, Warchoł J, Godelitsas A (2008) Chromium (VI) uptake bypolyhexamethylene-guanidine-modified natural zeolitic materials. Micropor.Mesopor.Mater. 108:162-167.

[58] Chutia P, Kato Sh, Kojima T, Satokawa S (2009) Adsorption of As(V) on surfactant-modified natural Zeolites. J.Hazard.Mater. 162:204-211.

[59] Widiastuti N, Wu H, Ming Ang H, Zhang D (2011) Removal of ammonium from greywater using natural zeolite. Desalination 277:15-23.

[60] Lina J, Zhan Y, Zhub Z (2011) Adsorption characteristics of copper (II) ions from aqueous solution onto humic acid-immobilized surfactant-modified zeolite. Colloids Surf. A: Physicoch. Eng. Aspect 384:1-3:9-16.

[61] Karatas M (2012) Removal of Pb(II) from water by natural zeolitic tuff: Kinetics and Thermodynamics. J.Hazard. Mater.199-200:383-389.

[62] Rajić N, Stojakovic Dj, Jevtić S, Zabukovec Logar N, Kovač J, Kaučič V (2009) Removal of aqueous manganese using the natural zeolitic tuff from the Vranjska Banja deposit in Serbia. J.Hazard.Mater. 172:1450-1457.

[63] Baskan MB, Pala A (2011) Removal of arsenic from drinking water using modified natural zeolite. Desalination 281:396-403.

[64] Malekian R, Abedi-Koupai J, Saeid Eslamian S, Farhad Mousavi S, Abbaspour KC, Afyuni M (2011) Ion exchange process for ammonium removal and release using natural Iranian zeolite. Appl. Clay Sci. 51:323-329.

[65] Kurniawan TA, Chan GYS, Lo WH, Babel S (2006) Physico–chemical treatment techniques for wastewater laden with heavy metals. Chem.Engin.J. 118:83-98.

[66] Fu F, Wang Q (2011) Removal of heavy metal ions from wastewaters: A review. J. Environ. Manage. 92:407-418.

[67] Choo KH, Lee H, Choi SJ (2005) Iron and manganese removal and membrane fouling during UF in conjunction with prechlorination for drinking water treatment. J.Memb.Sci. 267:18-26.

[68] Al-Anbera M, Al-Anberb ZA (2008) Utilization of natural zeolite as ion-exchange and sorbent material in the removal of iron. Desalination 225:70-81.

[69] Ćurković L, Cerjan Stefanović Š, Filipan T (1997) Metal ion exchange by natural and modified zeolites. Water Res. 31:6:1379-1382.

[70] Leyva-Ramos R, Jacobo-Azuara A, Diaz-Flores PE, Guerrero-Coronado RM, Mendoza-Barron J, Berber-Mendoza MS (2008) Adsorption of chromium (VI) from an aqueous solution on a surfactant-modified zeolite. Coll.Surf. A: Physicochem. Eng. Aspects 330:35-41.

[71] Woinarskia AZ, Snapeb I, Stevensa GW, Starkb SC (2003) The effects of cold temperature on copper ion exchange by natural zeolite for use in a permeable reactive barrier in Antarctica. Cold Reg. Sci.Technol. 37:159-168.

[72] Rajić N, Stojaković Dj, Jovanović M, Zabukovec Logar N, Mazaj M, Kaučič V (2010) Removal of nickel(II) ions from aqueous solutions using the natural clinoptilolite and preparation of nano-NiO on the exhausted clinoptilolite. Appl. Surf. Sci. 257:1524-1532.

[73] Shavandi MA, Haddadian Z, Ismail MHS, Abdullah N, Abidin ZZ (2012) Removal of Fe(III), Mn(II) and Zn(II) from palm oil mill effluent (POME) by natural zeolite. J. Taiwan Ins.Chem. Engin. (in press).

[74] Dal Bosco SM, Jimenez RS, Carvalho WA (2005) Removal of toxic metals from wastewater by Brazilian natural scolecite. J.Coll.Interf.Sci. 281:424-431.

[75] Sheta AS, Falatah AM, Al-Sewailem MS, Khaled EM, Sallam ASH (2003) Sorption characteristics of zinc and iron by natural zeolite and bentonite. Micropor. Mesopor. Mater. 61:127-136.

[76] Taffarel SR, Rubio J (2009) On the removal of Mn2+ ions by adsorption onto natural and activated Chilean zeolites. Mineral. Engin. 22:336-343.

[77] Katsou E, Malamis S, Tzanoudaki M, Haralambous KJ, Loizidou M (2011) Regeneration of natural zeolite polluted by lead and zinc in wastewater treatment systems. J.Hazard.Mater. 189:773-786.

[78] Farkaš A, Rožić M, Barbarić-Mikočević T (2005) Ammonium exchange in leakage waters dumps using natural zeolite from the Krapina region,Croatia, J.Hazard.Mater. 117:1:25-33.

[79] Perego C, Bagatin R, Tagliabue M, Vignola R (2012) Zeolites and Related Mesoporous Materials for Multi-talented Environmental Solutions, Micropor.Mesopor.Mater., doi: http://dx.doi.org/10.1016/j.micromeso.2012.04.048 (Accepted Manuscript).

[80] Vukojević Medvidović N, Perić J, Trgo M (2006) Column performance in lead removal from aqueous solution by fixed bed of natural zeolite-clinoptilolite. Sep. Purif. Tech. 49:3:237-244.

[81] Vukojević Medvidović N, Perić J, Trgo M, Mužek N (2007) Removal of lead ions by fixed bed of clinoptilolite – The effect of flow rate. Micropor.Mesopor.Mater. 105:3:298-304.

[82] Mercer BW, Ames LL, Touhill CJ, Van Slyke WJ, Dean RB (1970) Ammonia Removal from Secondary Effluents by Selective Ion Exchange. J.Wat.Pollut.Contr.Fed. 42:20:95-107.

[83] Koon JH, Kaufmann WJ (1975) Ammonia Removal from Municipal Wastewaters by Ion Exchange. J.Wat.Pollut.Contr.Fed. 47:3:448-465.

[84] El Akrami HA (1991) Column Studies of Ammonium Ion Exchange on Clinoptilolite, MS. Thesis, METU, Ankara.

[85] Abusafa A (1995) Column Studies of Ammonium and Cesium Exchanges on Bigadic Clinoptilolite, MS. Thesis, METU, Ankara.

[86] Milan Z, Sénchez E, Weiland P, de Las Pozas C, Borja R, Mayari R, Rovirosa N (1997) Ammonia Removal from Anaerobically Treated Piggery Manure by Ion Exchange in Columns Packed With Homoionic Zeolite. Chem.Engin.J. 66:1:65-71.

[87] Nguyen ML, Tanner CC (1998) Ammonium Removal from Wastewaters Using Natural New Zealand Zeolites. New Zealand. J.Agri.Res. 41:3:427-446.

[88] Roberto J, Hernández S, M. Andrés J, Ruiz C (2009) Ion exchange uptake of ammonium in wastewater from a Sewage Treatment Plant by zeolitic materials from fly ash. J.Hazard.Mater. 161:2-3:781-786.

[89] Lahav O, Green M (1998) Ammonium removal using ionexchange and biological regeneration. Water Res. 32:2019–2028.

[90] Petruzzelli D, Pagano M, Tiravanti G, Passino R (1999) Lead Removal and Recovery from Battery Wastewaters by Natural Zeolite Clinoptilolite. Sol.Extra.Ion Exchan.17:3:677-694.

[91] Abd El-Hady HM, Grünwald A, Vlčková K, Zeithammerová J (2001) Clinoptilolite in Drinking Water Treatment for Ammonia Removal. Acta Polytech. 41;1;41-45.

[92] Demir A, Günay A, Debik E (2002) Ammonium Removal from Aqueous Solution by Using Packed Bed Natural Zeolite. Water SA, 28:3:329-336.

[93] Inglezakis VJ, Loizidou MD, Grigoropoulou HP (2002) Equilibrium and Kinetic Ion Exchange Studies of Pb2+, Cr3+, Fe3+ and Cu2+ on Natural Clinoptilolite. Water Res. 36:11:2784-2792.

[94] Park JB, Lee SH, Lee JW, Lee CY (2002) Lab Scale Experiments for Permeable Reactive Barriers against Contaminated Groundwater with Ammonium and Heavy Metals Using Clinoptilolite (01-29B). J.Hazard.Mater. 95:1-2:65-79.

[95] Bolan NS, Mowatt C, Adriano DC, Blennerhassett JD (2003) Removal of Ammonium Ions from Fellmongery Effluent by Zeolite. Communi. Soil Sci. Plant Analy. 34:13-14:1861-1872.

[96] Turan M, Celik MS (2003) Regenerability of Turkish Clinoptilolite for Use in Ammonia Removal from Drinking Water. J.Water Sup. Res.Techn. – AQUA 52:10:59-66.

[97] Rahmani AR, Mahvi AH, Mesdaghinia AR, Nasseri N (2004) Investigation of Ammonia Removal from Polluted Waters by Clinoptilolite Zeolite. J. Environ.Sci.Tech. 1:2:125-133.

[98] Sarioglu M (2005) Removal of Ammonium from Municipal Wastewater Using Natural Turkish (Dogantepe) Zeolite. Sep.Purific.Tech., 41:1-11.

[99] Vukojević Medvidović N, Perić J, Trgo M (2008) Testing of Breakthrough Curves for Removal of Lead Ions from Aqueous Solutions by Natural Zeolite-Clinoptilolite According to the Clark Kinetic Equation. Sep. Sci. Technol. 43:4:944-959.

[100] Perić J, Trgo M, Vukojević Medvidović N, Nuić I (2009) The Effect of Zeolite Fixed Bed Depth on Lead Removal from Aqueous Solutions. Sep. Sci.Technol. 44:13:3113-3127.

[101] Miladinović N, Weatherley LR (2007) Intensification of Ammonia Removal in a Combined Ion-Exchange and Nitrification Column. Chem.Engin.J.135:1-20:15-24.

[102] Šiljeg M, Foglar L, Kukučka M (2010) The Ground Water Ammonium Sorption onto Croatian and Serbian Clinoptilolite. J.Hazard.Mater. 178:1-3:572-577.

[103] Tao YF, Qiu Y, Fang SY, Liu ZY, Wang Y, Zhu JH (2010) Trapping the Lead Ion in Multi-Components Aqueous Solution by Natural Clinoptilolite. J.Hazard.Mater. 180:1-3:282-288.

Chapter 5

Conventional Media Filtration with Biological Activities

Ivan X. Zhu and Brian J. Bates

Additional information is available at the end of the chapter

1. Introduction

Conventional gravity filtration takes advantage of gravity of water as a driving force, and is classified as slow media filtration or rapid media filtration. A slow sand filter is simple in design, construction, and operation. It is simply a filter box (usually made of concrete) containing sand media supported by a layer of gravel with appurtenances to deliver and remove water. The first recorded use of slow sand filters for a citywide water supply was in 1804 by John Gibbs in Paisley, Scotland (Barrett et al. 1991). Slow sand filters as their name implies, is accomplished with a relatively slow speed of filtration (typically 0.1 to 0.2 meters per hour) with 1 to 2 meters media depth. Because of the slow filtration rate, the head loss buildup is gradual and usually takes several months to achieve a significant level and form a condensed layer called *schmutzdecke* on media surface, which will be removed manually with media replenishment. Therefore the filter runtime is usually in the magnitude of months as opposed to 24-48 hours with rapid sand filters. A rapid sand filter is operated in a much higher speed (typically 2 to 10 meters per hour) with periodically backwashing the filter to recover headloss which builds up much faster due to a higher filtration speed. Backwashing is initiated normally by set time intervals, headloss across a filter media bed, or filter effluent turbidity. For both slow and rapid filters, filter run times are highly dependent on the freeboard on the top of the media. The freeboard is 1-3 meters, designed according to water qualities, especially turbidity and total suspended solids.

Sand, anthracite, and granular activated carbon or their combination was used as media with proper gradation. Some proprietary media as pumice, expanded clay, diatomaceous earth, and ceramic have also been applied. Characteristics of different media are shown in Table 1.

When media becomes clogged and dirty, the best way is to backwash the filter and flush dirt out. Backwash is classified as fluidized backwash and sub-fluidized backwash. Fluidized backwash requires a higher water rate to expand media bed usually by 20-30%, where the minimal fluidization water velocity is directly related to media type, media size, uniformity coefficient, water temperature, and salinity (important factor for sea water filtration) (Logsdon et al. 2002). However, a fluidization test is always recommended for precisely identifying the backwash rate to achieve the desired fluidization. Usually sub-fluidized backwash is applied to coarse media with 15-20 m/h, where media will only move or rotate locally and expand slightly as fluidization of larger media require an extremely high water velocity (Logsdon et al. 2002). Backwash can be water only backwash or air assisted backwash, the later of which has gained popularity because of water conservation and effectiveness of media cleaning. The air scouring (usually at 70-90 m/h) can be before water wash, after water wash, simultaneously with water, or combination thereof. Logsdon et al. (2000) summarized the typical water and air rates for backwashing filters with different types of media.

With concurrent air scouring and water wash, the filter bed undergoes a "collapse pulsing action" under optimal air and water rates, which can be predicted according to a set of empirical equations applicable to different media gradation and water temperature (Amirtharajah 1993). Obtaining and sustaining collapse pulse action within the backwashing process is optimum for the removal of particles from the media grains. The collapse pulsing action can be described as follows: the air bubble exits the air delivery device (orifice) and expands under the weight of the media. When the air bubble expands, the media expands slightly within the vicinity of the bubble, and the bubble collapses and reforms just above its original location. This collapsing is due to the weight of the media. The bubble reforms above its original location because the media is only partially expanded. Just prior to collapsing, high local water velocities occur at the perimeter of the bubble. Simultaneous to bubble collapse, media particles rush together and collide in a violent scouring action. This creates a "pulsation" in the bed. The bubble travels on upward, expands, collapses, and re-forms again, and repeats the process several times as it passes through the bed.

Media	Density (g/cm^3)	Major Constituents	Specific Surface Area (SSA) and its References	
			SSA (BET) (m^2/g)	References
Garnet	3.6-4.2	Nesosilicates		(Logsdon et al. 2002)
Sand	2.6	SiO$_2$	0.04 ±0.001 (0.25 and 0.5 mm size)	(Jerez et al. 2006)
Anthracite	1.6	Carbon	6-7 (0.2-0.4 mm size)	(Davidson et al. 1996)
GAC	1.3-1.5	Carbon	720 (<1 mm size) 900 (0.149 mm size) 928 (0.15 and 0.25 mm size)	(Gergova et al. 1993) (Oliveira et al. 2002) (Tang et al. 2004)
Diatomaceous earth	1.0-1.6	SiO$_2$, Al$_2$O$_3$, Fe$_2$O$_3$	27.8 (106–250 mm size)	(Al-Ghouti et al. 2003)
Expanded clay	1.0-1.6	SiO$_2$, Al$_2$O$_3$, Fe$_2$O$_3$	398 (10-100 µm size)	(Occelli et al. 2002)
Pumice	0.4	Highly vesicular texture glass	2.1-14.2 (<63-1000 µm size)	(Kitis et al. 2007)

Table 1. Characteristics of different media used for water filtration

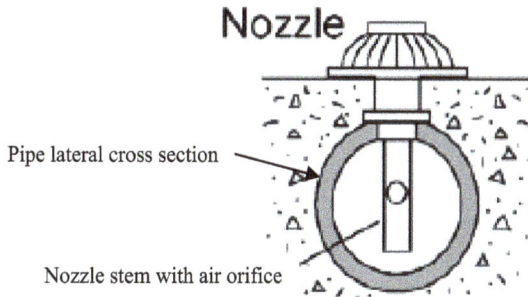

Figure 1. Sketch of typical nozzle design on a filter floor

To fully clean the filter media without forming dead zones, mud balls, media encrustation, and boiled media spots, even distribution of air and water during backwashing is critical. Fundamentally, there are two types of designs for simultaneous air and water distribution, nozzles and underdrain (with integrated dual channels). Nozzles (Figure 1) usually have slot sizes in the range 0.25–0.5 mm to minimize the risk of sand penetration. Nozzle arrangement density on the floor depends on the type of nozzles and is greater than about 35 nozzles/m² (Ratnayaka et al. 2009). However, the design of nozzle slots needs to be considered carefully to prevent fouling. There are several other underdrain systems, mostly of proprietary designs, successfully used in many parts of the world. An example is the design by Leopold (a Xylem brand) which comprises underdrain blocks (each block approximately 1 foot ×1 foot ×4 feet), formed from high-density polyethylene (HDPE), which snap lock together to form water resistant long laterals (Figure 2). The blocks incorporate a dual lateral design with a water recovery channel that ensures uniform distribution of concurrent air and water even over laterals up to 42 feet or 12.8 m. The blocks can be fitted with a porous HDPE IMS® Cap on top that helps to eliminate the need for support gravel (Figure 2).

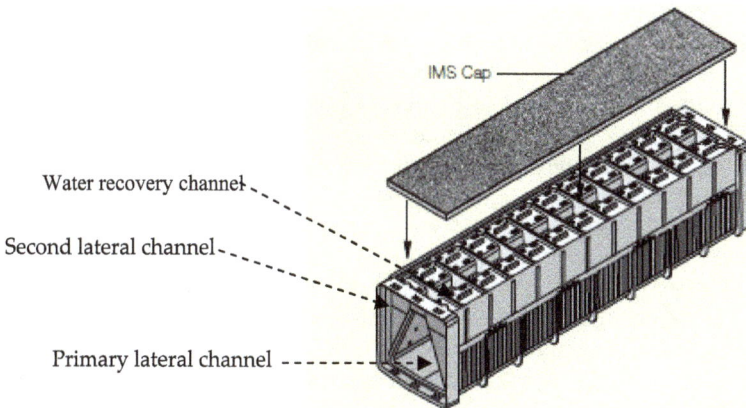

Figure 2. Sketch of typical underdrain design with dual channels (courtesy of Leopold, a Xylem brand)

2. Filtration with biological activity

Biological water and wastewater treatment processes are based on the growth of microbial communities capable of metabolizing contaminants through mediating oxidation-reduction reactions. The oxidants (electron acceptors) are normally oxygen, nitrate, perchlorate, sulfate, and Fe (III); the reductants (electron donors) are normally organic matter, trace organic compounds, ammonia, As (III), and iron (II) and Mn (II), etc. In a fixed-film biological process, biofilms are developed on media such as sand, anthracite, granular activated carbon (GAC), or membranes. A biofilm process mainly consists of two simultaneous steps, substrate diffusion and biological reaction, as illustrated in Figure 3. Electron donors and acceptors diffuse from bulk fluid into the biofilm and are metabolized by microbial cells in the biofilm, as a result of which the diffusion profiles are parabolic.

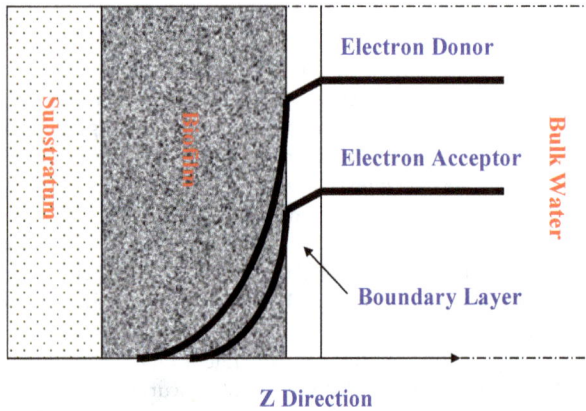

Figure 3. Schematic diagram of substrate diffusion in a biofilm attached to a solid substratum

Biologically active filters (BAFs) are essentially of the same physical structures as rapid gravity filters except BAFs are maximized with biological activities without backwashing the filters with chlorinated water or with no pre-chlorination. BAFs have been used for decades in North America and Europe in drinking water treatment, but have drawn more attention only recently. Regulatory and customer acceptance remain an issue because of the concern of microbial sloughing and breakthrough. A recent survey conducted by the AWWA Research Foundation indicated that 44% of the respondents believed biological processes in the drinking water industry were not accepted and 25% believed they were. Major operational concerns were breakthrough of pathogens and sloughing of bacteria (Evans et al. 2008). However, coliform bacteria were rarely observed in BAF effluents in laboratory studies, indicating that coliform organisms were eliminated by the microbial activity in the filters because of the competition for limited nutrients (Camper et al. 1985; Rollingger and Dott 1987). A pilot study demonstrated that biologically active filters reduced microbial activities in distribution systems (Characklis 1988). Furthermore, French experience indicated that removal of biodegradable materials resulted in a lesser amount of microbial presence in the distributions systems (Bourbigot et al. 1982). Comparison of physicochemical and biological treatments indicated that biological treatment limited mutagenic generation (Carraro et al. 2000).

3. Applications and design parameters

Perhaps, slow sand filters (SSFs) were the earliest application of a biological process in drinking water treatment. The major function of a filter occurs at the surface layer (Schmutzdecke) of the sand bed which contains a zoogeal jelly in which biological activities are highest (Babbitt and Doland 1939). Full scale experience indicated NOM removal was 15±5 mg/L by slow sand filters (Collins et al. 1992). Coliform reduction was 2-4 log (Barrett et al. 1991), and E Coli reduction was found strongly correlated with carbon dioxide respiration in the top 2.5 cm media and protistan abundance in the top 0.5 cm Schmutzdecke (Unger and Collins 2008). With over 150 years of history, river bank filtration (RBF) showed efficient organic substance removal in full scale plants in the Netherlands and Germany (Piet and Zoeteman 1980; Sontheimer 1980). River bank filtration removed TOC by 33-86% and disinfection by-product formation potential by 30-100% at the wells in several drinking water utilities in the US (Partinoudi and Collions 2007; Weiss et al. 2003).

A previous review of biological processes in drinking water treatment summarized that a wide range of contaminants can be removed through biological oxidation and reduction of dissolved constituents including natural organic matter (NOM), ammonia, nitrate, perchlorate, and iron and manganese, where operating parameters were discussed (Bouwer and Crowe 1988). Additionally, BAFs were reported to remove trace organic compounds, halogenated organics, perchlorate, and arsenic. A BAF usually does not require the addition of other chemicals for oxidizing and removing of contaminants. It does not require close monitoring of a breakthrough point, as in conventional column adsorption processes. Some organics adsorbed in activated carbon particles can be degraded by microorganisms attached on the activated carbon, or through enzymatic reaction during normal operation and hence create some active adsorption sites (Perrotti et al. 1974; Rice and Robson 1982; Rodman et al. 1978). This process is referred as biological regeneration. The service life of activated carbon can be extended by biologically regenerating exhausted carbon. The treated water from the BAF is unlikely to produce undesirable disinfection by-products and bacteria re-growth in the water distribution system (Dussert and Van Stone 1994; Scholz and Martin 1997).

3.1. Removal of natural organic matter (NOM)

Natural organic matter (NOM), consisting of humic acid, fulvic acid, carbohydrates, and other natural compounds, is present in natural water sources and is a precursor of disinfection by products (DBPs). DBPs are compounds formed when strong oxidants such as chlorine and ozone come into contact with NOM. Epidemiological evidences supported an association between chlorinated water or trihalomethanes and bladder cancer (Cantor et al. 1987; Cantor et al. 1999; Doyle et al. 1997). The most prevalent DPBs that form as a result of contact between organic carbon and chlorine include total trihalomethanes (TTHMs) and haloacetic acids (HAA5). To mitigate public exposure to these compounds, US Environment Protection Agency (EPA) has developed regulations restricting their concentrations at all points in the distribution system. The initial legislation formed for these restrictions is known as Stage 1 Disinfectants and Disinfection Byproducts Rule. A secondary stage for the DBP rule had also been promulgated by EPA in 2006. Stage 2 Disinfectants and Disinfections Byproducts Rule was applied as an addition to the continual improvement of safety in drinking water standards in the United States. Amendments to Stage 1 DBP rule include: (a) requiring annual averages at every point in the distribution system to adhere to the predefined maximum contamination levels (MCLs), (b) escalating the sampling frequency for communities with larger populations, and (c) the utility's distribution system must be evaluated to identify locations with elevated DBP concentrations.

To remove DBP precursors, biologically active filtration using different media was employed worldwide, usually assisted by pre-ozonation which increased assimilable organic carbon (AOC) or biodegradable organic matter (BOM) which was subsequently metabolized by biofilms in biofilters (Carlson and Amy 2001; Goel et al. 1995; Weiss et al. 2003). The design parameters are empty bed contact time (EBCT), media selection, media configuration, backwash regime selection, temperature in addition to pre-ozone doses. A summary of design parameters and media selection and observations in previous studies were provided in Table 2. Majority of the studies showed that an EBCT of 10 minutes should be used in process design to achieve 30-50% TOC removal with GAC. When combined with pre-ozonation, an EBCT of 5 minutes appeared sufficient (Hozalski et al. 1995; Laurent et al. 1999; Rittmann et al. 2002; Weiss et al. 2003).

The design of biologically active filters should be based on the achievement of one of the following criteria:

- Maximal removal of DOC to reduce the formation of DBPs;
- Maximal removal of BOM to minimize the risk of biological re-growth in the distribution system;
- Maximal removal of potential carcinogenic ozonated by-products (OBPs) such as formaldehyde.

Direct comparison showed that the GAC/sand filter produced better performance than the anthracite/sand filter (Rittmann et al. 2002) and the GAC/Sand filter out performed anthracite/sand filter by 11% for AOC removal (Weiss et al. 2003). GAC media showed better resistance to temperature at 1-3 °C in terms of oxalate and TOC removal compared with anthracite (Emelko et al. 2006). The BOM (10% acetate and 90% other organic matter with a maximal degradation rate less than one-tenth that of acetate) removal was reduced from 55% at 22.5 °C to 12% at 6 °C in a sand filter at EBCT 7.5 minutes (Hozalski et al. 1999).

Emelko et al. (2006) studied the effect of backwash and temperature on full scale biofiltration, and concluded that biodegradable organic material (BOM) removal was not influenced by backwash regimes even though some biomass expressed by phospholipid was lost during backwash with air scouring. Others also concluded that backwashing did not have noticeable impact on BOM removal because no more than 25% of biomass was washed out (Hozalski et al. 1999; Rittmann et al. 2002). However, microbial communities in the filters and during the operating condition shifts were not investigated in this study. In another study, it was found that backwashing caused changes in the relative compositions of microorganisms in a GAC biofilm in the top layer of the bed and reduced the attached bacterial abundance to 64% (Kasuga et al. 2007). The relative abundances of some terminal-restriction fragments (T-RFs) increased such as the *Planctomycetes*-derived fragment; however, some decreased, which included the *β-proteobacteria*-derived fragments (Kasuga et al. 2007).

Nutrient levels were also shown to influence the process efficiency. In a full scale study at Daugava water treatment plant in Riga, Latvia, the process including ozonation and biofiltration was not efficient for removal of dissolved organic carbon (DOC) from waters with a high amount of humic substances likely due to phosphorus limitation (Juhna and Rubulis 2004).

Phosphorus supplementation in a pilot study decreased biofilter terminal headloss by ~15 percent relative to the control likely result of reduced EPS formation in the filter, and decreased contaminant breakthrough relative to the control biofilter, including MIB (~75 percent less breakthrough), manganese (~90 percent less breakthrough), and DOC (~15 percent less breakthrough) (Lauderdale et al. 2011).

Study Scales	Parameters	Findings	References
Pilot using filters of granular activated carbon (GAC), GAC/sand, anthracite/sand	GAC/sand filter: EBCT 5 minutes with 29% TOC removal; EBCT 10 minutes with 33% TOC removal; EBCT 15 minutes with 42% TOC removal; EBCT 20 minutes with 51% TOC removal 0.3-1.0 mg O_3/mg TOC	Pre-ozonation increased Assimilable Organic Carbon (AOC) in influent, but also increased BAF effluent AOC relative to non-ozonated influent water. GAC/Sand filter was better than anthracite/sand filter by 11% for AOC removal.	(Weiss et al. 2003)
IRWD pilot facility in Santa Ana, California, including BAF following ozonation.	EBCT from 3.5 to 9 minutes 1.0-1.8 g O_3/g TOC	Up to 90% of color was removed and up to 38% DOC was removed; GAC biofilter gave better performance than anthracite.	(Rittmann et al. 2002)
Laboratory-scale batch degradation tests	0-7.3 mg O_3/mg TOC	Biodegradability of four NOM sources was improved by ozonation in the range of 0-7.3 mg	(Goel et al. 1995)

Study Scales	Parameters	Findings	References
		O_3/mg TOC. Degradation of high molecular weight organics were more influenced by ozonation	
Laboratory-scale biologically active sand (ES 0.5 mm) filters.	EBCT from 4 to 20 minutes 2-4 mg O_3/mg TOC	Ozonated NOM removal was significantly affected by the sources of the organic carbon independent of EBCT	(Hozalski et al. 1995)
Laboratory-scale biodegradability study with Ohio River	0.6-1.0 mg O_3/mg TOC (optimal for DBP reduction) 2.0 mg O_3/mg TOC (maximal for AOC)		(Shukairy et al. 1992)
Laboratory-scale biologically active glass beads and sand (ES 0.52 mm) filters.	EBCT 7.5 minutes 0.58±0.12 mg O_3/mg TOC	30% TOC removal was achieved. Perfromance was not impaired by backwash.	(Hozalski et al. 1999)
Pilot including biologically active filters (expanded clay 0.5-2.5 mm) following ozonation.	EBCT 11-54 minutes 1.0-1.7 mg O_3/mg TOC	EBCT did not have a significant impact; TOC removal 18-37% ; majority (80%) of BOM were removed;	(Melin and Odegaard 1999)
Full scale GAC filters at River dune Water Works	EBCT 20 minutes with two stages 0.35-0.45 mg O_3/mg TOC	50% TOC removal; successive reactivation of GAC was still effective	(van der Hoek et al. 1999)
Full scale BAC filters, St-Rose Treatment plant, Canada	EBCT 5-12 minutes	50% BDOC removal	(Laurent et al. 1999)

Study Scales	Parameters	Findings	References
Lab scale biodegradability study	2 mg O_3/mg TOC	40-50% DOC removal 40-60% THMFP removal 90-100% aldehydes removal	(Siddiqui et al. 1997)
Pilot including biologically active anthracite filters following ozonation.	0.6 mg O_3/mg DOC EBCT 2-11 minutes	Maximal 9% DOC was removed at EBCT 15 minutes with 5 m/h and EBCT 7 minutes with 9.7 m/h; 80% ozone by-products were removed at EBCT 3-5 minutes	
Lab scale fluidized GAC filter	EBCT 20 minutes 1 mg O_3/mg TOC	45% BDOC removal	(Yavich and Masten 2003)
Pilot including biologically active anthracite filters following ozonation.	0.6 mg O_3/mg DOC EBCT 6 minutes Temperature 6-10 °C	Up to 1.0 mg O_3/mg DOC produced maximal $BDOC_{rapid}$/DOC; and 0.4-0.6 mg O_3/mg DOC produced maximal $BDOC_{rapid}$/$BDOC_{total}$. Cumulative 90% DOC removal at EBCT 6 minutes.	(Carlson and Amy 2001)
Pilot including anthracite and GAC filters directly following ozonation.	EBCT 5 minutes 0.5-1.0 mg O_3/mg DOC	7-9% TOC (as UV254) removal; No difference was observed between GAC and anthracite filters due to the nature of the water	(Chaiket et al. 2002)
Full scale including GAC filters directly following ozonation at Sweeney WTP, Wilmington, NC.	EBCT 10-60 minutes 0.5-2 mg O_3/mg DOC	10-50% DOC removal; no significant difference between lignite and bituminous GAC.	(Najm et al. 2005)

Table 2. Design parameters and findings for the removal of natural organic matter removal (NOM) in previous studies

3.2. Removal of MIB and geosmin

The presence of tastes and odors in drinking water is an increasing and serious problem in the United States and the world. Some species of algae and bacteria naturally produce odorous chemicals inside their cells. Geosmin (trans-1, 10-dimethyl-trans-9-decalol) and MIB (2-methylisoborneol) are common odorous chemicals. There are no maximal contamination levels (MCLs) for MIB and Geosmin in drinking water systems according to US EPA. However, earthy and musty odors generated by Geosmin and MIB are detectable by individuals at the concentrations of 5 to 10 parts per trillion, and often result in customer complain. When large numbers of algae and bacteria flourish in a water body (an "algae bloom"), taste and odor-compound concentrations increase to levels above this threshold and cause taste and odor problems.

Biological active filtration was effectively used for the removal of Geosmin and MIB as summarized in Table 3. Since the concentrations of Geosmin and MIB encountered in drinking water systems are usually much less than that of TOC, a secondary utilization pathway was proposed as opposed to primary substrate utilization (Bouwer and Crowe 1988). Primary substrates support steady state biofilms which in turn metabolize secondary substrates such as Geosmin and MIB.

Unlike the removal of TOC, placing ozonation in front of GAC filters was not benefiting the removal of MIB and Geosmin likely due to the competition from increased AOC (Vik et al. 1988). An important design parameter is EBCT, which is usually in the range of 5-20 minutes depending on the required removal. GAC filters provided resistance to temperature variation while the removal was reduced by 24% for both Geosmin and MIB when temperature was reduced from 20 to 8°C in anthracite filters (Elhadi et al. 2006). At lower temperatures (6-12 °C), MIB and Geosmin removal was also reduced with expanded clay by 15% and 10%, respectively, compared to that at 15 °C (Persson et al. 2007). Biodegradation of both MIB and Geosmin was determined to be a pseudo-first-order reaction, with rates influenced by the initial amount of the biofilm biomass (Ho et al. 2007). As a result, sand with a well-established biofilm taken from a 26 years old filter was capable of removing MIB and Geosmin to below detection limits after 11 days of operation while sand without an established biofilm removed 60% Geosmin and 40% MIB after 154 days of operation (McDowall et al. 2007). Four bacteria, a *Pseudomonas sp.*, an *Alphaproteobacterium*, a *Sphingomonas sp.* and an *Acidobacteriaceae* member were identified as microorganisms most likely involved in the biodegradation of Geosmin within the sand filters (Ho et al. 2007).

Study Scales	Parameters	Findings	References
Full scale study at CLCJAWA Water Treatment Plant at Lake Bluff, Illinois, which included biologically active GAC filters following ozonation.	0.66-0.81 mg O_3/mg TOC EBCT 10-20 minutes	Ozonation removed 36-65% MIB and biofiltration removed 26-46% of MIB. The biodegradability of geosmin and MIB was confirmed by a bench scale study, where 55% and 44% removal was achieved for geosmin and MIB, respectively.	(Nerenberg et al. 2000)

Study Scales	Parameters	Findings	References
Pilot study of sand filters capped with biologically active GAC. Geosmin was in the range of about 70 to 110 ng/L,	EBCT 5 minutes	11 to 38%	(Ndiongue et al. 2006)
	EBCT 7.5 minutes	78% geosmin removal; Geosmin was better removed than MIB.	
Bench-scale two 2.0 m high glass GAC/sand filters	EBCT 5.6 minutes Temperature 12-16 °C	76 to 100% geosmin removal and 47% to 100% MIB removal. The exhausted GAC initially removed less geosmin and MIB, but the removals increased over time.	(Elhadi et al. 2004)
Bench-scale GAC/sand and anthracite/sand filters	EBCT 5.6 minutes Temperature 8 and 20 °C	60% geosmin and 40% MIB removal at 20 °C in GAC filters. 36% geosmin and 16% MIB removal at 8 °C in anthracite filters.	(Elhadi et al. 2006)
Bench-scale sand filters	EBCT 15 minutes	60% geosmin and 40% MIB with new sand after 154 days; reduced to below detection limit with sand from a 26 years old filter with a well-established biofilm	(McDowall et al. 2007)
Pilot study of GAC and expanded clay filters	EBCT 6, 15 and 30 minutes	Exhausted GAC had adsorption capability for MIB and Geosmin. At initial 20 ng/L MIB and 20 ng/L Geosmin: 97% removal at 30 minutes EBCT; 90% removal at 15 minutes EBCT; >40% remocal at 6 minutes EBCT	(Persson et al. 2007)
Bench-scale sand filters	EBCT 15 minutes 20±2 °C	95% removal of both MIB and geosmin with sand from an over 30 year facility	(Ho et al. 2007)
Pilot GAC and ozone plus GAC	EBCT 21 minutes 2-5 O_3 mg/L 1.5-4 TOC mg/L	TOC removal was better with ozone plus GAC; GAC was better for MIB and geosmin removal than ozone plus GAC because of the competition of TOC. GAC kept Geosmin and MIB below 10 ng/L	(Vik et al. 1988)

Table 3. Design parameters and findings for the removal of MIB and Geosmin in previous studies

3.3. Removal of iron, manganese, and arsenic

Appreciable amounts of iron and manganese usually exist in ground water or lake water experiencing low dissolved oxygen levels. The US EPA set secondary MCLs for iron and manganese at 0.3 mg/L and 0.05 mg/L, respectively.

There are two valences of iron and manganese, Fe (II) and Fe (III); Mn (II) and Mn (IV). Fe (II) and Mn (II) are quite soluble than Fe (III) and Mn (IV), respectively. As summarized in a previous study, the solubility product of ferrous hydroxide was in the range of 7×10^{-13} to 4.5×10^{-21} while the solubility product of ferric hydroxide was 3×10^{-38} to 4×10^{-36} (Gayer and Woontner 1956). The solubility product of manganous hydroxide was 9.0×10^{-14} and manganese dioxide was in equilibrium with $Mn(OH)_4$ (aq) with an equilibrium constant of 4.0×10^{-5} (Swain et al. 1975). $Mn(OH)_4$ (aq) will be prone to adsorption during filtration. Therefore, media filtration will not be effective to remove total iron and manganese if considerable portions are at the lower valance. Physicochemical removal requires a strong oxidant injected in front of media filtration to oxidize lower valance metals to a higher valance (Equations 1 and 2) and then filtered out.

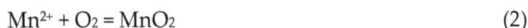

$$2Fe^{2+} + 3/2O_2 = Fe_2O_3 \qquad (1)$$

$$Mn^{2+} + O_2 = MnO_2 \qquad (2)$$

Fe (II) and Mn (II) can provide energy as electron donors for autotrophic biological reactions when oxygen is present. Biological filtration was demonstrated effective for iron and manganese removal assisted with aeration or ozonation in front of filtration (Table 4). It appeared that EBCT of 10 minutes reduced iron and manganese by 95-100% with coarse or fine sand media (Katsoyiannis and Zouboulis 2004; Lytle et al. 2007a; Štembal et al. 2005; Tekerlekopoulou et al. 2008). The use of ozone was beneficial (Pokhrel et al. 2005), but it appeared that aeration was sufficient for providing oxygen (Katsoyiannis and Zouboulis 2004; Lytle et al. 2007a; Štembal et al. 2005; Tekerlekopoulou et al. 2008).

Arsenic is a semi-metal element in the periodic table. It is odorless and tasteless. It enters drinking water supplies from natural deposits in the earth or from agricultural and industrial practices. Non-cancer effects can include thickening and discoloration of the skin, stomach pain, nausea, vomiting; diarrhea; numbness in hands and feet; partial paralysis; and blindness. Arsenic has been linked to cancer of the bladder, lungs, skin, kidney, nasal passages, liver, and prostate. US EPA set the arsenic standard for drinking water at 0.01 mg/L (10 parts per billion or ppb) to protect consumers served by public water systems subject to the effects of long-term, chronic exposure to arsenic.

Understanding the oxidation state is important for arsenic removal from drinking water. There are two oxidation states for arsenic: arsenite (As (III)) and arsenate (As (V)). Arsenite typically forms aqueous $As(OH)_3$, $As(OH)_4^-$, and AsO_2OH^{2-}, depending on pH; dissolved arsenate forms AsO_4^{3-}, $HAsO_4^{2-}$, or $H_2AsO_4^-$ (Edwards 1994; Katsoyiannis et al. 2002). At 6.9 < pH < 11.5, $HAsO_4^{2-}$ is the primary species; and at 2.2 < pH < 6.9, $H_2AsO_4^-$ is the primary arsenate species (Edwards 1994; Katsoyiannis et al. 2002). Arsenate adsorbs to soil minerals, particularly iron oxides and hydroxides. Arsenate sorption to iron oxides peaks around pH

5-7. Arsenite tends to adsorb less strongly than arsenate. Source water containing arsenite generally requires using a strong oxidant, e.g., chlorine, chlorine dioxide, and ozone, to oxidize arsenite to arsenate which can be removed by coagulation and filtration. Arsenic (V) removal by either ferric chloride or alum was relatively insensitive to variations in source water composition below pH 8 meanwhile arsenic (III) removal by ferric chloride was less efficient and more strongly influenced by source water composition than arsenic(V) removal (Hering et al. 1997). The presence of sulfate (at pH 4 and 5) and natural organic matter (at pH 4 through 9) adversely affected the efficiency of arsenic (III) removal by ferric chloride and arsenic(III) could not even be removed by coagulation with alum (Hering et al. 1997).

Study Scales	Water Qualities	Parameters	Findings	References
Pilot GAC column with 1740 mm (height) by 450 mm (diameter)	Fe 6.4-8.4 mg/L Mn 0.93-0.99 mg/L As 14.5-27.3 µg/L DOC 4.3-4.9 mg/L	EBCT 34.1 minutes 6.2-8.5 O_3 mg/L	The biologically active system removed 99.8% of Fe(II), and with the ozone pretreatment, the average removal increased to 99.9% for Fe(II).	(Pokhrel et al. 2005)
Pilot study of roughing filter with gravel and sand filter	Fe 0.09-0.44 mg/L Mn 0.18-1.83 mg/L	EBCT 30 and 60 minutes (roughing filter) EBCT 5 minutes (sand filter)	Iron and manganese removal efficiencies were between 85% and 95%.	(Pacini et al. 2005)
Pilot filter (1 m high polystyrene Beads 3-4 mm)	Fe 2.8 mg/L	EBCT 7.3 minutes	Fe(II) was microbiologically oxidized to Fe(III) precipitated on the filter bed.	(Katsoyiannis and Zouboulis 2004)
Bench-scale sand filter	Mn 0.86-1.83 mg/L		90% Mn removal	(Burger et al. 2008)
Pilot trickling filter (1.9 mm sand)	Mn 0.6–2.0 mg/L	EBCT 9 minutes	Close to 100% Mn removal	(Tekerlekopoulou et al. 2008)
Pilot sand filter (190 cm high, 0.5-2.0 mm sand)	Fe 0-2.45 mg/L Mn 0.1-1.06 mg/L NH4-N 0.02-2.62 mg/L	EBCT 4.75-10.4 minutes	95% Fe, Mn, and NH4-N removal	(Štembal et al. 2005)
Pilot dual media filter (20″ anthracite over 10″ sand)	As 37±2 µg/L NH4-N 1.15 mg/L Fe 2289 ±114 µg/L	EBCT 9.4 minutes	Reduced Fe to less than 25 µg/L	(Lytle et al. 2007a)

Table 4. Design parameters and findings for the removal of iron and manganese in previous studies

Study scale dual media filters at	Water Qualities	Parameters	Findings	References
Pilot dual media filter (20″ anthracite over 10″ sand)	As 37±2 µg/L NH₄-N 1.15 mg/L Fe 2289 ±114 µg/L	EBCT 9.4 minutes	Reduced to less than 10 µg/L; Majority of As in the effluent was particular.	(Lytle et al. 2007a)
Pilot anthracite/sand filters	As 20–46 µg/L	EBCT 10 minutes	Reduced to less than 10 µg/L	(Lytle et al. 2007b)
Pilot GAC column with 1740 mm (Height) by 450 mm (Diameter).	Fe 6.4-8.4 mg/L Mn 0.93-0.99 mg/L As 14.5-27.3 µg/L DOC 4.3-4.9 mg/L	EBCT 34.1 minutes 6.2-8.5 O₃ mg/L	97% without ozonation; 99% with pre-ozonation	(Pokhrel et al. 2005)
Pilot filter (1 m high polystyrene Beads 3-4 mm)	As 50–200 µg/L	EBCT 7.3 minutes	Reduced to less than 10 µg/L	(Katsoyiannis and Zouboulis 2004)
Pilot filter (1 m high polystyrene Beads 3-4 mm)	As 40–50 µg/L DO 2.7 mg/L	EBCT 7.3 minutes	80% As removal	(Katsoyiannis et al. 2002)
Pilot filter with at least 0. 66 m sand	As(III) 30-200 µg/L Fe(II) 0.5-1.5 mg/L Mn(II) 0.6-2.0 mg/L	EBCT 7.9 minutes	95% As removal	(Liu et al. 2010)
Pilot slow sand filter (sand 0.45-0.55 mm mixed with iron fillings)	As 50 µg/L	EBCT 4-5 hours 0.023 m/h	Columns containing filtration sand only removed As <11%; all iron/sand columns achieved greater than 92% removal.	(Gottinger 2010)
Pilot slow sand filter (90 cm, sand 0.25-2 mm)	As 10–35 µg/L	EBCT 4.25 hours	Reduced to less than 5 µg/L if Fe/As feed ratio was 40	(Pokhrel and Viraraghavan 2009)
Pilot GAC filter (100 cm, GAC 2-4 mm)	As(III) 25 mg/L Fe(II) 10 mg/L Mn(II) 2 mg/L	EBCT 6 hours	>99% As removal 80% Fe removal 95% Mn removal	(Mondal et al. 2008)

Table 5. Design parameters and findings for the removal of arsenic in previous studies

The current treatment options include activated alumina, iron oxide coated sand, greensand, reverse osmosis, ion exchange, and electrodialysis in addition to coagulation (Edwards 1994).

Similar to Fe (II) and Mn (II), arsenite can provide energy as an electron donor for autotrophic biological reactions when oxygen is present. Biological filtration was shown effective for arsenite removal assisted with aeration or ozonation in front of filtration (Table 5). Various media including sand, anthracite, GAC, and polystyrene beads were used for arsenic removal. Generally an EBCT of 10 minutes is required to reduce arsenic from up to 100 µg/L down to less than 10 µg/L. Due to the strong affinity of arsenate to ferric oxide, feeding ferric in the influent increased the removal efficiency with an increasing Fe/As ratio (Pokhrel and Viraraghavan 2009). A study found filtration columns containing mixture of sand and iron fillings improved removal and were capable of reducing arsenic from 50 to well below 10µg/L with an average of 92% removal (Gottinger 2010).

To improve the removal efficiency, immobilizing whole bacterial cells has attracted more research interest in recent years. *Ralstonia eutropha MTCC 2487* (this strain can produce ArsR protein and arsenate reductase enzyme) was immobilized on GAC bed in the column reactor (Mondal et al. 2008). *D. Desulfuricans, G. ferrigunea, L. ocracia, R. picketti, T. ynys1, Gallionella,* and *Leptothrix* were exploited to remove arsenic in biofiltration columns (Brunet et al. 2002; Elliot et al. 1998; Jong and Pany 2003; Katsoyiannis et al. 2002).

3.4. Removal of ammonia (nitrification)

Although there is no ammonia drinking water standard in the United States, the European community has established a maximum limit of approximately 0.5 mg/L and a guide level of 0.05 mg/L (EU Council 1980). Although there are no immediate indications that ammonia will become regulated within the United States, there are benefits for utilities to reduce the amount of ammonia that is able to enter a distribution system. The presence of ammonia in drinking water distribution systems has been correlated to increased biological activity, corrosion, formation of nitrite and nitrates, and adverse impacts on taste and odor (AWWA 2006). In addition, the presence of ammonia can interfere with the effectiveness of some water treatment processes including biological manganese removal as ammonia removal must be achieved before manganese removal due to the fact that the oxidation potential for nitrification is lower than manganese oxidation (McGovern and Nagy 2010).

Nitrification can be achieved in different ways, but may be most cost effectively accomplished by employing biofiltration (Table 6). The effectiveness of biological ammonia oxidation treatment to reduce source water ammonia levels is dependent on a number of source water and engineering design factors including temperature, dissolved oxygen, TOC, pH, biomass quantity and population, media type, and surface area, as well as hydraulic loading rate and contact time (Zhang et al. 2009). Factors affecting nitrification occurrence, nitrification impacts on water quality and corrosion, and nitrification monitoring and control methods were reviewed previously (Zhang et al. 2009). Arrhenius coefficient was

1.12 without acclimation and 1.06 with acclimation (Andersson et al. 2001). At temperature less than 4 °C, nitrification seemed un-sustained and feeding low temperature culture (psychrophiles) seemed necessary (Andersson et al. 2001). However, in fixed-film biofilters, the impact of temperature on nitrification rate was less significant than that predicted by the van't Hoff-Arrhenius equation, and a temperature increment at 20 °C resulted in a nitrification rate increase of 1.108% per degree and 4.275% per degree under DO and ammonia limiting conditions, respectively (Zhu and Chen 2002).

Study Scales	Water Qualities	Parameters	Findings	References
Pilot sand filter (190 cm high, 0.5-2.0 mm sand)	Fe 0-2.45 mg/L Mn 0.1-1.06 mg/L NH₄-N 0.02-2.62 mg/L	EBCT 4.75-10.4 minutes	95% Fe, Mn, and NH₄-N removal	(Štembal et al. 2005)
Full scale and pilot scale GAC filters at St Rose WTP at Laval, Canada	NH₄-N < 1 mg/L	EBCT up to 28 minutes	EBCT 5 minutes seems suitable for removing ammonia;	(Andersson et al. 2001)
Pilot gravel filter (180 cm high, 6-12 mm and 12-25 mm gravel)	NH₄-N < 1 mg/L 26 °C	EBCT 4-23 minutes	70-80% at EBCT 7.8 minutes; >90% at EBCT 10 minutes	(Forster 1974)
Pilot aerated GAC filter (85 cm high)	NH₄-N 2.88 mg/L	EBCT 10 minutes	95% ammonia removal	(Rogalla et al. 1990)
Pilot dual media filter (20" anthracite over 10" sand)	NH₄-N <1.7 mg/L	EBCT 9.4 minutes	Close to 100% ammonia removal	(Lytle et al. 2007a)

Table 6. Design parameters and findings for the removal of ammonia in previous studies

European experience on nitrification was reviewed for trickling filters, up-flow fluidized bed filters, rapid sand filters, and GAC filters (Rittmann and Snoeyink 1984). In a pilot trickling filter operated at 2.4 m/h with 2 m gravel media, ammonia removal was 80% at 20 °C; 78% at 15 °C; 67% at 10 °C; and 50% at 5 °C. In fluidized filters, nearly 100% removal was achieved as long as the fluidized solids were at least 30% by volume from 4-20 °C. Nearly 100% nitrification was achieved using rapid sand filters at Mulheim where raw water contained 1 mg/L ammonia nitrogen. Complete nitrification was achieved with GAC filters at 10 m/h with an EBCT of 10 minutes.

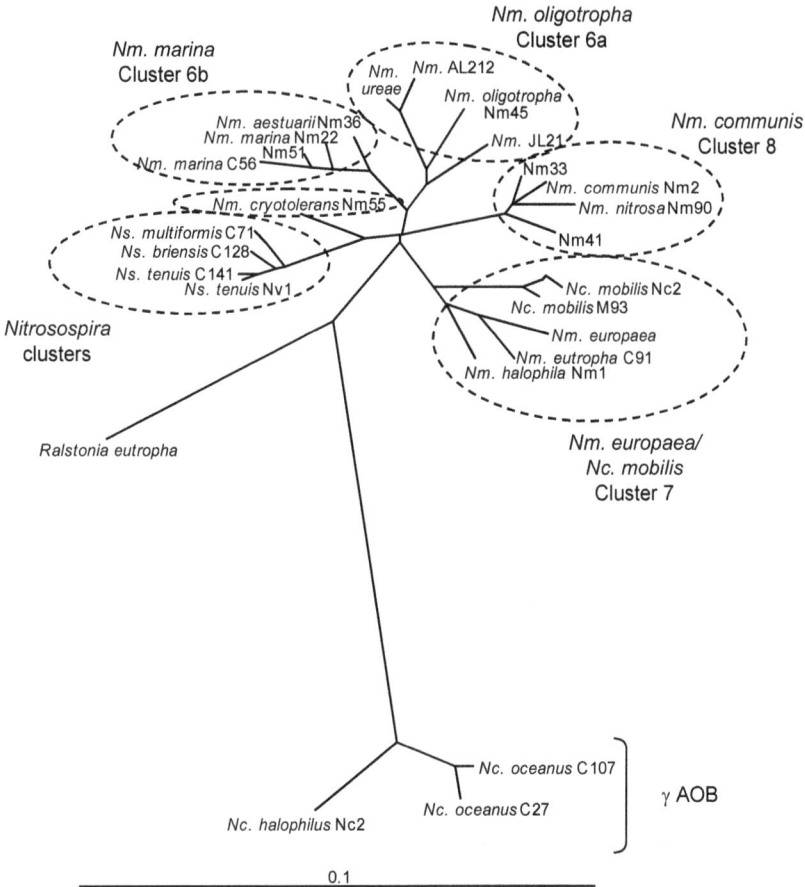

Figure 4. Phylogenetic tree of AOB based on multiple alignment of 55 nearly full-length AOB 16S rDNA sequences. Abbreviations are Nm for *Nitrosomonas*, Nc for *Nitrosococcus*, and Ns for *Nitrosospira*. *R. eutropha* is a non-AOB member of the *Betaproteobacteria* subphylum. Scale bar represents 10% sequence difference (Regan 2001)

Nitrification is a two-step process: ammonia oxidation and nitrite oxidation. The bacterial genera associated with ammonia oxidation are named as ammonia oxidizing bacteria (AOB) and the bacterial genera associated with nitrite oxidation are named as nitrite oxidizing bacteria (NOB). Both AOB and NOB are autotrophic bacteria using carbon dioxide for cellular synthesis under aerobic conditions. Phylogenetic trees for AOB and NOB were summarized elsewhere (Regan 2001) as shown in Figures 4 and 5, respectively. Kinetic parameters including specific substrate utilization rate, half saturation constant, yield, etc of *Nitrosomonas* and *Nitrobacter* were summarized at different temperatures (Rittmann and Snoeyink 1984).

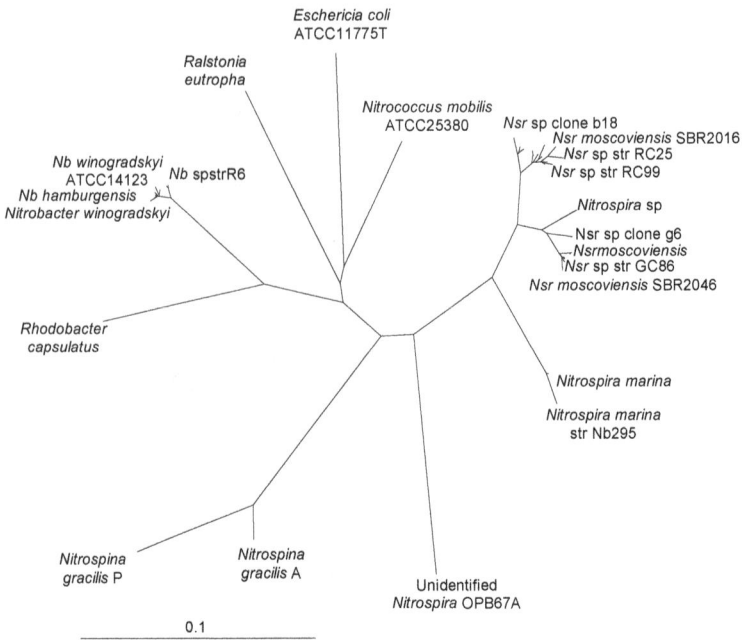

Figure 5. Phylogenetic tree of NOB based on a multiple alignment of 40 NOB 16S rDNA sequences. Abbreviations are Nb for *Nitrobacter* and Nsr for *Nitrospira*. *Rh. capsulatus* is in the *Alphaproteobacteria* class, *R. eutropha* is in the *Betaproteobacteria* class, and *E. coli* is in the *Gammaproteobacteria* class. Scale bar represents 10% sequence difference (Regan 2001)

3.5. Removal of nitrate (denitrification) and perchlorate

With the application of nitrate containing fertilizers and consumption of animal products, more nitrate is discharged into rivers and lakes, which may cause eutrophication and elevated levels of nitrate in ground water and surface water. Although nitrate was not identified as a carcinogen to laboratory animals, methaemoglobinaemia forms as a consequence of the reaction of nitrite (reduced from nitrate in human bodies) with haemoglobin in the human red blood cells to form methaemoglobin, which binds oxygen tightly and does not release it, therefore blocks oxygen transport (WHO 2008). The maximal contamination level in drinking water was 10 mg/L nitrate nitrogen in US, Japan, and Korea. The European Union countries set the standard for nitrate nitrogen at 11.3 mg/L. World Health Organization recommends 11.3 mg/L nitrate nitrogen to protect against methaemoglobinaemia in bottle-fed infants. Traditionally, nitrate removal was achieved by biological denitrification, ion-exchange, adsorption, chemical reduction, and membrane separation such as reverse osmosis. Ion exchange and membrane processes were often applied for high purity water treatment, which will generate concentrated nitrate reject (usually need additional treatment) from resin or membranes. Biological treatment processes were widely used for wastewater and drinking water treatment.

Denitrification filters are a unique type of biologically active filter where an external carbon source is usually added to the filter to provide a food source to anoxic biology and facilitate the reduction of nitrates in the filter. As a result, a dedicated anoxic biology is likely developed in the filter.

It is generally recognized that denitrification is carried out in the following steps with the aid of various enzymes produced during the process in the form of intracellular and extracellular polymeric substances, i.e., nitrate reductase (Nar), nitrite reductase (Nir), nitric oxide reductase (Nor), and nitrous oxide reductase (Nos) (Rittmann and McCarthy 2001).

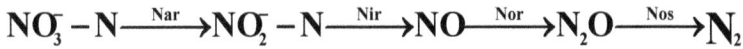

$$NO_3^- - N \xrightarrow{Nar} NO_2^- - N \xrightarrow{Nir} NO \xrightarrow{Nor} N_2O \xrightarrow{Nos} N_2$$

Denitrifiers are a group of heterotrophic bacteria and are phylogenetically diverse. They belong to over 50 genera and fall into all major physiological groups (Zumft 1992; Zumft 1997). Furthermore, microorganisms fed with different carbon sources showed distinct features. A report demonstrated that the metabolic profiles obtained from potential denitrification rates with 10 electron donors were altered with their preferences for certain compounds after supplementing methanol or ethanol, and that methanol had the greater impact (Hallin et al. 2006).

Denitrifying bacteria fed with methanol were recognized as methylotrophs. Fluorescence in situ hybridization (FISH) combined with microautoradiography (MAR) revealed that α-*Proteobacteria* metabolized ^{14}C methanol in the presence of nitrate, suggesting their involvement in denitrification in a methanol-fed fluidized marine denitrification reactor (Labbe et al. 2007). Using a molecular tool, *Paracoccus* sp. and *Hyphomicrobium* spp. were identified as denitrifiers in a denitrification sand filter fed with methanol (Neef et al. 1996). Research showed that methanol denitrifiers and acetate denitrifiers were distinctively different. When acetate was used as the carbon source, 16S rRNA gene sequences obtained from ^{13}C-labeled DNA were closely related to the 16S rRNA genes of *Comamonadaceae* and *Rhodocyclaceae* of the β-*Proteobacteria*, and *Rhodobacteraceae* of the α-*Proteobacteria*. When methanol was used as the carbon source, 16S rRNA gene sequences retrieved from ^{13}C-DNA were affiliated with *Methylophilaceae* and *Hyphomicrobiaceae* (Osaka et al. 2006).

A study showed that methanol-utilizing organisms can not use acetate or sugar (at least not immediately). Adding alternative carbon sources, i.e., acetate or sugar, will not result in an immediate improvement in denitrification (Dold et al. 2008). However, the substrate uptake rate (μ_{max}) and specific denitrification rate (SDNR), measured by feeding ethanol to methanol-utilizers, indicated that ethanol was also used essentially as easily and at a similar rate to methanol by the methanol-utilizers (Dold et al. 2008).

The stoichiometric and kinetic information for different carbon sources commonly used in a denitrification filter were studied previously and selected parameters are summarized elsewhere (Omnis-Hayden and Gu 2008). Table 7 summarized findings and operating parameters in previous studies. The important design parameters are EBCT, C/N ratio (the ratio of external carbon to nitrate nitrogen), pH and temperature.

Study Scales	Water Qualities	Parameters	Effluent Qualities, Efficiencies, and Findings	References
Pilot up-flow fluidized sand filter (1.2 m deep 0.6-0.8 mm sand)	NO_3-N 27 mg/L	EBCT 2.57 minutes C/N 1.3 (propionic acid)	Close to 100% NO_3-N removal	(Holló and Czakó 1987)
Pilot Rotating Biological Contactors (RBC) (75 m^2)	NO_3-N 40-250 mg/L pH 7 and 20±2 ºC	NO_3-N loading: 76 mg/$m^2 \bullet$h 490 mg/$m^2 \bullet$h Acetic acid as a carbon source	99% (for 76 mg/$m^2 \bullet$h) and 87% (490 mg/$m^2 \bullet$h) The optimum C/N ratio 4.3	(Mohseni-Bandpi et al. 1999)
Bench-scale membrane reactors.	NO_3-N 100-200 mg/L 20±1 ºC	C/N ratio 2-4 (ethanol)	99% removal with denitrification rates up to 1.23 g NO_3^--N/$m^2 \bullet$d	(Fuchs et al. 1997)
Bench-scale up-flow fixed-bed reactor	NO_3-N 20 mg/L 25±1 ºC	EBCT 48 minutes Molasses	85% NO_3-N removal with an optimum C/N ratio 2	(Ueda et al. 2006)
Bench-scale experiments. Calcium tartrate (4% w/w) co-immobilized in alginate beads with microorganisms	NO_3-N 110 mg/L		99% NO_3-N removal with NO_2-N residual (2-4 mg/L) higher than expected	(Liu et al. 2003)
Bench-scale carbon packed fixed-biofilm bed reactor, inoculated with *Paracoccus denitrificans* NRRL B-3784.	NO_3-N 200 mg/L	EBCT 78 minutes C/N ratio 2-4 (ethanol)	90% NO_3-N removal	(Pekdemir et al. 1998)
Bench scale packed Pall Rings (16 mm diameter and length) bed	NO_3-N 50 mg/L pH 7.5-7.8	EBCT 30 minutes COD/N ratio 4 (ethanol)	NO_3-N 0.9 mg/L; NO_2-N 0.7 mg/L; COD 34 mg/L	(Dahab and Kalagiri 1996)
Bench scale packed Pall Rings (16 mm diameter and length) bed	NO_3-N 50 mg/L pH 7 and 20 ºC	EBCT 30 minutes COD/N ratio 4 (ethanol)	NO_3-N 2.4 mg/L; NO_2-N 0.8 mg/L; COD 18 mg/L	(Woodbury and Dahab 2001)
Full scale packed polypropylene beads (3.8 mm and 5 mm diameter) bed	NO_3-N 20 mg/L pH 7.2 and 13-18 ºC	EBCT 66 minutes COD/N ratio 5.3 (corn syrup)	NO_3-N 5.0 mg/L; NO_2-N 1.7 mg/L; COD 20 mg/L	(Silverstein and Carlson 1999)

Study Scales	Water Qualities	Parameters	Effluent Qualities, Efficiencies, and Findings	References
Pilot ceramic media bed (1.5 m high and 0.3 m diameter)	NO_3-N 68 mg/L pH 7-7.5 and 15-20 °C	EBCT 72 minutes COD/N ratio 4.3 (ethanol)	NO_3-N <4 mg/L; NO_2-N<3 mg/L; COD 10 mg/L	(Moreno et al. 2005)
Bench scale PVC/GAC beads bed (88 cm high and 12 cm diameter)	NO_3-N 45 mg/L and 20 °C	EBCT 306 minutes COD/N ratio 5.5 (acetate)	NO_3-N 5 mg/L; NO_2-N<0.5 mg/L; COD 60 mg/L	(Vrtovsek and Ros 2006)
Bench scale moving bed (Kaldnes K1)	NO_3-N 60 mg/L pH >7 and 20 °C	EBCT 54 minutes COD/N ratio 13.1 (acetate)	NO_3-N 4.7 mg/L; NO_2-N<0.25 mg/L; COD 400 mg/L	(Welander and Mattiasson 2003)
Pilot 900 L moving bed reactor (carrier, Natrix 6/6C, ANOX AB, Lund)	NO_3-N 800 mg/L pH 7.8 and 17 °C	EBCT 17 hours COD/N ratio 4 (acetate)	NO_3-N ~0 mg/L; NO_2-N ~0 mg/L	(Welander et al. 1998)
Pilot moving plastic media bed	NO_3-N 13 mg/L NO_2-N 0.5 mg/L 7-10 °C	EBCT 26 minutes COD/N ratio 4 (acetate)	NO_3-N 2.0 mg/L; NO_2-N 0.9 mg/L; COD 50 mg/L	(Rusten et al. 1995)
Membrane bioreactor with hollow-fibers (1.1 mm inner diameter, 1.4 mm external diameter, 0.38m length)	NO_3-N 200 mg/L pH 7.2	EBCT 26 minutes COD/N ratio 3 (methanol)	NO_3-N 5.7 mg/L; NO_2-N 0.02 mg/L; COD 70 mg/L	(Ergas and Rheinheimer 2004)
Pilot-scale fixed-bed bioreactors packed with sand or plastics.	ClO_4- 77 µg/L	EBCT 15 minutes Acetic acid as a carbon source	Reduced to <4 µg/L	(Min et al. 2004)
Six-month pilot at the Castaic Lake Water Agency, Santa Clarita, CA, using fixed-bed bioreactors	18-20 mg NO_3-N/L 17-20 µg ClO_4-/L	EBCT 15 minutes Acetic acid as a carbon source	Reduced to less than detection limit	(Brown et al. 2005)

Study Scales	Water Qualities	Parameters	Effluent Qualities, Efficiencies, and Findings	References
Bench-scale Granular Activated Carbon (GAC) bioreactor	ClO_4- 50 µg/L	EBCT 25 minutes Acetic acid as a carbon source	Reduced to less than detection limit	(Brown et al. 2002)
Bench-scale hydrogen permeable membrane fixed film reactor	1,000 mg NO_3- N/L and/or 500 mg ClO_4-/L	EBCT 48 hours 72 cm^2 membrane surface	80% nitrate removal and 60% ClO_4- removal (no nitrate); the presence of nitrate reduced ClO_4- removal.	(Chung et al. 2007)
Bench-scale hydrogen permeable membrane fixed film reactor	12.5 mg NO_3- N/L 9.4 µg ClO_4-/L	EBCT 55 minutes 83.6 cm^2 membrane surface	99.5% reduction of 0.21 mg NO_3- $N/cm^2 \bullet d$ and 3.4 µg $ClO_4/cm^2 \bullet d$	(Ziv-El and Rittmann 2009)
Pilot up-flow packed bed reactors (plastic media)	NO_3-N 325 mg/L ClO_4- 6.37 mg/L	EBCT 15 h pH of 7.0	Completely removed perchlorate and nitrate with up to 10% salinity	(Chung et al. 2010)
Pilot anthracite filter (0.31 m deep and 1.0 mm ES)	NO_3-N 2 mg/L ClO_4- 50 µg/L	EBCT 47 minutes	Reduced to <2 µg/L with temperature as low as 10 ºC	(Dugan et al. 2009)

Table 7. Design parameters and findings for the removal of nitrate and perchlorate in previous studies

Perchlorate occurs in water due to natural presence or manufacturing for ammonia perchlorate (Srinivasan and Sorial 2009). Being a strong oxidant, perchlorate was used as solids propellants for rockets and missiles or used for fireworks. The US EPA is not currently regulating perchlorate in drinking water but already placed it in the contaminant candidate list.

Table 7 also summarized findings and operating parameters in previous studies for pechlorate removal. Usually perchlorate is removed simultaneously with nitrate. Because the standard oxidation potential of the perchlorate/chloride pair (1.28 V) is much higher than nitrate/N_2 pair (0.75 V), perchlorate is reduced preferentially, and can be reduced down to less than 4 µg/L with an EBCT at least 15 minutes. High reduction rates of nitrate and perchlorate occurred in a synthetic high-strength salt medium 20 g/L (~2%) NaCl, while 40 g/L NaCl slowed reduction by 40% or more (Chung et al. 2010).

Similar to nitrate removal, biological processes are still cost effective for perchlorate removal. In an anoxic environment, perchlorate is reduced to chloride at the expense of an external carbon source.

4. Conclusions and perspectives

While BAFs are playing an important role in contaminant removal from water sources, understanding the process design parameters such as EBCT, media selection, backwash velocity, pH, temperature, oxygen demand, pre-oxidation requirement, inhibiting metal elements, etc, is important in that it will provide insights on treatment process control, pathogenic impacts, disinfection by-product control, and the potential to improve treatment efficiencies.

An EBCT of 10 minutes is generally recommended for the removal of TOC, MIB and geosmin, iron and manganese, arsenic, and ammonia. At least 15 minutes are generally required for the removal of nitrate and perchlorate. Backwash was found not influencing the process performance. A GAC/sand filter produced better performance than an anthracite/sand filter for the removal of NOM and taste and odor compounds, and the GAC/Sand filter out performed the anthracite/sand filter by 11% for AOC removal and showed better resistance to temperature at 1-3 ºC. Unlike the situation with NOM, placing ozonation in front of GAC filters was not benefitting the removal of MIB and geosmin. Simple aeration is sufficient for providing oxygen for the removal of ammonia, and iron and manganese.

Effective microbial adhesion and immobilization is essential for biofilm activities, and still drives further research on physicochemical (for example roughing, grafting, coating, etc) and biological (inoculated with selected species) approach in BAFs. To improve the process efficiency, dedicated microbial species targeting specific contaminants are usually desired. However, it is challenging and presents significant scientific and engineering opportunities to select microbial communities in biofilms specifically adapted to targeted contaminants. Besides currently employed media, alternative cost effective media are always interested, especially the ones from waste materials and engineered with specific surface properties.

Author details

Ivan X. Zhu and Brian J. Bates
Xylem Water Solutions Zelienople LLC, Zelienople, PA, USA

5. References

Al-Ghouti MA, Khraisheh MAM, Allen SJ, Ahmad MN. 2003. The removal of dyes from textile wastewater: a study of the physical characteristics and adsorption mechanisms of diatomaceous earth. J. Environ. Manage. 69:229-238.

Amirtharajah A. 1993. Optimum backwashing of filters with air scour: a review. Wat. Sci. Tech. 27(10):195-211.

Andersson A, Laurent P, Kihn A, Prévost M, Servais P. 2001. Impact of temperature on nitrification in biological activated carbon (BAC) filters used for drinking water treatment. Water Res. 35(12): 2923–2934.

AWWA. 2006. Manual of Water Supply Practices—M56. Kozyra MK, editor. Denver: Glacier Publishing Services, Inc.

Babbitt HE, Doland JJ. 1939. Water Supply Engineering. New York: McGraw Hill.

Barrett JM, Bryck J, Collins MR, Janonis BA, Logsdon GS. 1991. Manual of Design of Slow Sand Filtration. Hendricks D, editor. Denver: AWWA RF and AWWA.

Bourbigot MM, Dodin A, Lheritier R. Limiting bacterial aftergrowth in distribution systems by removing biodegradable organics; 1982; Miami.

Bouwer EJ, Crowe PB. 1988. Biological processes in drinking water treatment. J AWWA September:82-93.

Brown JC, Anderson RD, Min JH, Boulos L, Prasifka D, Juby GJG. 2005. Fixed-bed biological treatment of perchlorate-contaminated drinking water. J. AWWA 97(9):70-81.

Brown JC, Snoeyink VL, Kirisits MJ. 2002. Abiotic and biotic perchlorate removal in an activated carbon filter. J. AWWA 94(2):70-79.

Brunet B, Dictor MC, Carrido F, Crouzet C, Morin D, Dekeyser K, Clarens M, Baranger P. 2002. An arsenic(III) oxidizing bacterial population: selection, characterization, and performance in reactors. J. Appl. Microbiol. 93:656–667.

Burger MS, Mercer SS, Shupe GD, Gagnon GA. 2008. Manganese removal during bench-scale biofiltration. Water Res. 42:4733-4742.

Camper AK, Lechevallier MW, Broadaway SC, Mcfeters GA. 1985. Growth and persistence of pathogens on granular activated carbon filters. Appl. Environ. Microbiol. 50(6):1378-1382.

Cantor KP, Hoover R, Hartge P, Mason TJ, Silverman DT, Altman R, Austin DF, Child MA, Key CR, Marrett LD and others. 1987. Bladder cancer, drinking water source, and tap water consumption: a case-control study. J. Natl Cancer Institute 79(6):1269-1279.

Cantor KP, Lynch CF, Hildesheim ME, Dosemeci M, Lubin J, Alavanja M, Craun G. 1999. Drinking water source and chlorination byproducts in Iowa. III. Risk of brain cancer. Am J Epidemiol. 150(6):552-560.

Carlson KH, Amy GL. 2001. Ozone and biofiltration optimization for multiple objectives. J. AWWA 93(1):88-98.

Carraro E, Bugliosi EH, Meucci L, Baiocchi C, Gilli G. 2000. Biological drinking water treatment processes, with special reference to mutagenicity. Water Res. 34(11):3042–3054.

Chaiket T, Singer PC, Miles A, Morgan M, Pallotta C. 2002. Effectiveness of coagulation, ozonation, and biofiltration in controlling DBPs. J. AWWA 94(12):81-95.

Characklis WG. 1988. Bacterial Regrowth in Distribution Systems. Denver: AWWA RF and AWWA.

Chung J, Nerenberg R, Rittmann BE. 2007. Evaluation for biological reduction of nitrate and perchlorate in brine water using the hydrogen-based membrane biofilm reactor. J. Environ. Eng. 133(2):157-164.

Chung J, Shin S, Oh J. 2010. Biological reduction of nitrate and perchlorate in brine water using up-flow packed bed reactors. J. Environ. Sci. Health Part A 45:1109-1118.

Collins MR, Eighmy TR, Fenstermacher JM, Spanos SK. 1992. Removing natural organic matter by conventional slow sand filtration. J. AWWA 84(5):80-90.

Dahab MF, Kalagiri J. 1996. Nitrateremoval from water using cyclicallyoperated fixedfilm bio-denitrification reactors. Wat. Sci. Technol. 34(1-2):331-338.

Davidson MI, Bryant R, Williams DJA. 1996. Characterization of anthracite. Geological Society, London, Special Publications 109:213-225.

Dold P, Takács I, Mokhayeri Y, Nichols A, Hinojosa J, Riffat R, Bott C, Bailey W, Murthy S. 2008. Denitrification with carbon addition—kinetic considerations Water Environ. Res. 80(5):417-427.

Doyle TJ, Sheng W, Cerhan JR, Hong CP, Sellers TA, Kushi LH, Folsom AR. 1997. The association of drinking water source and chlorination by-products with cancer incidence among postmenopausal women in Iowa: a prospective cohort study. Am. J. Public Health 87(7):1168-1176.

Dugan NR, Williams DJ, Meyer M, Schneider RS, Speth TF, Metz DH. 2009. The impact of temperature on the performance of anaerobic biological treatment of perchlorate in drinking water. Water Res. 43:1867-1878.

Dussert BW, Van Stone GR. 1994. The biological activated carbon process for water purification. Water Eng. Manage. 141(12):22-24.

Edwards M. 1994. Chemistry of arsenic removal duirng coagulation and Fe-Mn oxidation. J. AWWA 86(9):64-78.

Elhadi SLN, Huck PM, Slawson RM. 2004. Removal of geosmin and 2-methylisoborneol by biological filtration. Water. Sci. Technol. 49(9):273-280.

Elhadi SLN, Huck PM, Slawson RM. 2006. Factors afecting the removal of geosmin and MIB in drinking water biofilters. J. AWWA 98(8):108-119.

Elliot P, Ragusa S, Catcheside D. 1998. Growth of sulphate reducing bacteria under acid conditions in an up-flow anaerobic bioreactor as a treatment system for acid mine drainage. Water Res. 32:3724–3730.

Emelko MB, Huck PM, Coffey BM, Smith EF. 2006. Effects of media, backwash, and temperature on full scale biological filtration. J. AWWA 98(12):61-73.

Ergas SJ, Rheinheimer DE. 2004. Drinking water denitrification using amembranebioreactor. Water Res. 38(14-15):3225-3232.

EU Council. 1980. Council Directive 98/83/EC of 3 November 1998 on the Quality of Water Intended for Human Consumption. EC Drinking Water Directive.

Evans P, Opitz E, Daniel P, Shultz C, Skerly A, Shivakumar S. Preliminary results of a survey on the use of biological processes for drinking water treatment; 2008; Cincinnati, Ohio. AWWA.

Forster JRM. 1974. Studies on nitroification in marine biological filters. Aquaculture 4:387-397.

Fuchs W, Schatzmayr G, Braun R. 1997. Nitrate removal from drinking water using a membrane-fixed biofilm reactor. Appl Microbiol Biotechnol. 48(2):267-274.

Gayer KH, Woontner L. 1956. The solubility of ferrous hydroxide and ferric hydroxide in acidic and basic media at 25 °C. J. Phys. Chem. 60:1569-1571.

Gergova K, Eser S, Schobert HH. 1993. Preparation and characterization of activated carbons from anthracite. Energy & Feuls 7:661-668.

Goel S, Hozalski RM, Bouwer EJ. 1995. Biodegradation of NOM: effect of NOM source and ozone dose. J. AWWA 87(1):90-105.

Gottinger AM. 2010. Chemical-free arsenic removal from potable water with a zvi-amended biofilter. Regina: University of Regina.

Hallin S, Throbäck IN, Dicksved J, Pell M. 2006. Metabolic profiles and genetic diversity of denitrifying communities in activated sludge after addition of methanol or ethanol. Appl. Environ. Microbiol. 72(8):5445-5452.

Hering JG, Chen PY, Wilkie JA, Elimelech M. 1997. Arsenic removal from drinking water during coagulation. J. Environ. Eng. 123:800-807.

Ho L, Hoefel D, Bock F, Saint CP, Newcombe G. 2007. Biodegradation rates of 2-methylisoborneol (MIB) and geosmin through sand filters and in bioreactors. Chemosphere 66:2210-2218.

Holló J, Czakó L. 1987. Nitrate removal from drinking water in a fluidized-bed biological denitrification bioreactor. Acta Biotechnologica 7(5):417-423.

Hozalski RM, Bouwer EJ, Goel S. 1999. Removal of natural organic matter (NOM) from drinking water supplies by ozone biofiltration. Wat. Sci. Tech. 40(9):157-163.

Hozalski RM, Goel S, Bouwer EJ. 1995. TOC removal in biological filters. J. AWWA 87(12):40-54.

Jerez J, Flury M, Shang J, Deng Y. 2006. Coating of silica sand with aluminosilicate clay. J. Colloid Interface Sci. 294:155-164.

Jong T, Pany DL. 2003. Removal of sulfate and heavy metals by sulfate reducing bacteria in short term bench scale up-flow anaerobic packed bed reactor runs. Water Res. 37:3379–3389.

Juhna T, Rubulis J. 2004. Problem of DOC removal during biological treatment of surface water with a high amount of humic substances. Water Sci. & Technol.: Water Supply 4(4):183-197.

Kasuga I, Shimazaki D, Kunikane S. 2007. Influence of backwashing on the microbial community in a biofilm developed on biological activated carbon used in a drinking water treatment plant. Water Sci. & Technol. 55(8-9):173-180.

Katsoyiannis I, Zouboulis A, Althoff H, Bartel H. 2002. As(III) removal from groundwaters using fixed-bed upflow bioreactors. Chemosphere 47:325-332.

Katsoyiannis IA, Zouboulis AI. 2004. Application of biological processes for the removal of arsenic from groundwaters. Water Res. 38:17-26.

Kitis M, Kaplan SS, Karakaya E, Yigit NO, Civelekoglu G. 2007. Adsorption of natural organic matter from waters by iron coatedpumice. Chemosphere 66(1):130-138.

Labbe N, Laurin V, Juteau P, Parent S, Villemur R. 2007. Microbiological community structure of the biofilm of a methanol-fed, marine denitrification system, and identification of the methanol-utilizing microorganisms. Microb. Ecol. 53(4):621-630.

Lauderdale CV, Brown JC, chadik PA, Kirisits MJ. 2011. Engineered Biofiltration for Enhanced Hydraulic and Water Treatment Performance. Denver: Water Research Foundation.

Laurent P, Prevost M, Cigana J, Niquette P, Servais P. 1999. Biodegradable organic matter removal in biological filters: evaluation of the chabrol model. Water Res. 33(6):1387-1398.

Liu SX, Hermanowicz SW, Peng M. 2003. Nitrate removal from drinking water through the use of encapsulated microorganisms in alginate beads. Environ. Technol. 24(9):1129-1134.

Liu XT, Li D, Zeng HP, Zhang J. 2010. Arsenic(III) removal by biofilter for iron and manganese removal. Journal of Harbin Institute of Technology 42(6):873-875.

Logsdon GS, Hess AF, Chipps MJ, Rachwal AJ. 2002. Filter Maintenance and Operations Guidance Manual. Denver: AWWA RF and AWWA.

Lytle DA, Chen AS, Sorg TJ, Phillips S, French K. 2007a. Microbial As(III) oxidation in water treatment plant filters. J. AWWA 99(12):72-86.

Lytle DA, Sorg TJ, Lili W, Muhlen C, Rahrig M, French K. 2007b. Biological nitrification in a full-scale and pilot-scale iron removal drinking water treatment plant. J. Water Supply : Res. Technol. AQUA 56(2):125-136.

McDowall B, Ho L, Saint C, Newcombe G. 2007. Removal of geosmin and 2-methylisoborneol through biologically active sand filters. Int J. Environ. Waste Manage. 1(4):311-320.

McGovern B, Nagy R. Biological removal of iron, ammonia, and manganese simultaneously utilizing pressurized downflow filtration; 2010 November 14–18; Savannah, GA. AWWA.

Melin ES, Odegaard H. 1999. Biofiltration of ozonated humic water in expanded clay aggregate filters. Wat. Sci. Tech. 40(9):165-172.

Min B, Evans PJ, Chu AK, Logan BE. 2004. Perchlorate removal in sand and plastic media bioreactors. Water Res. 38(1):47-60.

Mohseni-Bandpi A, Elliott DJ, Momeny-Mazdeh A. 1999. Denitrification of groundwater using aceticacid as a carbon source. Water Sci. Technol. 40(2):53-59.

Mondal P, Majumder CB, B Mohanty B. 2008. Treatment of arsenic contaminated water in a laboratory scale up-flow bio-column reactor. J. Hazard. Mat. 153:136-145.

Moreno B, Gómez MA, González-López J, Hontoria E. 2005. Inoculation of a submerged filter for biological denitrification of nitrate polluted groundwater: a comparative study. J. Hazard Mater. 117(2-3):141-174.

Najm I, Kennedy M, Naylor W. 2005. Lignite versus bituminous GAC for biofiltration - a case study. J. AWWA 97(1):94-101.

Ndiongue S, Anderson WB, Tadwalkar A, Rudnickas J, Lin M, Huck PM. 2006. Using pilot-scale investigations to estimate the remaining geosmin and MIB removal capacity of full-scale GAC-capped drinking water filters. Water Qual. Res. J. Canada 41(3):296-306.

Neef A, Zaglauer A, Meier H, Amann R, Lemmer H, Schleifer KH. 1996. Population analysis in a denitrifying sand filter: conventional and In situ identification of Paracoccus spp. in methanol-fed biofilms. Appl. Environ. Microbiol. 62(12):4329–4339.

Nerenberg R, Rittmann BE, Soucie Wj. 2000. Ozone/biofiltration for removing MIB and Geosmin. J. AWWA 92(12):85-95.

Occelli ML, Olivier JP, Perdigon-Melon JA, Auroux A. 2002. Surface area, pore volume distribution, and acidity in mesoporous expanded clay catalysts from hybrid density functional theory (DFT) and adsorption microcalorimetry methods. Langmuir 18(9816-9823).

Oliveira LCA, Rios RVRA, Fabris JD, Garg V, Sapag K, Lago RM. 2002. Activated carbon/iron oxide magnetic composites for the adsorption of contaminants in water. Carbon 40:2177-2183.

Omnis-Hayden A, Gu AZ. Comparison of organic sources for denitrification: biodegradability, denitrification rates, kinetic constants and practical implication for their application in WWTPs; 2008 October 20-22; Chicago. WEF. p 253-273.

Osaka T, Yoshie S, Tsuneda S, Hirata A, Iwami N, Inamori Y. 2006. Identification of acetate- or methanol-assimilating bacteria under nitrate-reducing conditions by stable-isotope probing. Microb. Ecol. 52(2):253-266.

Pacini VA, Ingallinella AM, Sanguinetti G. 2005. Removal of iron and manganese using biological roughing up flow filtration technology. Water Res. 39:4463-4475.

Partinoudi V, Collions MR. 2007. Assessing RBF reduction/removal mechanisms for microbial and organic DBP precursors. J. AWWA 97(12):61-71.

Pekdemir T, Kacmazoglu EK, Keskinler B, Algur OF. 1998. Drinking water denitrification in a fixed bed packed biofilm reactor. Turk. J. Eng. Environ. Sci. 22(1):39-45.

Perrotti AE, Martins SAM, Mazzola PG. 1974. Factors involved with biological regeneration of activated carbon. AIChE Sym. Series 144:317–325

Persson F, Heinicke G, Hedberg T, Hermansson M, Uhl W. 2007. Removal of geosmin and mib by biofiltration - an investigation discriminating between adsorption and biodegradation. Environ. Technol. 28:95-104.

Piet GJ, Zoeteman BCJ. 1980. Organic water quality changes during sand bank and dune filtration of surface waters in Netherlands. J. AWWA 72(7):400-404.

Pokhrel D, Viraraghavan T. 2009. Biological filtration for removal of arsenic from drinking water. J. Environ. Manage. 90:1956-1961.

Pokhrel D, Viraraghavan T, Braul L. 2005. Evaluation of treatment systems for the removal of arsenic from groundwater. Pract. Period. Hazard. Toxic Radioact. Waste Manage. 9:152-157.

Ratnayaka DD, Brandt MJ, Johnson KM. 2009. Water Supply. Oxford, UK: Elsevier Ltd.

Regan JM. 2001. Microbial Ecology of Nitrification in Chloraminated Drinking Water Distribution Systems. Madison: University of Wisconsin.

Rice RG, Robson RG. 1982. Biological Activated Carbon. Ann Arbor, Michigan: Ann Arbor Science.

Rittmann BE, McCarthy PL. 2001. Environmental Biotechnology: Principles and Applications. New York: McGraw-Hill.

Rittmann BE, Snoeyink VL. 1984. Achieving biologically stable drinking water. J. AWWA 10:106-114.

Rittmann BE, Stilwell D, Garside JC, Amy GL, Spangenberg C, Kalinsky A, Akiyoshi E. 2002. Treatment of a colored groundwater by ozone-biofiltration: pilot studies and modeling interpretation. Water Res. 36:3387-3397.

Rodman CA, Shunney EL, Perrotti AE. 1978. Biological Regenration of Activated Carbon. Cheremisinoff PN, Ellerbursch F, editors. Ann Arbor, Michigan: Ann Arbor Science.

Rogalla F, Ravarini P, De Larminat G, Couttelle J. 1990. Large-scale biological nitrate and ammonia removal. J. IWEM 4(4):319-328.

Rollingger Y, Dott W. 1987. Survival of selected bacterial species in sterilized activated carbon filters and biological activated carbon filters. Appl. Environ. Microbiol. 53(4):777-781.

Rusten B, Hem LJ, Ødegaard H. 1995. Nitrogen removal from dilute wastewater in cold climate using moving-bed biofilm reactors. Water Environ. Res. 67(1):65-74.

Scholz M, Martin J. 1997. Ecological equilibrium on biological activated carbon. Water Res. 31(12):2959-2968.

Shukairy HM, Miltner RJ, Summers RS. 1992. Control of disinfection by-products and biodegradable organic matter through biological treatment. Revue des sciences de l'eau 5:1-15.

Siddiqui MS, Amy GL, Murphy BD. 1997. Ozone enhanced removal of natural organic matter from drinking water sources. Water Res. 31(12):3098-3106.

Silverstein JS, Carlson GL. 1999. Biological Denitrification of Drinking Water for Rural Communities. Final Project Report. NRECA (National Rural Electric Cooperative Association) and EPRI (Electric Power Research Institute).

Sontheimer H. 1980. Experience with riverbank filtration along the Rhine River. J. AWWA 72(7):386-390.

Srinivasan R, Sorial GA. 2009. Treatment of perchlorate in drinking water: a critical review. Separation and Purification Technol. 69:7-21.

Štembal T, Markic M, Ribicic N, Briski F, Sipos L. 2005. Removal of ammonia, iron and manganese from groundwaters of northern Croatia—pilot plant studies. Process Biochem. 40:327-335.

Swain HA, Lee C, Rozelle RB. 1975. Determination of the solubility of manganese hydroxide and manganese dioxide at 25 °C by atomic absorption spectrometry. Anal. Chem. 47(7):1135-1137.

Tang Z, Maroto-Valer MM, Zhang Y. 2004. CO2 capture using anthracite based sorbents. Prepr. Pap.-Am. Chem. Soc., Div. Fuel Chem. 49(1):298-299.

Tekerlekopoulou AG, Vasiliadou IA, Vayenas DV. 2008. Biological manganese removal from potable water using trickling filters. Biochem. Eng. J. 38:292-301.

Ueda T, Shinogi Y, Yamaoka M. 2006. Biological nitrate removal using sugar-industry wastes. Paddy Water Environ. 4(3):139-144.

Unger M, Collins MR. 2008. Assessing *Escherichia coli* removal in the *Schmutzdecke* of slow rate biofilters. J. AWWA 100(12):60-73.

van der Hoek JP, Hofman JAMH, Graveland A. 1999. The use of biological activated carbon filtration for the removal of natural organic matter and organic micropollutants from water. Wat. Sci. Tech. 40(9):257-264.

Vik EA, Storhaug R, Naes H, Utkilen HC. 1988. Pilot scale studies of Geosmin and 2-Methylisoborneol removal. Water Sci. & Technol. 20(8-9):229–236

Vrtovsek J, Ros M. 2006. Denitrification of groundwater in the biofilm reactor with a specific biomass support material. Acta Chim. Slov. 53:396-400.

Weiss J, Bouwer EJ, Ball WP, O'Melia CR, LeChevallier MW, Arora H, Speth TF. 2003. Riverbank filtration - fate of DBP precursors and selected microorganisms. J. AWWA 95(10):68-81.

Welander U, Henrysson T, Welander T. 1998. Biologicalnitrogenremoval from municipal landfill leachate in apilotscalesuspendedcarrierbiofilmprocess. Water Res. 32(5):1564-1570.

Welander U, Mattiasson B. 2003. Denitrification at low temperatures using a suspended carrier biofilm process. Water Res. 37(10):2394-8.

WHO. 2008. Guidelines for Drinking-water Quality. Geneva, Switzerland: WHO Press.

Woodbury BL, Dahab MF. 2001. Comparison of conventional and two-stage reversible flow, static-bed biodenitrification reactors. Water Res. 35(6):1563-1571.

Yavich AA, Masten SJ. 2003. Use of ozonation and FBT to control THM precursors. J. AWWA 95(4):159-171.

Zhang Y, Love N, Edwards M. 2009. Nitrification in drinking water systems. Crit. Rev. Environ. Sci. and Technol. 39:153-208.

Zhu S, Chen S. 2002. The impact of temperature on nitrification rate in fixed film biofilters. Aquacultural Eng. 26(4):222-237.

Ziv-El MC, Rittmann BE. 2009. Systematic evaluation of nitrate and perchlorate bioreduction kinetics in groundwater using a hydrogen-based membrane biofilm reactor. Water Res. 43:173-181.

Zumft WG. 1992. The denitrifying prokaryotes. In: Balows A, Trüper HG, Dworking M, Harder W, Schleifer KH, editors. The Prokaryotes. 2 ed. New York, Berlin, Heidelberg: Springer-Verlag. p 554–582.

Zumft WG. 1997. Cell biology and molecular basis of denitrification. Microbiol. Mol. Biol. Rev. 61(4):533-616.

Chapter 6

Application of Hybrid Process of Coagulation/Flocculation and Membrane Filtration for the Removal of Protozoan Parasites from Water

Letícia Nishi, Angélica Marquetotti Salcedo Vieira,
Ana Lúcia Falavigna Guilherme, Milene Carvalho Bongiovani,
Gabriel Francisco da Silva and Rosângela Bergamasco

Additional information is available at the end of the chapter

1. Introduction

Contamination of water resources, especially in areas with inadequate sanitation and water supply, has become a risk factor for health problems (Fundação Nacional de Saúde [FUNASA], 2003), with water playing a role as a vehicle for transmission of biological agents (viruses, bacteria, and parasites) as well as a source of contamination by chemicals (industrial effluents).

Among the waterborne diseases, enteric diseases are most frequent. Approximately 19% of waterborne gastroenteritis outbreaks in the United States are attributed to parasitic protozoans (Lindquist, 1999), particularly *Giardia* and *Cryptosporidium* species, due to their wide distribution in the environment, high incidence in the population, and resistance to conventional water treatment (Iacovski et al., 2004).

Despite regulations and control measures turning to be more and more stringent, outbreaks of waterborne *Cryptosporidium* spp. and *Giardia* spp. have been reported worldwide (United States Environmental Protection Agency [USEPA], 1996; Centers for Disease Control and Prevention [CDC], 2006). Therefore, the treatment applied to the collected water must ensure that it is free of pathogens and chemicals that pose health risks, when distributed by the water supply system. Furthermore, physicochemical parameters must meet the drinking water standards required by the laws of each country (Bergamasco et al., 2011). Thus, there is great importance in either the development of more sophisticated treatments or the improvement of the current ones.

Due to its small size and resistance to chlorine disinfection, conventional processes used in water treatment systems are unable to remove or inactivate efficiently all (oo)cysts of these protozoans, depending, among other factors, on concentration of (oo)cysts in water and the integrity of the water treatment plants. In addition to these microorganisms, many other impurities can harm human health if not reduced or eliminated. These impurities do not approach each other, it is necessary to add a coagulant.

Chemical coagulants are the most used and among them the most common is aluminum sulfate, since it is cheap and easily obtained. However, chemical coagulants have certain disadvantages, as they require tight control over their residual concentration in treated water for human consumption as well as in industrial food production. The possibility of trace aluminum contamination in food, as well as undesirable damage to the human body, especially the nervous system, are scientifically proven facts and subject of constant and innovative medical research worldwide, which increasingly requires a rigorous control of the presence of metals in both drinking water and groundwater (Ndabigengesere & Narasiah, 1998; Rondeau et al. 2000).

One alternative that arises in this context is the use of natural coagulants that have advantages over chemical coagulants since they are biodegradable and non-toxic, and produce sludge in less quantity and with lower metal content. Seeds of *Moringa oleifera* Lam (moringa), which contain active agents with coagulant properties, are an example of natural coagulants.

Moringa oleifera is a tropical plant belonging to the family of Moringaceae (Katayon et al. 2006). This plant is native to India, but is now found in other tropical regions (Bhatia et al. 2007), and it is drought tolerant. It has nutritional, medicinal and water coagulant properties. The seeds have an active compound that acts in colloidal systems, neutralizing charges and forming bridges between the particles. This process is responsible for floc formation and subsequent sedimentation (Ndabigengesere et al., 1995; Nkurunziza et al., 2009).

Studies have shown that moringa seeds have coagulation properties for treating effluents, as well as water. The seeds can be prepared either in water or in saline solution, removing color, turbidity, and total and thermotolerant coliforms (Nkurunziza et al., 2009; Madrona et al., 2010).

The use of coagulants for drinking water treatment, in spite of being efficient in the removal of most contaminants, is not able to generate water of high potability standards, which leads to the necessity of the simultaneous use of other techniques. Membrane filtration technique is already widely recognized and can be implemented in combination with coagulation processes.

The NOM found in the liquid leads to membrane fouling, flux reduction and inferior effluent quality. Therefore, the application of coagulants for the raw water pretreatment may bring about an improvement in permeates quality. This is very important, especially in the case of drinking water production. Conjunctive use of coagulation and membranes is becoming more attractive for water treatment because the coagulation is an opportunity to join NOM with other particles present in water before NOM reaches the membrane surface.

This way, this chapter will look at the use of alternative techniques for water treatment based on the use of natural coagulant (moringa seeds) associated with the membrane filtration process (microfiltration) to obtain *Giardia* spp. cysts, *Cryptosporidium* spp. oocysts, color and turbidity removal of surface water.

1.1. *Cryptosporidium* spp. and cryptosporidiosis

The oocyst is the stage transmitted from an infected host to a susceptible host by the faecal-oral route. Routes of transmission can be (1) person-to-person through direct or indirect contact, possibly including sexual activities, (2) animal-to-animal, (3) animal-to-human, (4) water-borne through drinking water or recreational water, (5) food-borne, and (6) possibly airborne (Fayer et al., 2000). The water-borne transmission depends by the level of environmental contamination, survival of the oocysts to environmental conditions (Robertson et al., 1992), and oocyst resistance to a variety of methods used to water treatment (Korich et al., 1990) as chlorination, ozonation or incomplete removal of the oocysts by filtration methods. The disease is caused by protozoan parasites of genus *Cryptosporidium* spp., intracellular coccidian parasite.

According to Neves (2005), the genus *Cryptosporidium* has been recognized in 1907 by Tyzzer, to designate a small coccidia found in the gastric glands of mice, with specific name of *C. muris*. Subsequently, in 1911, the same author found another species, lower than the first, located in the small intestine of mice, and described as *C. parvum*. Other species have been described in various animals and humans, but studies about the biology, morphology and low specificity that this coccidian has with respect to the hosts, led most researchers to consider them as synonyms of *C. muris* e *C. parvum*. In recent years this concept has undergone changes due to the use of molecular biology in the study of these parasites and some species previously considered synonymous came to be regarded as valid and others were described. The use of molecular techniques allowed to distinguish differences in the structure of some sporozoites components within the oocysts as enzymes and nucleic acids, which allowed the separation of different genotypes *C. parvum*, the most pathogenic species and increase the frequency in human infections.

Until now, it is recognized 15 species of *Cryptosporidium* (Fayer, 2004), with several subtypes (Xiao, 2010), *C. andersoni, C. baileyi, C. canis, C. felis, C. galli, C. hominis, C. meleagridis, C. molnari, C. muris, C. nasorum, C. parvum, C. saurophilum, C. serpentis, C. varanii* and *C. wrairi*. These are recognized to date seven species that can cause diarrheal disease in human, *C. hominis, C. parvum, C. meleagridis, C. felis, C. canis, C. suis* and *C. muris; C.* parvum and C. hominis are responsible... (Cacciò et al., 2005; Bouzid et al., 2008).

Although *Cryptosporidium* species are described since the early 20th century, it was only at the end of that century which the protozoan was recognized as a pathogen of cattle and horses, pets and wildlife, and a problem for public health. It was recognized that drinking or recreational water contaminated is a major source of transmission, following the discovery of resistant oocyst stage in the environment that transmits the infection of host to host via contaminated feces. The development of molecular tools to identify morphologically indistinguishable species allowed the researchers to define relationships between parasite species, potential hosts and routes of transmission (Fayer, 2004).

The diversity of reservoirs and the means of *Cryptosporidium* spp. transmission associated with their ability to survive in the environment generate an extensive network of transmission and sources of infection for humans and animals (Furtado et al., 1998). The individual's susceptibility to infection as well as the severity and duration of that vary considerably from individual to individual, depending on immune health, nutritional status and previous exposure. Similarly, an acquired immune response may limit the duration and severity of the infection. The period between the ingestion of oocysts and the development of symptoms is 7-10 days, ranging from 5-28 days (Fayer & Ungar, 1986). Infected hosts may excrete between 10^9 and 10^{10} oocysts per gram of feces, with the intake of 30 oocysts may result in human infections, which after the third day of exposure to the agent can cause profuse diarrhea with severe dehydration, intense abdominal pain, nausea, vomiting and even fever (Smith & Rose, 1998).

The rapid lifetime, the auto-infective cycles of *Cryptosporidium* can lead to large numbers of infected cells in the small intestine, resulting in secondary infection of the duodenum and large intestine. In immunocompromised patients the parasite has been found in the stomach, pancreatic and bile duct, and respiratory tract (Butler & Mayfield, 1996).

In the immunocompetent patients, the disease is self-limiting, the symptoms are often sudden with diarrhea over about 10-14 days, although there are cases described as long as three to five weeks. The symptoms disappear spontaneously and within a few months get the parasitological cure. The excretion of oocysts in the feces is intermittent and usually longer than the symptomatic phase of disease (Andrade Neto & Assef, 1996).

In immunocompromised individuals, infection is chronic and the severity and persistent diarrhea are greater than in immunocompetent individuals, it can persist indefinitely and evolve into intense dehydration and death. There is no data to indicate that immunocompromised individuals are more susceptible to infection than immunocompetent individuals. However, those with compromised immune systems are more severely affected by cryptosporidiosis (Butler & Mayfield, 1996).

The control of cryptosporidiosis is often limited by the high resistance of the oocysts to disinfectants commonly used, such as ammonia, sodium hypochlorite and chlorine. The absence of an effective treatment to combat infection increases the magnitude and enhances the importance of preventing cryptosporidiosis. The oocysts control in public water supplies, in the face of their resistance to conventional treatment is a concern worldwide, and numerous studies have been developed for this purpose (Assavasilavasukul et al., 2008; Brown & Emelko, 2009).

The small size of the oocysts, as well as apparently existing flexibility facilitate the passage of *Cryptosporidium* in filtration processes performed during the water treatment for supply. Rose (1990) report that 57% of oocysts were able to cross a membrane filter with a 3μm pore diameter. On the other hand, in analysis of washing water from rapid filters was observed high concentrations (2,906 to 24,306 oocysts/L), indicating that the filters were effective in removing oocysts (Rose, 1990). Sterling (1990) analyzing the number of oocysts detected in the raw water and in the same post-treatment, noted a removal efficiency of 91%. Filtration without prior coagulation achieved oocyst removal of 54.6% and 91.4% from effluent and surface water (rivers), respectively. Moreover, oocysts were detected in conventional water after treatment (coagulation, sedimentation, filtration and chlorination) and removal was estimated at 93.3%, with no operational problem identified (Muller, 1999).

Further studies are being conducted to improve the treatment processes, optimizing filtration processes, operating conditions, including the coagulant dosage, mixing and monitoring of filters (Rose, 1990; Brown & Emelko, 2009). Problems such as coagulant dosages, inadequate processes of mixing, inadequate monitoring of filters, turbidity, using saturated filters, among others were analyzed in Pennsylvania by Consonery et al. (1997), with optimization of processes, positive samples were reduced from 35% to values lower than 5%.

Oocysts and other microorganisms that are not effectively removed by filtration should therefore be inactivated by the disinfection process. However, the *Cryptosporidium* spp. oocysts have shown considerably resistant to the effects of most commercial disinfectants (Muller, 1999). Although chlorine and related components could reduce the ability of the oocyst to infect, the concentration relatively high and long exposure time necessary limit the practical implementation. For the majority of chemicals studied, effective concentrations are unusual for disinfection outside the laboratory, and high concentrations to reduce infectivity are expensive and toxic. Exposure to ammonia in liquid or gaseous phase and hydrogen peroxide has been effective in reducing and eliminating oocysts (Fayer, 1997).

Campbell et al. (1982) demonstrated that *Cryptosporidium* is able to remain viable, when exposed to sodium hypochlorite and chlorine. Thus, disinfection by chlorine in the concentrations commonly used in conventional water treatment is not effective in the inactivation of the parasite. Due to high doses and long period of exposure required for disinfecting, ultraviolet radiation is not considered a viable practice for the disinfection of oocysts (Fayer, 1997). Ozone has been appointed by some authors as one of the most important chemical disinfectants in inactivating oocysts (Fayer, 1997). Peeters et al. (1989) and Korich et al. (1990) reported that at a concentration of disinfectant (TC) from 3.5 – 10 mg/min/L, 99 to 99.99% of oocysts were inactivated. Laboratory studies have been conducted to assess the limits of oocysts survival exposed to heat, cold, desiccation and ultraviolet radiation. As documented in the scientific literature, temperatures above 64.2 °C for 5 minutes and 72.4 ° C for 1 minute, inactivate oocysts. On the other hand, oocysts suspended in water or milk remain infective when exposed to 71.7 °C for 5, 10 and 15 seconds. The oocysts can resist freezing at -20°C for long periods, but do not survive to -70 °C (Butler & Mayfield, 1996; Fayer, 1997).

1.2. *Giardia* spp. and giardiosis

The genus *Giardia* spp. includes flagellate parasites presented in small intestine of mammals, birds, reptiles and amphibians, and possibly the first human intestinal protozoan to be known. The first description of the trophozoite has been attributed to Anton van Leeuwenhoek (1681), who noted *Giardia* trophozoites in their own feces, but it was Larnbl in 1859, who described it in more detail. The genus was created by Kunstler (1882) when observing a flagellate present in the intestines of tadpoles of anuran amphibians (Neves, 2005).

Giardia duodenalis (syn. *G. lamblia*, *G. intestinalis*) causes giardiosis, gastroenteritis caused by unicellular protozoan. This protozoan is probably the most widespread cause of diarrhea with 200 million of symptomatic individuals in the world. The majority of infections by *G. duodenalis* are asymptomatic, and the prevalence is 2-5% in industrialized countries and 20-30% in developing countries. Regarding the symptoms may be asymptomatic or present a wide spectrum of symptoms, from self-limiting enteritis to chronic conditions with debilitating diarrhea, steatorrhea and weight loss (Rey, 2001).

Giardia and giardiosis have been extensively studied and, despite the efforts, many fundamental questions still remain unanswered. The own taxonomy is still controversial and the determination of *Giardia* species has been made considering the host of origin and morphological characteristics. According to some authors, consider the host of origin is not a valid criterion, once, by the DNA analysis, *Giardia* species from different hosts are similar, while those from the same host can be quite different (Neves, 2005). The difficulty in determining accurately the species of isolated *Giardia* from different hosts has been a limiting factor for the establishment of the zoonotic potential of giardiosis and to clarify the possibility of the existence of animals that can participate as reservoirs of *Giardia* spp.

Currently, the classification has been adopted for *G. duodenalis* is the categorization of the species in 7 genotypes or "assemblages": A, B, C, D, E, F and G. Only the A and B genotypes have been detected in humans, but both can infect other mammals, i.e, they have potential zoonotic (Bouzid et al., 2008). Subgroup AI is a mixture of isolated human and animals, since the sub-group AII is strictly isolated from humans. The genotype B comprises two subgroups III and IV and this last one is probably specific human (Thompson, 2004). The other five assemblages (CG) comprise species *G. duodenalis* restricted to host animals. Molecular analysis of *G duodenalis* isolates and other *Giardia* species conclude that the first is a complex of species and shows the need for a revision in the taxonomy of *Giardia* (Bertrand & Schwartzbrod, 2007).

The cysts are released for an extended period and in large numbers (1.4×10^{10} daily) in the feces of humans and certain animals. Can survive for several months in cold water, are relatively resistant to chlorination and ultraviolet light. The boiling is very effective in destroying the cysts, but some may resist freezing for a few days. The infection can be caused by low doses of up to 10 cysts, can occur by direct contact, very common among children in day care center and through ingestion of contaminated food or water (Ortega et al., 1997).

Due to several reasons, nowadays giardiosis is considered a reemerging parasitic. The first of these reasons relates to the large number of diarrhea outbreaks in day care centers. In developed countries, the increasing use of these institutions is one reason for the reemergence of the disease. Another reason pointed out as a cause of re-emergence of this disease relates to high rates of *Giardia* infection in domestic animals (Thompson, 2000).

The filtration processes are important barrier for removal of cysts in the water treatment. The main concern in water treatment plants for cysts removal should focus on the filtration processes, as well as optimizing operations and improvements in processes for cysts removal (Plutzer et al., 2010).

Disinfection with chlorine and derivatives is an important barrier to waterborne pathogens, however they are less effective against *Giardia* cysts (Khalifa et al., 2001). As previously mentioned, chlorine can cause the inactivation of cysts, but requires concentration and contact time high (Medema et al., 2006). Ozone is the powerful agent against protozoa, however the contact time required is still high (Haas & Kaymak, 2003). Furthermore, chemical disinfecting agents cannot be used at very high doses due to the formation of disinfection byproducts formed by the reaction with organic matter of water, such as trihalomethanes (Von Gunten, 2003).

There is also the risk of post-contamination (Karanis et al., 2007) when the water distribution system or water reservoirs are contaminated. This post-contamination can occur because of leakage of contaminants along the distribution system, on distribution reservoirs opened or due to inadequate disinfection after construction or repair in the network. Improper connections can inject water sources contaminated with domestic sewage on the distribution network. Public or domestic water tanks can be contaminated due to animals' access (Robertson et al., 2009).

Biofilms formation in the pipeline distribution network can also contribute as a potential source for water contamination by *Giardia* cysts, serving as a means of accumulation during periods of slow flow (Helmi et al., 2008).

1.3. Moringa oleifera Lam

Moringa oleifera (moringa) is a tropical plant belonging to the family Moringaceae (Katayon et al., 2006), a single family of shrubs with 14 known species. Moringa is native of India but is now found throughout the tropics (Bhatia et al., 2007). Moringa seeds contain a non-toxic natural organic polymer which is an active agent with excellent activity and coagulating properties. The tree is generally known in the developing world as a vegetable, a medicinal plant, and a source of vegetable oil (Katayon et al., 2006). Its leaves, flowers, fruits, and roots are used locally as food ingredients. The medicinal and therapeutic properties of moringa have led to its utilization as a cure for different ailments and diseases, physiological disorders, and in Eastern allopathic medicine (Akhtar et al., 2007). Additionally, the coagulant is obtained at extremely low or zero net cost (Ghebremichael et al., 2005).

If moringa is proven to be active, safe, and inexpensive, it is possible to use it widely for drinking water and wastewater treatment. Besides, moringa may yet have financial advantages bringing more economic benefits for the developing countries (Okuda et al., 1999).

The moringa seed has a protein that when solubilized in water is able to promote coagulation and flocculation of compounds that cause color and turbidity in highly turbid water. Several studies have also shown their effective antimicrobial and antifungal capacity, thereby contributing to good water quality at low cost (Chuang et al., 2007; Coelho et al., 2009). Several studies have shown the effectiveness of moringa on color, turbidity and other compounds removal present in the water.

Moringa has been found to be effective as a natural coagulant for high turbidity water in previous studies (Okuda et al., 2001). This was verified in a study by Nishi (2011). The authors obtained values of color and turbidity removal over 90% when the water to be treated showed high values of initial turbidity, between 350 and 450 NTU. A moringa concentration of 150 mg/L would have been sufficient to achieve this level of removal. The coagulant derived from moringa seeds, as it contains a certain amount of organic matter, can give color and turbidity to the treated water.

An important point to be considered when using moringa as a coagulant is related to the pH of the water to be treated. For chemical coagulants, water pH adjustment is necessary for the flakes to be properly formed. In the case of the moringa, there is no need for this adjustment, and this parameter is not changed after treatment, as evidenced by Vieira et al. (2010). Moringa is an efficient coagulant in a wide pH range (6-8), which is an advantage compared with other coagulants, as the pH adjustment step can be eliminated in the coagulation/flocculation processes.

2. Materials and methods

2.1. Water samples

Surface water used in the tests was collected in the river basin Pirapó, which supplies the Maringa city, Paraná, Brazil. Samples of the water with high and low turbidity were mixed in order to obtain different initial turbidity in the range 50 - 450 NTU. Prepared samples were artificially contaminated with 10^6 cysts/L for *Giardia* spp. and 10^6 oocysts of *Cryptosporidium* spp. obtained from the positive control (suspension of cysts and oocysts) present in the commercial kit Merifluor® (Meridian Bioscience, Cincinnati, OH, USA).

After sample preparation, these were submitted to processes of (1) coagulation/flocculation with moringa seeds (CFM), (2) microfiltration (MF) and (3) combined coagulation/flocculation with moringa seeds followed by microfiltration (CFM-MF).

2.2. Coagulation/flocculation with moringa (CFM)

The coagulant solution of moringa was prepared and used on the same day. 1 g of mature *Moringa oleifera* Lam seeds, from the Federal University of Sergipe (UFS), manually removed from the dried pods and shelled were crushed in a blender with 100 mL of distilled water. After grinding, the solution was stirred for 30 min and vacuum filtered, obtaining a solution 1% of moringa seeds (moringa concentration in solution = 10.000 mg/L) (Cardoso et al., 2008, Madrona et al., 2010). From this solution, it was used 12 concentration levels ranging from: 25 to 350 mg/L (Table 1).

Initial turbidity of water samples (NTU)	50	150	250	350	450							
Moringa solution concentration (mg/L)	25	50	75	100	125	150	175	200	225	250	275	300

Table 1. Initial turbidity and e concentrations of moringa solution used in experiments.

CFM tests were conducted in a Jar-Test equipment, Nova Ética - Model 218 LDB in six buckets, with rotation regulator of mixing rods. The experimental conditions for the CFM process were: rapid mixing gradient (100 rpm), rapid mixing time (3 min), slow mixing gradient (10 rpm), slow mixing time (15 min) and settling time (60 min) (Cardoso et al., 2008, Madrona et al., 2010).

In this process, the measured parameters in experiments were color, turbidity, pH, *Giardia* spp. and *Cryptosporidium* spp. removal.

2.3. Microfiltration process (MF) and coagulation/flocculation/ microfiltration sequence (CFM/MF)

The membrane filtration tests were performed on a microfiltration bench unit (PAM-Membranas Seletivas®) (Figure 1). This module is made of stainless steel, with polymeric membrane (1). The system is composed by manometers (2) and flow meter (3) to control the transmembrane pressure and flow rate, feed tank (4) with volume capacity of 5 liters. The permeate output (5) was collected with valve opening (6) and the return of the concentrate to the feed tank was conducted through the pipe (7).

The MF membrane employed had form of hollow fibers made up of poly (imide), with 0.40 μm pore diameter, fiber external diameter between 0.8 and 0.9 mm, fibers with external selective layer. The filtration is cross flow and the pressure used was 1.0 bar. In order to have uniformity in the tests set up the feed initial volume of 5 liters and assay time of 60 minutes.

In this part of the study, MF without pretreatment and coagulation/flocculation/ microfiltration sequence with moringa (MF-CFM) were performed. These processes were performed to observe if MF with pre-treatment (coagulation/flocculation with moringa) showed differences between the MF processes without pre-treatment. The membrane processes, in addition to analysis of *Giardia* spp., *Cryptosporidium* spp., color and turbidity removal, it was also measured permeate flux and membrane fouling.

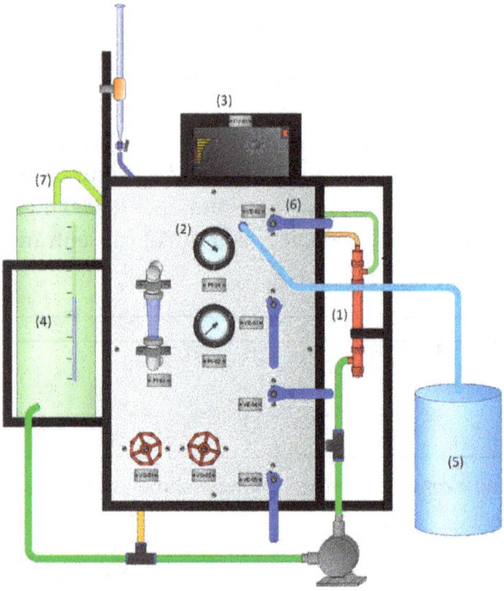

Figure 1. Scheme of front view of MF module.

2.3.1. Evaluation of permeate flux and fouling percentage of membranes for MF and CFM-MF processes

The membranes were first compacted and were then stabilized with deionized water until achieving a steady permeate flux. Permeate samples were collected at predetermined times, by known time intervals, and analytical balance were used to measure the permeate flow rate indirectly based on weight increase, according to the Equation (1).

$$f = \frac{m}{\rho.\Delta t.A_m} \tag{1}$$

where f is the permeate flux, m is permeate mass (g), ρ is the water density (25°C), Δt is the time interval during which the sample was collected (s) and A_m is the surface area.

The permeate samples were collected at shorter time intervals at the beginning of filtration, such intervals, increased subsequently to determine the curve of permeate flux versus time. The membrane filtration processes were performed using raw water (RW) without pre-treatment and after coagulation/flocculation with moringa (CFM).

The removal efficiency for each parameter analyzed using different treatment processes was calculated from Equation (2), where C_i and C_f are the initial and final concentrations, respectively, to each parameters:

$$\% \text{ removal efficiency } = \left(\frac{C_i - C_f}{C_i}\right) x100 \tag{2}$$

The water flux of deionized water (DW) were determined before each experiment ($J_{initial}$) and after the MF of RW and CFM solutions (J_{final}) for determining the fouling of the membrane. The percentage of fouling (% F) was calculated according to Equation (3), proposed by Balakrishnan et al. (2001), using stable flow values, which assume that the flow tends to constant values. This %F represents a reduction of deionized water flow after tests with contaminated water. In Equation (3), F% is the percentage of fouling, J_i is the initial flow of water obtained in the first filtration with deionized water and J_f is the final flow of water obtained with the filtration of deionized water after filtration of surface water.

$$\%F = \frac{\left(J_i - J_f\right)}{J_i} x100 \tag{3}$$

2.4. Parameters evaluated

The parameters evaluated were apparent color, turbidity, pH, *Giardia* spp. e *Cryptosporidium* spp. These parameters were analyzed according to the procedures described below.

2.4.1. Apparent color

The color was measured on a spectrophotometer HACH DR 2010, method 8025, program 120, wavelength of 455 nm, by visual comparison with platinum-cobalt standard, according to the procedure recommended by the Standard Methods (APHA, 1995). The result of color was given in uH = Hazen unit (mgPt-Co/L).

2.4.2. Turbidity

The turbidity was determined in Turbidimeter portable HACH - Model 2100P, according to the procedure recommended by the Standard Methods (APHA, 1995). The result of turbidity was expressed in NTU (Nephelometric Turbidity Units).

2.4.3. Giardia spp. e Cryptosporidium spp. analisys

The treated samples were evaluated for the presence of *Giardia* spp. e *Cryptosporidium* spp. by membrane-filtration technique, with mechanical extraction and elution (Aldom & Chagla 1995; Dawson et al. 1993; Franco et al., 2001) (Figure 2). This technique was performed as follows:

1) Filtration in cellulose acetate membrane (Millipore ®), with 47 mm pore diameter and 1.2 µm porosity, 2) mechanical extraction of the material retained on the membrane with the aid of plastic spatulas by alternating washes of the membrane surface during 10 minutes in a Petri dish with a solution of elution Tween 80 1%, 3) Repeat of procedure (2) to extract the greatest possible amount of material retained; 4) the material was centrifuged at 600 x g for

15 minutes. The supernatant was discarded with the aid of glass pipettes until the volume of 3 ml and the sediment was resuspended in distilled water by supplementing 15ml of centrifuge tube and then centrifuged again at 600 xg for 15 minutes, 5) The supernatant was discarded to a volume of 1 ml of centrifuge tube and the sediment was resuspended in this volume.

Of the sediment resuspended in 1 ml of distilled water, 5 µl were used for the direct immunofluorescence technique, using the Merifluor commercial kit (Meridian Bioscience,

Cincinnati, OH, USA). Simultaneously, a confirmatory test was performed with inclusion of the fluorogenic vital stain DAPI (4¢, 6¢-diamidine-2-phenylindole; Sigma Chemicals Co., St Louis, MO, USA) to reveal the morphological characters (nucleus, axoneme and suture) (Cantusio Neto & Franco, 2004). To read the preparations, an Olympus BX51 epifluorescent microscope was used (with excitation filter: 450–490 nm and emission filter: 520 nm for the Merifluor; excitation filter: 365–400 nm and emission filter: 395 nm for the DAPI). The numbers of oocysts and cysts per litre in the positive samples were estimated by the following calculation (EPA, 1999; Cantusio Neto & Franco, 2004):

$$X = \left(\frac{n°\text{oocysts} \times 10^6 \times \text{pellet volume (1mL)}}{\text{volume used/blade} \times \text{sample volume (mL)}} \right) \tag{4}$$

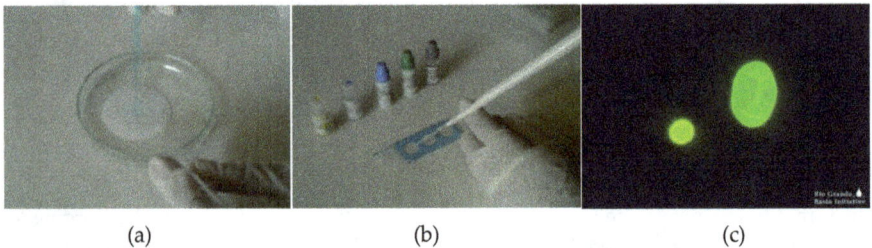

(a)   (b)   (c)

Figure 2. Membrane-filtration technique, with mechanical extraction and elution (a), analysis by direct immunofluorescence with Merifluor Kit (b) and *Cryptosporidium* spp. (left) e *Giardia* spp. (right) visualized in immunofluorescence.

3. Results

The following are the results obtained in the processes of coagulation/flocculation with moringa (CFM), microfiltration (MF) and coagulation/flocculation/microfiltration sequence with moringa (CFM-MF) for protozoan parasites, color, and turbidity removal and pH values of the of treated water samples.

3.1. Results obtained in coagulation/flocculation process with moringa (CFM)

The initial characteristics of water samples used in the study are presented in Table 2.

Turbidity (NTU)	Color (uH)	pH
50	350	7.80
150	902	7.81
250	1000	7.50
350	1849	7.64
450	1885	7.70

Table 2. Water sample parameters before treatment processes.

Table 3 presents the results of turbidity, color and parasitic protozoa removal efficiency and pH values of water samples after coagulation/flocculation with moringa.

Initial turbidity (NTU)	Removal efficiency (%)	Moringa solution concentration 1% (mg/L)											
		25	50	75	100	125	150	175	200	225	250	275	300
50	Turbidity	27.7	23.0	20.7	33.4	33.2	35.6	44.4	45.6	41.6	27.6	10.0	3.0
	Color	0.35	0.11	1.65	3.07	4.00	6.60	22.8	27.6	30.0	26.0	15.4	16.7
	Giardia spp.	6.00	42.3	38.4	69.2	84.6	76.9	82.0	80.0	80.0	76.9	85.0	69.0
	Cryptosporidium	76.0	91.0	90.0	98.0	93.0	91.0	98.0	86.0	91.0	88.0	89.0	83.0
	pH	7.90	8.20	8.10	8.20	8.20	8.10	8.20	7.90	8.00	8.10	8.00	7.90
150	Turbidity	42.0	52.4	69.8	71.0	74.0	75.8	67.7	65.3	69.0	73.6	76.0	72.0
	Color	10.0	47.5	67.0	68.8	70.8	73.5	73.5	71.4	65.0	63.4	66.4	61.6
	Giardia spp.	74.0	97.0	85.0	98.0	94.0	98.0	98.0	98.0	97.0	97.0	91.0	82.0
	Cryptosporidium	42.0	50.0	77.0	81.0	81.0	92.0	92.0	92.0	85.0	92.0	81.0	58.0
	pH	7.93	7.96	7.85	7.74	7.8	7.76	7.70	7.74	7.81	7.74	7.72	7.68
250	Turbidity	68.9	74.8	80.6	93.4	90.1	94.4	93.9	94.2	90.9	94.6	91.8	92.7
	Color	21.8	46.5	46.2	68.4	64.4	80.7	79.0	81.3	78.0	77.8	88.4	81.6
	Giardia spp.	80.0	65.0	65.0	80.0	80.0	95.0	95.0	95.0	95.0	90.0	92.5	90.0
	Cryptosporidium	67.0	61.0	75.0	74.0	86.0	96.0	95.0	92.0	94.0	90.0	87.0	78.0
	pH	7.60	7.70	7.80	7.80	7.60	7.60	7.70	7.80	7.80	7.80	7.70	7.80
350	Turbidity	49.4	62.8	70.8	75.0	82.0	90.0	93.7	95.0	96.0	95.8	96.4	92.5
	Color	47.0	82.3	76.4	94.0	94.0	88.2	97.0	97.0	94.0	88.2	94.0	94.0
	Giardia spp.	47.0	82.3	76.4	94.0	94.0	88.2	97.0	97.0	94.0	88.2	94.0	94.0
	Cryptosporidium	22.0	81.0	68.0	95.0	86.0	81.0	97.0	96.0	92.0	86.0	96.0	92.0
	pH	7.78	7.87	7.75	7.77	7.81	7.74	7.82	7.75	7.73	7.76	7.73	7.78
450	Turbidity	63.0	61.0	75.0	79.9	92	94.0	97.2	97.4	97.2	97.0	96.7	94.7
	Color	39.0	47.8	61.6	68.5	88.3	91.5	96.1	96.1	96.4	96.0	95.6	92.8
	Giardia spp.	63.0	92.0	96.0	96.0	92	96.0	90.0	97.0	94.0	97.0	93.0	89.2
	Cryptosporidium	45.0	51.0	94.0	85.0	80	94.0	86.0	93.0	95.0	98.0	90.0	76.0
	pH	7.80	7.70	7.60	7.60	7.60	7.60	7.60	7.60	7.60	7.60	7.50	7.60

Table 3. Percentage removal efficiency of turbidity, color, *Giardia* e *Cryptosporidium* and pH values after coagulation/floccullation process with moringa.

Using moringa as coagulant, turbidity ramoval was in the range of 3 – 97,4%. The lower removal efficiencies were in the range of 3-45,6% for water with low initial turbidity (50 NTU), removal below 90% were observed for water with turbidity 250, 350 e 450 NTU. The decrease in efficiency of turbidity removal in water with initial turbidity 50 NTU, after the moringa addition, can be explained by an increase in organic load. This is justifiable, considering moringa is an oilseed rich in organic substances such as oil, protein, fat, vitamins, etc.. This parameter increase in color and turbidity in water treated with moringa is observed in other studies, especially when the water has initial color and turbidity relatively low (Ramos, 2005).

Nkurunziza et al. (2009), using solution of moringa seeds 3%, prepared in saline solution, in water from rivers of province Rwanda, observed removal efficiency of 83.2% in the samples with turbidity of 50 NTU and higher removals (99.8 %) in water with turbidity of 450 NTU. The best concentrations found in this study were 150 mg/L for 50 NTU and 125mg/L for other levels of turbidity tested by researchers. The results of turbidity removal for water with low initial turbidity (50 NTU) were higher than those obtained in this study (45.6%) and for water with high initial turbidity (450 NTU) were similar (97.4%). The differences may be due to the way of moringa solution preparation by extracting aqueous or saline, and different concentrations used of the moringa stock solution (1% in this study). In both studies, it was noted that the coagulating property of the moringa is shown more effectively in water with high initial turbidity, in agreement with published studies (Ndabigengesere et al., 1995; Madrona et al, 2010).

Ndabigengesere et al. (1995) using solution of moringa seeds 5% on synthetic turbid water (kaolin) with initial turbidity of 426 NTU, obtained removal of 80-90% and reached the optimum concentration of 500 mg/L coagulant solution. This concentration is higher than optimal concentration of 450 NTU water obtained in this study, which was 275 mg/L. This difference between the optimal concentrations of the moringa solution may have occurred due to the water used in the study. Ndabigengesere et al. (1995) used synthetic water prepared with kaolin and the present study used surface water. Cardoso et al. (2008), however, used similar conditions to the present study, obtained turbidity removal of 91.4% in surface water from the river Pirapó with 247 NTU, with aqueous moringa, removal efficiency with a value close to that observed in this study (96% rof turbidity removal for water with initial turbidity of 250 NTU). Thus, the similarities or differences between the removal efficiencies of turbidity and optimal concentrations can be explained by the different types of water samples used by the works (raw water, artificial water), as well as the method of moringa solution preparation (salt extraction , aqueous) concentrations evaluated, seed quality, among other factors.

With respect to color, removal ranged from 0.11 to 30% for water with initial turbidity of 50 NTU, high removals in this sample were within the concentration range 175-250 mg/L. Water with higher initial turbidity (150-450 NTU), the removal efficiency varied from 10 to 97%, with the major removals from the concentration of 150 mg/L of the coagulant. It is observed that the color removal by Moringa is similar to its behavior with respect to turbidity, the lowest values of the color parameter are obtained in waters with high initial turbidity, which agrees with literature data (Cardoso et al., 2008; Nkurunziza et al., 2009; Madrona et al., 2010). For pH values of water samples after the coagulation process with different concentrations of moringa, it was observed that the pH is maintained, on average,

7.6, ranging from about 10%. There was little variation between samples regardless of the amount of moringa solution added, which consists of one of the advantages of moringa as a coagulating agent, i.e., its addition does not significantly alter the pH of water (Ndabigengesere et al., 1995; Nkurunziza et al., 2009), unlike treatment with aluminum sulfate in which it is necessary to adjust the pH of the water to improve its coagulating action, increasing the amount and cost of chemicals for water treatment.

Regarding the removal of *Giardia* cysts and *Cryptosporidium* oocysts were observed similar behavior between the samples. The best removal of both *Giardia* and *Cryptosporidium* occurred since the concentration of 150mg/L of moringa solution for all treated water samples, with an average removal efficiency of 93% (1.2 log removal) and 90% (1 log removal), respectively. In the literature no studies were found regarding the removal of these protozoan parasites using Moringa as a coagulating agent. The obtained high removal can be explained by the moringa coagulant action, based on the presence of cationic proteins in seeds. These proteins are heavily loaded cationic dimers with molecular weight of about 13kDa, and the charge neutralization and adsorption the main mechanisms of moringa coagulation (Ndabigengesere et al., 1995). As the calculated zeta potential of *Giardia* and *Cryptosporidium* oocysts in water at neutral pH are, on average, -17 and -38 mV, respectively (Hsu & Huang, 2002), the mechanism for charge neutralization of proteins natural coagulant could act in the removal of protozoan parasites.

The removal of protozoan parasites obtained in this study is close to the results of other studies, using chemical coagulants such as aluminum sulfate and ferric chloride, for these microorganisms removal of these microorganisms (Bustamante *et al* 2001; Xagoraraki & Harrington, 2004), and also charge neutralization the main mechanism of coagulation using aluminum sulfate in that frame. Brown & Emelko (2009) analyzed other natural coagulant, chitosan, for the removal of *Cryptosporidium parvum* in pilot scale treatment with artificial water (kaolin), using concentrations of 0.1, 0.5 and 1.0 mg/L of chitosan solution. The authors achieved excellent reductions in turbidity, but not observed good results in *C. parvum* removal with average less than 10%. A possible explanation for this difference, since the chitosan is a cationic polymer, also is the possibility that during the coagulation /flocculation process, the oocysts are also removed by engagement with the flakes, and this is another mechanism participating in the removal of the protozoan (Bustamante et al., 2001). Whereas the flakes formed depending on the characteristics of particles in the water, it can be said that the removal of microorganism will also depend on these characteristics, since Brown & Emelko (2009) used artificial water and in this study it was used natural raw water.

3.2. Results obtained in microfiltration process (MF) and coagulation/flocculation/microfiltration sequence (CFM-MF)

The results obtained in the processes of microfiltration (MF) and coagulation/flocculation with moringa followed by microfiltration (CFM-MF) are presented below. These results are presented together to show if the pretreatment (coagulation/flocculation with moringa) had differences in relation to the MF process without pretreatment. The removal efficiencies and the pH of the water treated by the MF and CFM-MF processes are presented in Table 4.

Treatment process	Removal efficiency (%)	Initial turbidity (NTU)			
		150	250	350	450
MF	Turbidity	81.09	84.16	76.82	76.33
	Color	78.28	83.45	74.27	72.56
	Giardia	ND	ND	ND	ND
	Cryptosporidium	ND	ND	ND	ND
	pH	7.38	7.85	7.36	7.81
CFM-MF	Turbidity	93.54	92.28	84.78	99.39
	Color	96.15	92.19	88.96	100.0
	Giardia	ND	ND	ND	ND
	Cryptosporidium	ND	ND	ND	ND
	pH	7.33	7.72	7.34	7.51

ND – not detected.

Table 4. Removal efficiencies of turbidity, color, *Giardia* and *Cryptosporidium*, and pH values of the water treated by the MF and CFM-MF processes.

It can be observed that the largest color and turbidity removals occurred with the combined CFM-MF process, compared with the MF process without pretreatment. There were no changes in the pH of the treated water. It is clear that the use of coagulation/flocculation with moringa prior to microfiltration improves the quality of treated water.

Few studies were found in the literature regarding the CF/MF process using moringa as a coagulant for surface water treatment. Madrona (2010) evaluated the combined process of coagulation/flocculation with moringa and MF with ceramic membranes, and obtained 97 to 100% removal of turbidity and color in the treatment of surface water from the Pirapó River, in Maringá, Paraná. These results were similar to those obtained in the present study, which used a polymer membrane for the MF process. Parker et al. (1999), using hollow fiber MF membranes with 0.2 μm pores for the treatment of water that had been previously treated in settling tanks, obtained water with turbidity below 0.1 NTU, with average removal of 99.46%, similar to those obtained in this study.

Neither in the microfiltration (MF) process alone, nor in the combined (CFM-MF) processes, (oo)cysts of *Giardia* and *Cryptosporidium* were detected in the filtered water, being below the detection limit (<1 cyst or oocyst/L) (approximately 6 log removal), in agreement with literature data. Jacangelo et al. (1995), studying the application of three MF membranes with pore sizes between 0.08 and 0.22 μm for the treatment of water contaminated with *Giardia* and *Cryptosporidium*, found that the protozoa concentration was below detectable levels in the filtered water (<1 cyst or oocyst/L) from two of the membranes (corresponding to log removal> 4.7 to> 7.0 for *Giardia* and > 4.4 to> 6.9 for *Cryptosporidium*). They also concluded that the level of removal depends on the concentration of protozoa in the water to be treated and on membrane integrity. In another study, MF membranes with average pore size of 0.2 μm resulted in significant removal of particles that were the same size as *Giardia* cysts (5-15 μm). Log removal was, on average, 3.3 to 4.4. The removal of particles that were the same size as *Cryptosporidium* oocysts (2-5 μm) was lower, 2.3 to 3.5 log removal. These removals were obtained according to the concentration of (oo)cysts used for artificial contamination of water and proved to be independent of the membrane flux (114-170 L/hm^2) (Karimi et al., 1999).

Thus, one can say that MF may act as a barrier against protozoan (oo)cysts. The coagulation/flocculation with moringa associated with microfiltration resulted in high levels of removal of the evaluated parameters.

Figure 3 shows the permeate flux versus time for the microfiltration of deionized water (DW), raw water without coagulant (SW), and pretreated water (CFM).

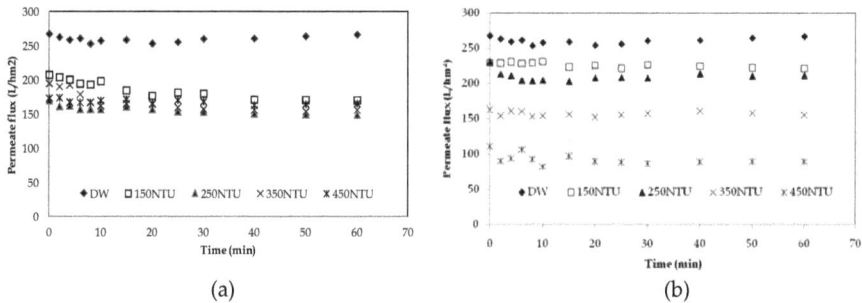

(a)   (b)

Source: Nishi, 2011.

Figure 3. Permeate flux with deionized water (DW) and raw water with initial turbidity from 150 to 450 NTU in the MF (a) and CFM-MF (b) processes.

For the MF process with raw water, that is, without previous treatment (coagulation/ flocculation), permeate flux ranged from 157 to 187 L/hm^2 for water samples of turbidity from 150 to 450 NTU. In the combined process (CFM-MF), permeate flux ranged from 157 to 226 L/hm^2 for water samples with initial turbidity of 150 to 350 NTU. Samples of 450 NTU presented the lowest permeate flux, 91 L/hm^2, on average (Nishi, 2011). This may be due to the presence of a greater number of particles that can cause the process of concentration polarization and due to superposition of various fouling mechanisms in the membrane, which may cause the decrease of the permeate flux (Stopka et al., 2001).

The combined processes of coagulation/flocculation/microfiltration showed slightly higher fluxes when compared with the microfiltration process alone. The improvement in permeate flux using coagulation/flocculation prior to microfiltration was also observed in other studies (Katayon et al., 2007; Horčičková et al., 2009).

The percentage of fouling (%F) for the MF process with raw water (SW) and water coagulated/flocculated with moringa (CFM) with initial turbidity from 150 to 450 NTU is shown in Figure 4.

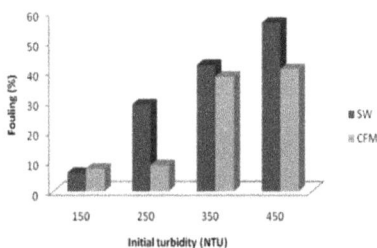

Figure 4. Percentage of fouling for the MF process with superficial water (SW) and water coagulated/flocculated with moringa (CFM) with initial turbidity from 150 to 450 NTU.

It is observed that the MF process with raw water showed higher percentages of fouling, ranging from 6.13 to 56.32% when compared with the combined process of coagulation/flocculation with moringa followed by MF, which presented percentages of fouling from 7.48 to 40.9% (Nishi , 2011). This reduction in membrane fouling when using the process of coagulation/flocculation as pretreatment was also observed in other studies. Madrona (2010) used coagulation/ flocculation with moringa, followed by MF with ceramic membranes with porosity of 0.1 and 0.2 μm, for the treatment of surface water and observed fouling percentages of around 94% during the filtration of raw water and slightly lower values, around 88%, when water previously coagulated/flocculated with moringa was filtered. Carroll et al. (2000) used polypropylene hollow fiber MF membrane to filter surface water from the Moorabool River, Australia, and observed fouling percentages of 80% for water without pretreatment and 50% for water pretreated by coagulation with alum.

According to Cheryan (1998), the type and extent of fouling depend on the chemical nature of the membrane, the solute, and the solute-membrane interactions, as well as on the porosity of the membrane and the working pressure used in the process.

4. Conclusion

It was possible to conclude that the coagulation with moringa performed satisfactory results, reducing the number of (oo)cysts of protozoan parasites in the study, 1.2 log removal for *Giardia* and 1.0 log removal for *Cryptosporidium*, which is in accordance with the WHO recommendations for the coagulation / flocculation process.

Furthermore, there was a reduction on turbidity and color approximately 97% and the pH remained stable in all water samples treated with moringa.

The combined process, coagulation/flocculation/microfiltration sequence, removed almost 100% of *Giardia* and *Cryptosporidium* (oo)cysts, besides the high color and turbidity removal (above 90%) of the water samples treated by the proposed process, with the advantage of improving the quality of filtered water and to improve the characteristics of the MF process, noting that the use of pretreatment with the coagulant moringa before the MF reduces the membrane fouling and slightly increases the permeate flow.

The use of *M. oleifera* Lam seeds can be considered advantageous and a promising step towards improving the processes of water coagulation/flocculation followed by microfiltration.

List of abbreviations

Coagulation/flocculation = CF
Coagulation/flocculation using moringa as coagulant = CFM
Coagulation/flocculation using moringa as coagulant followed by microfiltration = CFM-MF
Microfiltration = MF
Moringa oleifera = moringa
Natural organic matter = NOM

Author details

Letícia Nishi, Angélica Marquetotti Salcedo Vieira, Ana Lúcia Falavigna Guilherme, Milene Carvalho Bongiovani, Rosângela Bergamasco
Universidade Estadual de Maringá, Brazil

Gabriel Francisco da Silva
Universidade Federal de Sergipe, Brazil

Acknowledgement

The authors acknowledge the financial support provided by the research agency National Counsel of Technological and Scientific Development (CNPq).

5. References

Akhtar, M., Moosa Hasany, S., Bhanger, M. I. & Iqbal, S. (2007). Sorption potential of *Moringa oleifera* pods for the removal of organic pollutants from aqueous solutions. *Journal of Hazardous Materials*, Vol. 141, No. 3, pp. 546–556.

Aldom, J.E. & Chagla, A.H. (1995). Recovery of *Cryptosporidium* oocysts from water by a membrane filter dissolution method. *Letters in Applied Microbiology*, Vol. 20, pp. 186-187.

American Public Health Association [APHA]. [1995]. *Standard Methods for the Examination for Water and Wastewater* (19th edition). Byrd Prepess Springfield, ISBN 0875532233, Washington.

Andrade Neto, J.L. & Assef, M.C.V. (1996). Criptoporidiose e Microsporidiose. In: *Tratado de Infectologia*. Focaccia, R. & Veronesi, R. (ed.), p.1170, Atheneu, ISBN 8538801015, São Paulo.

Assavasilavasukula, P., Lau, B.L.T., Harrington, G.W., Hoffman, R.M. & Borchardt, M.A. (2008). Effect of pathogen concentrations on removal of *Cryptosporidium* and *Giardia* by conventional drinking water treatment. *Water Research*, Vol. 42, pp. 2678–2690.

Balakrishnan, M., Dua, M. & Khairnar, P.N. (2001). Significance of membrane type and feed stream in the ultrafiltration of sugarcane juice. *Journal of Separation Science and Technology*, Vol. 36, No. 4, pp. 619-637.

Bhatia, S., Othman, Z. & Ahmad, A.L. (2007). Pretreatment of palm oil mill effluent (POME) using Moringa oleifera seeds as natural coagulant. *Journal of Hazardous Materials*, Vol. 145, No. 1-2, pp. 120-126.

Bergamasco, R., Konradt-Moraes, L.C., Vieira, M.F., Fagundes-Klen, M.R. & Vieira, A.M.S. (2011). Performance of a coagulation–ultrafiltration hybrid process for water supply treatment. *Chemical Engineering Journal*, Vol. 166, pp. 483–489.

Bertrand, I. & Schwartzbrod, J. (2007). Detection and genotyping of *Giardia duodenalis* in wastewater: relation between assemblages and faecal contamination origin. *Water Research*, Vol. 41, No. 16, pp. 3675-3682.

Bouzid, M., Steverding, D. & Tyler, K.M. (2008). Detection and surveillance of waterborne protozoan parasites. *Current Opinion in Biotechnology*, Vol. 19, pp. 302–306.

Brown, T.J. & Emelko, M.B. (2009). Chitosan and metal salt coagulant impacts on *Cryptosporidium* and microsphere removal by filtration. *Water Research*, Vol. 43, pp. 331–338.

Bustamante, H.A., Shanker, S.R., Pashley, R.M. & Karaman, M.E. (2001). Interaction between *Cryptosporidium* oocysts and water treatment coagulants. *Water Research*, Vol. 35, pp. 3179-3189.

Butler, B.J.& Mayfield, C.I. (1996). *Cryptosporidium* spp. – A review of the organism, the disease, and implications for managing water resources. Waterloo Centre for Groundwater Research, Waterloo., Ontario: Canada.

Cacciò, S., Thompson, R.C., Mclauchlin, J. & Smith, H.V. (2005). Unravelling *Cryptosporidium* and *Giardia* epidemiology. *Trends in Parasitology*, Vol. 21, pp. 430-437.

Campbell, I., Tzipori, S., Hutchison, G.E. & Angus, K.W. (1982). Effect of disinfectants on survival of *Cryptosporidium* oocysts. *Veterinary Record*, Vol. 3, pp. 414-415.

Cantusio Neto, R. & Franco, R.M.B. (2004). Ocorrência de oocistos de *Cryptosporidium* spp. e cistos de *Giardia* spp. em diferentes pontos do processo de tratamento de água, em Campinas, São Paulo, Brasil. *Revista Higiene Alimentar*, Vol. 118, pp. 52-59.

Cardoso, K.C., Bergamasco, R., Cossich, E.S. & Konradt-Moraes, L.C. (2008). Otimização dos tempos de mistura e decantação no processo de coagulação/floculação da água bruta por meio da *Moringa oleifera* Lam. *Acta Scientiarum – Technology*, Vol. 30, pp. 193-198.

Carroll, T., King, S., Gray, S. R., Bolto, B. A. & Booker, N. A. (2000). The fouling of microfiltration membranes by nom after coagulation treatment. *Water Research*, Vol. 34, No. 11, pp. 2861 – 2868.

Centers for Disease Control and Prevention [CDC]. (2006). Surveillance Summaries, December 22. MMWR, 55 (No SS-12).

Cheryan, M. (1998). *Ultrafiltration and microfiltration handbook*. Technomic Publishing CO, Illinois, Lancaster, USA.

Chuang, P.H., Lee, C.W., Chou, J.Y., Murugan, M., Shieh, B.J. & Chen, H.M. (2007). Anti-fungal activity of crude extracts and essential oil of *Moringa oleifera* Lam. *Bioresource Technology*, Vol. 98, pp. 232–236.

Coelho, J.S., Santos, N.D.L., Napoleão, T.H., Gomes, F.S., Ferreira, R.S., Zingali, R.B., Coelho, L.C.B.B., Leite, S.P., Navarro, D.M.A.F. & Paiva P.M.G. (2009). Effect of *Moringa oleifera* lectin on development and mortality of *Aedes aegypti* larvae. *Chemosphere*, Vol. 77, No. 7, pp. 934-938.

Consonery, P.J., Greenfield, D.N. & Lee, J.J. (1997). Pennsylvania's filtration evaluation programme. *Journal of the American Water Works Association*, Vol. 89, pp. 67-77.

Dawson, D.J., Maddocks, M., Roberts, J. & Vidler, J.S. (1993). Evaluation of recovery of *Cryptosporidium parvum* oocysts using membrane filtration. *Letters in Applied Microbiology*, Vol. 17, pp. 276-279.

Fayer, R. & Ungar, B.L.P. (1986). *Cryptosporidium* spp and cryptosporidiosis. *Microbiology Reviews*, Vol. 50, pp. 458-484.

Fayer, R., (1997). *Cryptosporidium* and cryptosporidiosis. Boca Raton: CRC Press, p. 251.

Fayer, R., Morgan, U. & Upton, S.J. (2000). Epidemiology of Cryptosporidium: transmission, detection and identication. *International Journal for Parasitology*, Vol. 30, pp. 1305-1322.

Fayer, R. (2004). *Cryptosporidium*: a water-borne zoonotic parasite. *Veterinary Parasitology*, Vol. 126, pp. 37–56.

Franco, R.M.B., Rocha-Eberhardt, R. & Cantusio Neto, R. (2001). Occurrence of *Cryptosporidium* oocysts and *Giardia* cysts in raw water from the Atibaia river, Campinas, Brazil. *Rev. Inst. Med. Trop. S. Paulo*, Vol. 43, No. 2, pp. 109-111.

Fundação Nacional de Saúde (National Health Foundation) [FUNASA]. *Toxic cyanobacterias in water for human consumption in public health, and process for their removal from water for human consumption*. Ministry of Health, Brasília, (2003) 56 pp. (In Portuguese).

Furtado, C., Adak, G.K., Stuart, J.M., Wall, P.G., Evans, H.S. & Casemore, D.P. (1998). Outbreaks of waterborne infectious intestinal disease in England and Wales, 1992-5. *Epidemiology and Infection*, Vol. 121, pp. 109-119.

Ghebremichael, K.A., Gunaratna, K.R., Henriksson, H., Brumer, H. & Dalhammar, G. (2005). A simple purification and activity assay of the coagulant protein from Moringa *oleifera* seed. *Water Research*, Vol. 39, No. 11, pp. 2338-2344.

Haas, C.N. & Kaymak, B. (2003). Effect of initial microbial density on inactivation of *Giardia muris* by ozone.*Water Research*, Vol. 37, pp. 2980–2988

Helmi, K., Skraber, S., Gantzer, C., Willame, R., Hoffmann, L. & Cauchie, H.M. (2008). Interactions of *Cryptosporidium parvum*, *Giardia lamblia*, vaccinal poliovirus type 1, and bacteriophages phiX174 and MS2 with a drinking water biofilm and a wastewater biofilm. *Applied and Environmental Microbiology*, Vol. 74, pp. 2079–2088.

Horčičková, J., Mikulášek, P. & Dvořáková, J. (2009). The effect of pre-treatment on crossflow microfiltration of titanium dioxide dispersions. *Desalination*, Vol. 240, pp. 257-261.

Hsu, B.M. & Huang, C. (2002). Influence of ionic strength and pH on hydrophobicity and zeta potential of *Giardia* and *Cryptosporidium*. *Colloids and Surfaces A: Physicochemical and Engineering Aspects*, Vol. 201, pp. 201-206.

Iacovski, R.B., Barardi, C.R.M. & Simões, C.M.O. (2004). Detection and enumeration of Cryptosporidium sp. oocysts in sewage sludge samples from the city of Florianópolis

(Brazil) by using immunomagnetic separation combined with indirect immunofluorescence assay. *Waste Management & Research*, Vol. 22, pp. 171–176.

Jacangelo, J. G., Adham, S. S. & Laîné, J-M. (1995). Mechanism of *Cryptosporidium, Giardia* and MS2 virus removal by MF and UF. *Journal of the American Water Works Association*, Vol. 87, No. 9, pp. 107–121.

Karimi, A.A., Vickers, J.C. & Harasick, R.F. (1999). Microfiltration goes Hollywood: the Los Angles experience. *Journal of the American Water Works Association*, Vol. 91, No. 6, pp. 90–103.

Karanis, P., Kourenti, C. & Smith, H. (2007). Water-borne transmission of protozoan parasites: a review of world-wide outbreaks and lessons learnt. *Journal of Water and Health*. 5: 1–38.

Katayon, S., Noor, M.J.M.M., Asma, M., Ghani, L.A.A., Thamer, A.M., Azni, I., Ahmad, J., Khor, B.C. & Suleyman, A.M. (2006). Effects of storage conditions of *Moringa oleifera* seeds on its performance in coagulation. *Bioresource Technology*, Vol. 97, No. 13, pp. 1455–1460.

Khalifa, A.M., El Temsahy, M.M. & Abou El Naga, I.F. (2001). Effect of ozone on the viability of some protozoa in drinking water. *Journal of the Egyptian Society of Parasitology*, Vol. 31, pp. 603–616.

Korich, D.G., Mead, J.R., Madore, M.S., Sinclair, N.A. & Sterling, C.R. (1990). Effects of ozone, chlorine dioxide, chloride, and monochloramine on *Cryptosporidium parvum* oocyst viability. Applied and Environmental Microbiology, Vol. 56, pp. 1423-1428.

Lindquist, A. (1999). *Emerging pathogens of concern in drinking water*. United States Environmental Protection Agency, EPA, 600/R-99/070.

Madrona, G.S., Serpelloni, G.B., Vieira, A.M.S., Nishi, L., Cardoso, K.C & Bergamasco, R. (2010). Study of the effect of saline solution on the extraction of the *Moringa oleifera* seed's active component for water treatment. *Water, Air, & Soil Pollution*, Vol. 211, pp. 409–415.

Medema, G., Teunis, P., Blokker,M., Deere, D., Davison, A., Charles, P. & Loret, J.F. (2006). WHO Guidelines for Drinking Water Quality, Environmental Health Criteria, *Cryptosporidium*, Draft 2.

Muller, A.P.B., 1999, Detecção de oocistos de *Cryptosporidium* spp em águas de abastecimento superficiais e tratadas da região metropolitana de São Paulo. Dissertação de Mestrado, Instituto de Ciências Biomédicas da USP – São Paulo (in portuguese).

Ndabigengesere A., Narasiah, S.K. & Talbot, B.G. (1995). Active agents and mechanism of coagulation of turbid waters using *Moringa oleifera, Water Research*, Vol. 29, No. 2, pp. 703-710.

Ndabigengesere, A. & Narasiah, K.S. (1998). Quality of water treated by coagulation using *Moringa oleifera* seeds. *Water Research*, Vol. 32, pp. 781–791.

Neves, D.P. (2005). *Parasitologia humana*. Atheneu. ISBN 8573797371. São Paulo.

Nishi, L. (2011). Estudo dos processos de coagulação/floculação seguido de filtração com membranas para remoção de protozoários parasitas e células de cianobactérias. Doctoral Thesis, Universidade Estadual de Maringá – Maringá, PR, Brasil, 203 pp. (in Portuguese).

Nkurunziza, T., Nduwayezu, J.B., Banadda, E.N. & Nhapi, I. (2009). The effect of turbidity levels and *Moringa oleifera* concentration on the effectiveness of coagulation in water treatment. *Water Science and Technology*, Vol. 59, pp. 1551–1558.

Okuda, T., Baes, A.U., Nishijima, W. & Okada, M. (1999). Improvement of extraction method of coagulation active components from *Moringa oleifera* seed. *Water Research*, Vol. 33, No. 15, pp. 3373–3378.

Okuda, T., Baes, A.U., Nishijima, W. & Okada, M. (2001). Isolation and characterization of coagulant extracted from *Moringa oleifera* seed by salt solution. *Water Research*, Vol. 35, No. 2, pp. 405-410.

Ortega, Y.R., Roxas, C.R., Gilman, R.H., Miller, N.J., Cabrera, L., Taquiri, C. & Sterling, C.R. (1997). Isolation of *Cryptosporidium parvum* and *Cyclospora cayetanensis* from vegetables collected in markets of a endemic region in Peru. The American Journal of Tropical Medicine and Hygiene, Vol. 57, No. 6, pp. 683-686.

Parker, D. Y., Leonard, M. J., Barber, P., Bonic, G., Jones W. & Leavell, K. L. (1999). Microfiltration treatment of filter backwash recycle water from a drinking water treatment facility. Proceedings of the American Water Works Association Water Quality Technology Conference. Denver, CO, American Water Works Association.

Peeters, J.E., Ares Mazás, E., Masschelein, W.J., Villacorta-Martinez De Maturana, I. & Debacker, E. (1989). Effect of disinfection of drinking water with ozone or chlorine dioxide on survival of *Cryptosporidium parvum* oocysts. Applied and Environmental Microbiology, Vol. 55, pp. 1519-1522.

Plutzer, J., Ongerth, J. & Karanis, P. (2010). *Giardia* taxonomy, phylogeny and epidemiology: facts and open questions. *International Journal of Hygiene and Environmental Health*, Vol. 213, pp. 321–333.

Ramos, R.O. (2005). Clarification of water with low turbulence and moderate color using seeds of *Moringa oleifera*. State University of Campinas, Campinas-SP, Brazil. 276 pages. (Doctoral Thesis; in Portuguese).

Rey, L. (2001). *Parasitologia*. Guanabara Koogan. ISBN 9788527714068 .Rio de Janeiro.

Robertson, L.J., Campbell, A.T. & Smith, H.V. (1992). Survival of *Cryptosporidium parvum* oocysts under various environmental pressures. *Applied and Environmental Microbiology*, Vol. 58, pp. 3494-3500.

Robertson, L., Gjerde, B., Hansen, E.F. & Stachurska-Hagen, T. (2009). A water contamination incident in Oslo. Norway during October 2007; a basis for discussion of boil-water notices and the potential for post-treatment contamination of drinking water supplies. *Journal of Water and Health*, Vol. 7, pp. 55–66.

Rondeau, V., Commenges, D., Jacqmin-Gadda, H. & Dartigues, J.F. (2000). Relation between aluminum concentrations in drinking water and Alzheimer's disease: an 8-year follow-up study. *American Journal of Epidemiology*, Vol. 152, pp. 59-66.

Rose, J.B. (1990). Occurrence and control of *Cryptosporidium* in drinking water. In: *Drinking Water Microbiology*. p. 294-321.Springer-Verlag, New York.

Smith, H.V. & Rose, J.B. (1998). Waterbone cryptosporidiosis: current status. *Parasitology Today*, Vol. 14, pp. 14-22.

Sterling, C.R. (1990). Waterborne Cryptosporidiosis. In: *Cryptosporidiosis in Man and Animals.* Dubey, J.P., Speer, C.A. & Fayer, R. Ed. CRC Press, Boca Raton, FL. p. 51-59.

Stopka, J., Bugan, S. G. & Broussous, L. (2001). Microfiltration of beer yeast suspensions through stamped ceramic membranes. *Separation and Purification Technology,* Vol. 25, pp. 535-543.

Thompson, R.C.A. (2000). Giardiasis as a re-emerging infectious disease and its zoonotic potential. *International Journal of Parasitology,* Vol. 30, pp. 1259-1267.

Thompson, R.C.A. (2004). The zoonotic significance and molecular epidemiology of *Giardia* and giardiasis. *Veterinary Parasitology,* Vol. 126, pp. 15–35.

United States Environmental Protection Agency [USEPA]. 1996. *National Primary Drinking Water Regulations: Monitoring Requirements for Public Drinking Water Supplies; Final Rule.* 40CFR Part 141.

Vieira, A.M.S., Vieira, M.F., Silva, G.F., Araújo, A.A., Fagundes-Klen, M.R., Veit, M.T. & Bergamasco, R. (2010). Use of *Moringa oleifera* Seed as a Natural Adsorbent for Wastewater Treatment. *Water, Air, & Soil Pollution,* Vol. 206, pp. 273–281.

Von Gunten, U. (2003). Ozonation of drinking water. Part II. Disinfection and by-product formation in presence of bromide, iodide or chlorine. *Water Research,* Vol. 37, pp. 1469–1487.

Xagoraraki, I. & Harrington, G.W. (2004). Zeta potential, dissolved organic carbon, and removal of *Cryptosporidium* oocysts by coagulation and sedimentation. *Journal of Environmental Engineering,* Vol. 130, pp. 1424-1432.

Xiao, L. (2010). Molecular epidemiology of cryptosporidiosis: an update. *Experimental Parasitology,* Vol. 124, pp. 80–89.

Chapter 7

Waste Water Treatment Methods

Adina Elena Segneanu, Cristina Orbeci, Carmen Lazau,
Paula Sfirloaga, Paulina Vlazan, Cornelia Bandas and Ioan Grozescu

Additional information is available at the end of the chapter

1. Introduction

The last decades have shown a reevaluation of the issue of environmental pollution, under all aspects, both at regional and at international level. The progressive accumulation of more and more organic compounds in natural waters is mostly due to the development and extension of chemical technologies for organic synthesis and processing.

Population explosion, expansion of urban areas increased adverse impacts on water resources, particularly in regions in which natural resources are still limited. Currently, water use and reuse has become a major concern. Population growth leads to significant increases in default volumes of waste water, which makes it an urgent imperative to develop effective and affordable technologies for wastewater treatment.

The physico-chemical processes common treatment (coagulation and flocculation) using various chemical reagents (aluminum chloride or ferric chloride, polyelectrolytes, etc.) and generates large amounts of sludge. Increasing demands for water quality indicators and drastic change regulations on wastewater disposal require the emergence and development of processes more efficient and more effective (ion exchange, ultrafiltration, reverse osmosis and chemical precipitation, electrochemical technologies). Each of these treatment methods has advantages and disadvantages.

Water resources management exercises ever more pressing demands on wastewater treatment technologies to reduce industrial negative impact on natural water sources. Thus, the new regulations and emission limits are imposed and industrial activities are required to seek new methods and technologies capable of effective removal of heavy metal pollution loads and reduction of wastewater volume, closing the water cycle, or by reusing and recycling water waste.

Advanced technologies for wastewater treatment are required to eliminate pollution and may also increase pollutant destruction or separation processes, such as advanced oxidation methods (catalytic and photocatalytic oxidation), chemical precipitation, adsorption on various media, etc.. These technologies can be applied successfully to remove pollutants that are partially removed by conventional methods, e.g. biodegradable organic compounds, suspended solids, colloidal substances, phosphorus and nitrogen compounds, heavy metals, dissolved compounds, microorganisms that thus enabling recycling of residual water. (Zhou, 2002) Special attention was paid to electrochemical technologies, because they have advantages: versatility, safety, selectivity, possibility of automation, environmentally friendly and requires low investment costs (Chen, 2004; Hansen et. al., 2007).

The technologies for treating wastewater containing organic compounds fall within one of the following categories:

- Non-destructive procedures – based on physical processes of adsorption, removal, stripping etc.
- Biological destructive procedures – based on biological processes using active mud.
- Oxidative destructive processes – based on oxidative chemical processes which, in their turn, can fall within one of the following categories:
  - Incineration;
  - WO - "Wet Oxidation", operating in conditions of high temperature and pressure, with the versions:
    - WAO - "Wet Air Oxidation" (wet oxidation with O_2 air oxidative agent);
    - CWAO - "Catalytic Wet Air Oxidation" (catalytic wet oxidation with O_2 air oxidative agent);
    - SWA - "Supercritical Water Oxidation" (oxidation with O_2 air oxidative agent in supercritical conditions).
  - Liquid oxidation: AOPs - "Advanced Oxidation Processes", operating in conditions of temperature and pressure and use as oxidative agents O_3, H_2O_2 and even O_2, catalysts and/ or UV radiations.

2. Advanced oxidation processes

Advanced oxidation processes (AOPs) are widely used for the removal of recalcitrant organic constituents from industrial and municipal wastewater. In this sense, AOP type procedures can become very promising technologies for treating wastewater containing non-biodegradable or hardly biodegradable organic compounds with high toxicity. These procedures are based on generating highly oxidative HO radicals in the reaction medium.

- H_2O_2

  H_2O_2 + UV (direct photolysis)

  H_2O_2 + $Fe^{2+/3+}$ (classic, homogeneous Fenton)

  H_2O_2 + Fe/support (heterogeneous Fenton)

  H_2O_2 + $Fe^{2+/3+}$ + UV (VIS) (Photo-Fenton)

- O_3

  O_3 (direct ozone feeding)

  O_3 + UV (photo-ozone feeding)

  O_3 + catalysts (catalytic ozone feeding)

- $H_2O_2 + O_3$

  TiO_2 (heterogeneous catalysis)

  TiO_2 + UV (photo-catalysis)

The preferential use of H_2O_2 as oxidative agent and HO radicals generator is justified by the fact that the hydrogen peroxide is easy to store, transported and used, and the procedure is safe and efficient.

The technologies developed so far indicate the use of zeolites, active coal, structured clay, silica textures, Nafion membranes or Fe under the form of goethit (α-FeOOH), as support materials for the catalytic component.

The AOPs (Advanced Oxidation Processes) can be successfully used in wastewater treatment to degrade the persistent organic pollutants, the oxidation process being determined by the very high oxidative potential of the HO$\cdot$ radicals generated into the reaction medium by different mechanisms *(Pera-Titus et al., 2004)*.

AOPs can be applied to fully or partially oxidize pollutants, usually using a combination of oxidants. Photo-chemical and photocatalytic advanced oxidation processes including UV/H_2O_2, UV/O_3, UV/H_2O_2/O_3, UV/H_2O_2/Fe^{2+}(Fe^{3+}), UV/TiO_2 and UV/H_2O_2/TiO_2 can be used for oxidative degradation of organic contaminants. A complete mineralization of the organic pollutants is not necessary, being more worthwhile to transform them into biodegradable aliphatic carboxylic acids followed by a biological process *(Wang and Wang, 2007)*.

The oxidation process is determined by the very high oxidative potential of the HO$\cdot$ radicals generated into reaction medium by different mechanisms. In the case of the AOPs Fenton-type procedure (hydrogen peroxide and Fe^{2+} as catalyst), the generation of hydroxyl radicals takes place through a catalytic mechanism in which the iron ions play a very important role *(Andreozzi et al., 1999; Esplugas, et al., 2002)* the main reactions involved being presented in equations (1) – (4):

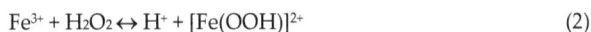

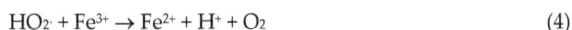

$$Fe^{2+} + H_2O_2 \rightarrow Fe^{3+} + HO^- + HO \tag{1}$$

$$Fe^{3+} + H_2O_2 \leftrightarrow H^+ + [Fe(OOH)]^{2+} \tag{2}$$

$$[Fe(OOH)]^{2+} \rightarrow Fe^{2+} + HO_2\cdot \tag{3}$$

$$HO_2\cdot + Fe^{3+} \rightarrow Fe^{2+} + H^+ + O_2 \tag{4}$$

In the presence of UV radiations (photo-Fenton process), an additional number of HO· radicals are produced both through direct H_2O_2 photolysis and through UV radiations interaction with the iron species in aqueous solutions (eq. 5-7) *(Spacek et al., 1995; Pignatello, J.J., 1992):*

$$H_2O_2 + UV \rightarrow 2HO \tag{5}$$

$$Fe^{3+} + H_2O + UV \rightarrow Fe^{2+} + H^+ + HO· \tag{6}$$

$$[Fe(OH)]^{2+} + UV \rightarrow Fe^{2+} + HO· \tag{7}$$

The main parameters which determine the efficiency of the oxidation process are: the structure of the organic compounds, the hydrogen peroxide and the catalyst concentrations, the wave length and intensity of UV radiations, the initial solution ph and the reaction contact time.

As recalcitrant organic pollutants continue to increase in air and wastewater streams, environmental laws and regulations become more stringent (Gayaa et al, 2008). The main causes of surface and groundwater contamination are industrial effluents (even in small amounts), excessive use of pesticides, fertilizers (agrochemicals) and domestic waste landfills. Wastewater treatment is usually based on physical and biological processes. After elimination of particles in suspension, the usual process is biological treatment (natural decontamination), but unfortunately, some organic pollutants, classified as bio-recalcitrant, are not biodegradable. In this way advanced oxidation processes (AOPs) may become the most widely used water treatment technologies for organic pollutants not treatable by conventional techniques due to their high chemical stability and/or low biodegradability (Munoz et al.2005). Advanced oxidation processes are indicated for removal of organic contaminants such as halogenated hydrocarbons (trichloroethane, trichlorethylene), aromatics (benzene, toluene, and xylene), pentachlorophenol (PCP), nitrophenol, detergents, pesticides, etc. These processes can also be applied to oxidation of inorganic contaminants such as cyanides, sulfides and nitrites (Munter, 2001). A general classification of advanced oxidation processes based on source allowing radicals. This classification is presented in Figure 1.

Heterogeneous photocatalysis has proved to be of real interest as efficient tool for degrading both aquatic and atmospheric organic contaminants because this technique involved the acceleration of photoreaction in presence of semiconductor photocatalyst (Guillard, 1999). Thus these processes can be classified in: advanced oxidation processes based on ozone based advanced oxidation processes H_2O_2, photocatalysis, POA "hot" technologies based on ultrasound, electrochemical oxidation process, oxidation processes with electron beam. These processes involve generation and subsequent reaction of hydroxyl radicals (_OH), which are one of the most powerful oxidizing species. Photocatalytic reaction is initiated when a photoexcited electron is promoted from the filled valence band of semiconductor photocatalyst to the empty conduction band as the absorbed photon energy, h_, equals or exceeds the band gap of the semiconductor photocatalyst leaving behind a hole in the valence band. Thus in concert, electron and hole pair (e—h+) is generated (Horvath, 2003). An ideal photocatalyst for photocatalytic oxidation is characterized by the following attributes: photo-stability, chemically and biologically inert nature, availability and low cost

(Carp et. al., 2004). Many semiconductors such as TiO_2 (Lazau, 2011), ZnO (Daneshvar et al., 2007), ZrO_2 (Lopez et al. 2007), CdS (Yingchun, 2011), MoS_2 (Kun Hong, 2011), Fe_2O_3 (Seiji, 2009) and WO_3 (Yuji, 2011) have been examined and used as photocatalysts for the degradation of organic contaminants. TiO_2 is most preferred one due to its chemical and biological inertness, high photocatalytic activity, photodurability, mechanical robustness and cheapness. Thus, these materials were used in the degradation of phenol, 1,4-dichlorobenzene (Papp et al., 1993),methanol (Nobuaki et. al., 2007), azo dye (Daneshvar, 2003), trichloromethane, hexachloro cyclohexane (Byrappa et. al., 2002), trichloroethylene and dichloropropionic acid (Nikola, 2001). To avoid the problem of filtration, many methods were proposed to immobilize the photocatalysts, but in these conditions the photocatalyst is expected to be used for a relatively long time, especially for industrial applications (Venkata, 2004). Various substrates have been used as a catalyst support for the photocatalytic degradation of polluted water. For example glass materials: glass mesh, glass fabric, glass wool, glass beads and glass reactors were very commonly used as a support for titania. Other uncommon materials such as microporous cellulosic membranes, alumina clays, ceramic membranes, monoliths , zeolites , and even stainless steel were also experimented as a support for TiO2 (Gianluca, 2008).

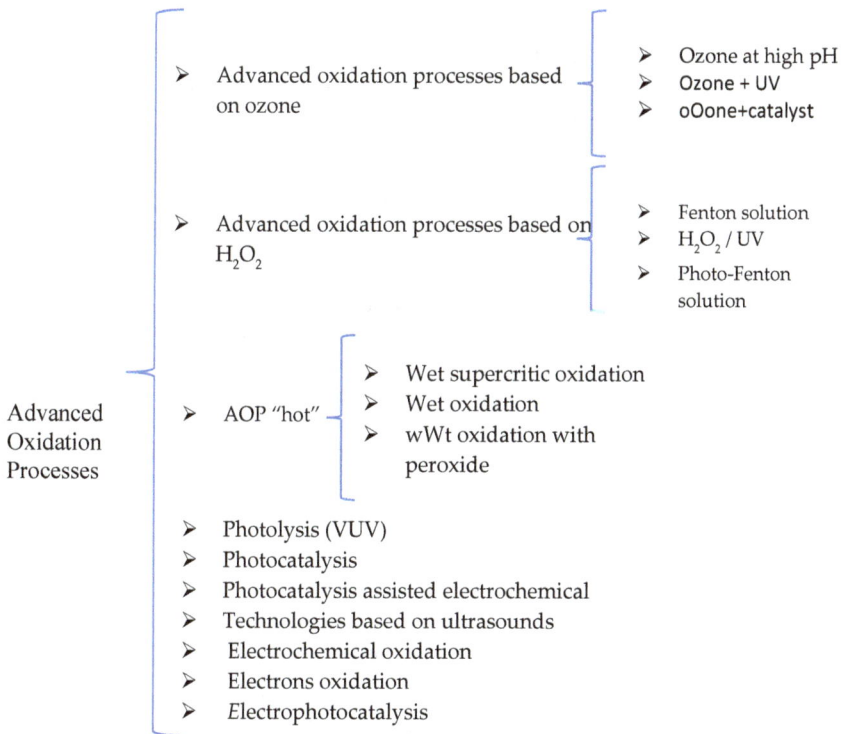

Figure 1. Classification of advanced oxidation processes

Advanced oxidation processes (AOPs) and electrochemical oxidation is based on the in-situ generation of OH radicals, which allow its non-selective reaction with organics allowing organics mineralization by its conversion into CO_2. The electrochemical methods are very promising alternatives for organics degradation because of their environmental compatibility, versatility, simplicity, and easy possibility of automation. The electrochemical oxidation performance depends strongly on the electrode material. To generate OH radicals by electrooxidation, several types of anodes with high overpotential for oxygen potential are suitable, i.e., DSA-type, PbO_2, boron-doped diamond (BDD) electrodes etc. Recently, electrochemical oxidation with a boron-doped diamond electrode is one of the most promising technologies in the treatment of the industrial effluents containing organics. BDD electrode exhibited a very good chemical stability and its application in the electrooxidation of organics led to complete mineralization into CO_2 in relation with applied potential or current density. A major drawback of the electrochemical oxidation consists of the high energy consumption to the mineralization. The presence of a catalyst in the electrical field or combined and direct photoelectrochemical application can enhance the treatment efficiency with lower energy consumption (Ratiu et. al., 2010).

Electrochemical and photochemical technologies may offer an efficient means of controlling pollution. Their effectiveness is based on the generation of highly reactive and non-selective hydroxyl radicals, which are able to degrade many organic pollutants. Electrolysis, heterogeneous photocatalysis, or photo-assisted electrolysis may be regarded as advanced oxidation processes (AOPs) and used in the supplementary treatment of wastewaters. The efficiency of the electrochemical oxidation depends on the anode material and the operating conditions, e.g., current density or potential. In general, in most applications of photoelectrocatalysis in the degradation of organics, the applied anodic bias potential is lower than the oxidation potential of organics on the electrode, due to direct electro-oxidation does not complicate the photocatalytic mechanism (Ratiu et. al.,2011).

The efficiency of photoelectrochemical degradation for organic pollutants depends not only on the selection of a suitable supporting electrolyte and pH values, but also on the electrode potential and preparation conditions of the semiconductors involved. In a photoelectrochemical system, photoelectrons and photoholes can be separated under the influence of an applied electric field. The problem of the separation of semiconductor particles from the treated solution, so persistent in heterogeneous photolysis, is not an issue in photoelectrochemical systems. There are numerous semiconductors which can be used as photoelectrocatalytic materials, such as TiO_2, WO_3, SnO_2, ZnO, CdS, diamond, and others (Hepel 2005).

2.1. Particular aspects

In the case of the chlorinated phenols, the number and the position on aromatic ring of the chlorine atoms modifies the oxidation efficiency *(Pera-Titus et al., 2004)*.

The oxidation rate constant decreases linearly with increasing number of chlorine content on the aromatic ring. Also, the increase of chlorine content will block some favorable positions susceptible to hydroxyl radical attack.

The oxidation process is also controlled by the presence of another species in reaction medium (intermediate products) in the sense that they interact with the catalyst component in a different manner. The species of reductive character accelerate the oxidation process because they reduce Fe^{3+} (inactive) to Fe^{2+} (active) and thus the generation of OH· radicals intensifies *(Du et al. 2006; Riga, A., et al, 2007)*. The acid type species lower the pH of the reaction medium and can form stable complexes with Fe^{3+} or Fe^{2+} ions, strongly slowing the oxidation process. The presence into reaction medium of the inorganic ionic species (Cl^-, ClO_4^-, NO_3^-, HCO_3^-, CO_3^{2-}, SO_4^{2-}, $H_2PO_4^-$) modifies the rate of the oxidation of the organic compounds as function of their nature and concentration. The inorganic anions can change the overall efficiency of the system by different ways. The influence of ClO_4^- and NO_3^- ions is less pronounced than another anions because they do not form complexes with Fe(II) and Fe(III) and do not react with HO· *(Lu et al., 1997; Siedlecka, E. M., 2007)*. Cl^-, SO_4^{2-} and $H_2PO_4^-$ anions decrease the rate of decomposition of H_2O_2 by forming ferric unreactive complexes and react with hydroxyl radicals forming Cl_2^-, SO_4^- and H_2PO_4 radicals who are less or much less reactive than HO· *(Siedlecka, E. M., 2007; De Laat et al., 2004)*. The influence of Cl^- is in correlation with the solution pH and its concentration, being insignificant at low concentration (<5 mM) but becomes very important at higher concentration values (>28 mM) *(Kwon et al., 1999)*. As function of the nature of the inorganic anions, at higher concentration of 0.1 M, the inhibition order of the oxidation rate is the following: $H_2PO_4^- > Cl^- > HCO_3^- > CO_3^{2-} > SO_4^{2-} > NO_3^-$ *(Riga, A., et al., 2007)*.

The presence of the inorganic species inside the reaction medium influences the rate of the oxidation process as function of their nature and concentration. The inorganic anionic species reduce the 4-CP oxidation efficiency by Fe(II) and Fe(III) complexes forming, HO· radicals scavenging or iron precipitate forming.

NO_3^- induces a small influence on 4-CP oxidation efficiency. This may be explained by the absence of the interactions between NO_3^- and the catalyst ($Fe^{2+/3+}$) and hydroxyl radicals. The anions Cl^-, SO_4^{2-} and PO_4^{3-} modify drastically the 4-CP oxidation efficiency, especially at high concentration into reaction medium. They interact with Fe^{2+} and Fe^{3+} forming chloro-, sulfato- and phosphate-iron complexes which are inactive in HO· generation mechanism. Also, Cl^-, SO_4^{2-} and $H_2PO_4^-$ anions interact with hydroxyl radicals (scavenging process), forming less reactive species (Cl_2^-, SO_4^- and H_2PO_4) into reaction medium.

The decrease of the 4-CP oxidation degree by the photo-Fenton procedure is correlated with the nature of the anions as following: $Cl^- > PO_4^{3-} > SO_4^{2-} \gg NO_3^-$ *(Orbeci et al., 2008)*.

The presence of the insoluble inorganic species (bentonite) modifies the 4-CP oxidation efficiency in different manner. Into reaction medium, 4-CP can be adsorbed by the bentonite substratum or can be destruct by oxidation, both processes increasing the 4-CP removal degree from the solution. The presence of the insoluble inorganic species (bentonite) modifies the oxidation efficiency by additional 4-CP and UV sorption processes, especially at high solution turbidity values *(Orbeci et al., 2008)*.

The efficiency of the various AOPs depends both on the rate of generating the free radicals and the extent of contact between the radicals and the organic compound. Also, the pH has a significant role in determining the efficiency of Fenton and photo-Fenton oxidation processes *(Gogate and Pandit, 2004)*. Limitations due to the use of homogeneous catalysts, such as limited pH range, production of Fe containing sludge, and deactivation could be overcome by heterogeneous catalysts.

The optimum pH range in the case of homogeneous photo-Fenton process is 2.5-4, a correction of solution pH being necessary. Also, at the end of the oxidation process, iron precipitation and catalyst separation and recovery are necessary. These disadvantages can be avoided using the heterogeneous photo-Fenton procedure by immobilization of active iron species on small particulate solid supports. In this case, different iron-containing catalysts can be used, such as the iron bulk catalysts (iron oxy-hydroxyl compounds: hematite, goethite, magnetite) or iron supported catalysts (zeolites, clays, bentonite, glass, active carbon, polymers etc.) *(Duarte and Madeira, 2010; Feng et al., 2005; He et al., 2005; Leland and Bard, 1987; Nie et al., 2008; Ortiz de la Plata et al., 2010; Vinita et al. 2010)*.

The use of the heterogeneous photo-Fenton procedure in the catalytic component version (Fe in various oxidation states) precipitated on solid support presents several drawbacks:

- catalyst's relatively high cost – associated with the cost of the so-called support, with the cost of the Fe compounds and with the operations necessary for Fe compounds to fix on the support;
- decrease in the efficiency of the UV radiations due to their partial adoption on the solid support;
- progressive solubility of the catalytic component (Fe) during oxidative processes and as a result of a progressive loss of catalytic activity.

In the case of the heterogeneous photo-Fenton process, a relevant fraction of the incident UV radiation can be lost via scattering, due to particulate solid support suspended into the reaction medium. As a consequence, the photo-Fenton process may be seriously affected *(Herney-Ramirez et al., 2010)*. Also, the solution pH affects the iron leaching from the support, at pH values less than 3 a higher amount of iron being released into the solution *(Duarte and Madeira, 2010)*. By using a zero-valent iron with iron oxide composite catalysts, the oxidation process proceeds via hydroxyl radicals generated from Fe^{2+}(surf) species and H_2O_2 in a Fenton like mechanism. The Fe^{2+}(surf) species are formed by electron transfer from Fe^0 to Fe^{3+} at the interface metal/oxide *(Moura et al., 2005, 2006)*. The experimental data *(Nie et al., 2008)* indicate that the hydrogen peroxide provides a driving force in the electron transfer from Fe^{2+} to Fe^{3+}, while the degradation of organic pollutants increases the electron transfer at the interface of Fe^0/iron oxide due to their reaction with hydroxyl radicals.

The degradation of organic pollutants using photo-Fenton processes occurs by intermediate oxidation products formation. In the case of phenol oxidation by Fenton reagent, a series of intermediates were identified, corresponding mainly to ring compounds and short-chain organic acids *(Zazo et al., 2005)*. Most significant among the former were catechol, hydroquinone, and p-benzoquinone; the main organic acids were maleic, acetic, oxalic, and formic, with substantially lower amounts of muconic, fumaric, and malonic acids. Oxalic and acetic acid appeared to be fairly refractory to the Fenton oxidation process. In the Fenton process, carboxylic acids like acetic and oxalic acid may be formed as end products during the degradation of phenol while in photo-Fenton process, both these acids were identified during the early stages of phenol degradation and were oxidized almost completely at the end of the process *(Kavitha and Palanivelu, 2004)*. The chlorophenols are common persistent organic contaminants, which show low biodegradability, posing serious risks to the environment once discharged into natural water *(Du et al., 2006)*.

Studying the degradation of 4-chlorophenol by an electrochemical advanced oxidation process, several authors *(Wang and Wang, 2007)* have proposed the following possible pathways: (a) 4-chlorophenol dechlorination to phenol; (b) hydroxylation of phenol to hydroquinone; (c) dehydrogenation of hydroquinone to benzoquinone; (d) oxidation of benzoquinone (with aromatic ring cleavage) to aliphatic carboxylic acids such as maleic acid, fumaric acid, malonic acid; (e) oxidation of maleic and fumaric acids to oxalic acid, formic acid and finally, to carbon dioxide and water.

The main intermediate products detected by HPLC analyses were chlorocatechol and benzoquinone after 60 min reaction time and aliphatic carboxylic acids after 120 min reaction time. Benzoquinone and hydroquinone-like intermediates such as catechol, hydroquinone and 4-chlorocatechol can reduce the ferric ion to ferrous ion and the oxidation process becomes faster *(Du et al., 2006)*. It is not necessary to degrade 4-chlorophenol to the final products of CO_2 and H_2O, being more worthwhile to treat to the biodegradable stage-aliphatic carboxylic acids followed by a biological process *(Wang and Wang, 2007)*. The photocatalytic processes may be used as a pre-treatment of toxic chemicals, including chlorophenols, in order to convert them into fully biodegradable compounds.

Recently, a series of pharmaceuticals such as analgesics, antibiotics, steroids etc. have been detected in the water feeding systems of several countries in Europe, the USA and Australia *(Bound and Voulvoulis, 2004; Kaniou et al., 2005)*. Unless antibiotics are removed from wastewaters through specific purification processes, they can affect the microbial communities in filtering systems using active sludge and, in general, the bacteria found in water, and, as a result, they can disturb the natural elementary cycles. The accumulation of antibiotics in surface waters represents a potential danger in the case in which they are used as sources of drinking water. Photocatalytic oxidation of antibiotics in aqueous solution is based on the oxidative potential of the HO radicals (2.80 V) generated in the reaction medium though photocatalytic mechanisms, in the presence of H_2O_2 and UV radiations. Through the photo-Fenton procedure, the efficiency of the oxidation is controlled by the nature and the structure of the organic substrate, the initial pH of the solution, the concentration of H_2O_2 and of the catalytic component (Fe^{2+}) as well as by the time the reaction medium stands in the area where UV radiations act.

The kinetic assessment of the oxidative degradation process applied to antibiotics of the type amoxicillin, ampicillin and streptomycin (pseudo 1st degree Lagergren kinetic model) suggests that the oxidative process occurs in two successive steps, with the formation of reaction intermediates. The ratio of the 1st degree kinetic constant values corresponding to the two oxidation stages depends on the structure of the antibiotics and indicates a marked decrease in the oxidation rate in the second stage. This decrease can be attributed to the formation of reaction intermediates such as inferior organic acids with a high stability in regard to oxidation and/or blocking active catalytic centers through the formation of compounds of the $Fe^{2+/3+}$ species with the reaction intermediates, compounds which are inactive in the process of generating HO·radicals *(Orbeci et al., 2010)*.

Advanced oxidation processes of Fenton and photo-Fenton type can be used for antibiotics degradation from wastewater *(Orbeci et al., 2010)* or for increasing their biodegradability in biological wastewater treatment *(Elmolla and Chaudhuri, 2009)*. Unlike complete amoxicillin degradation, the mineralization of the organic compounds from solution is not complete in the Fenton oxidation process due to formation of refractory intermediates *(Ay and Kargi, 2010)*.

The photo-Fenton process degradation of amoxicillin by using iron species as catalyst ($FeSO_4$ and potassium ferrioxalate complex) and solar radiation reduces the bactericide effect of amoxicillin but the toxicity may persist due to intermediates formed during the oxidation process. The toxicity decreases significantly when these intermediates are converted into short chain carboxylic acids, allowing further conventional treatment *(Trovó et al., 2011)*. The homogeneous photo-Fenton process is limited by the narrow working pH range (2.5-4) and requires the correction of solution pH for iron precipitation and catalyst separation and recovery. Otherwise, high amounts of metal-containing sludge can be formed and the catalytic metals are lost in these sludge. Because of these disadvantages, several attempts have been made to develop heterogeneous photo-Fenton procedure by immobilization of active iron species on solid supports. Since iron is relatively inexpensive and nontoxic, it has been widely used in different environmental treatment processes *(Herney-Ramirez et al., 2010; Nie et al., 2008)*. In the heterogeneous photo-Fenton process, different iron-containing catalysts can be used, such as the iron bulk catalysts (iron oxy-hydroxyl compounds: hematite, goethite, magnetite) or iron supported catalysts (zeolites, clays, bentonite, glass, active carbon, polymers etc.) *(Feng et al., 2005; He et al., 2005; Leland and Bard, 1987; Nie et al., 2008; Ortiz de la Plata et al., 2010; Vinita et al. 2010)*.

Antibiotics can be more or less extensively metabolized by humans and animals. Depending on the quantities used and their rate of excretion, they are released in effluents and reach sewage treatment plants *(Alexy et al., 2004; Bound and Voulvoulis, 2004; Kümmerer, 2009)*.

Available data on antibiotics (ampicillin, erythromycin, tetracycline and penicilloyl groups) indicate their capability to exert toxic effects to living organisms (bacteria, algae etc.), even at very low concentration. These antibiotics are practically non-biodegradable having the potential to survive sewage treatment, leading to a persistence of these compounds in the environment and a potential for bio-accumulation *(Arslan-Alaton et al., 2004)*. The presence of antibiotics in the environment has favored the emergence of antibiotic-resistant bacteria, increasing the likelihood of infections as well as the need to find new and more powerful antibiotics. As expected, antibiotic-contaminated water is incompatible with conventional biological water treatment technologies *(Rozas et al., 2010)*. Antibiotics have the potential to affect the microbial community in sewage systems and can affect bacteria in the environment and thus disturb natural elementary cycles *(Kümmerer, 2009)*. If they are not eliminated during the purification process, they pass through the sewage system and may end up in the environment, mainly in the surface water.

This is of special importance, since surface water is a possible source of drinking water *(Kaniou et al., 2005)*. The antibiotics degradation by advanced oxidation processes has proven to be reasonably suited and quite feasible for application as a pre-treatment method by combining with biological treatment *(Arslan-Alaton et al., 2004)*. The pre-treatment of penicillin formulation effluent by advanced oxidation processes based on O_3 and H_2O_2/O_3 did not completely remove the toxicity of procaine penicillin G from the effluents, leading to serious inhibition of the treatment of activated sludge *(Arslan-Alaton et al., 2006)*. One of the novel technologies for treating polluted sources of industrial wastewater and drinking water is the photo-Fenton process by which hydroxyl radicals are generated in the presence of H_2O_2, Fe^{2+} catalyst and UV radiation.

The advanced oxidation processes or even the hybrid methods may not be useful in degrading large quantity of the effluent with economic efficiency and hence it is advisable to use these methods for reducing the toxicity of the pollutant stream to a certain level beyond which biological oxidation can ensure the complete mineralization of the biodegradable products *(Gogate and Pandit, 2004)*.

Removal of organic compounds in wastewater is a very important subject of research in the field of environmental chemistry. In this sense, photocatalysis is a handy promising technology, very attractive for wastewater treatment and water potabilization *(Nikolaki et al., 2006; Lim et al., 2009)*. Using titanium dioxide for water splitting after UV irradiation, it has been shown that this can encompass a wide range of reactions, especially the oxidation of organic compounds. The study of the photodegradation for a large series of substances such as halogenated hydrocarbons, aromatics, nitrogenated heterocycles, hydrogen sulfide, surfactants and herbicides, and toxic metallicions, among others has clearly shown that the majority of organic pollutants present in waters can be mineralized or at least partially destroyed. The photocatalytic treatment of many organic compounds has been successfully achieved. The photocatalytic activity is dependent on the surface and structural properties of the semiconductor such as crystal composition, surface area, particle size distribution, porosity, band gap and surface hydroxyl density *(Ahmed et al., 2010)*.

A variety of semiconductor powders (oxides, sulfides etc.) acting as photocatalysts have been used. Most attention has been given to TiO_2 because of its high photocatalytic activity having a maximum quantum yields, its resistance to photo-corrosion, its biological immunity and low cost. There are two types of reactors: reactors with TiO_2 suspended in the reaction medium and reactors with TiO_2 fixed on a carrier material *(Lim et al.; 2009, Mozia, 2010; Li et al., 2009)*. A very promising method for solving problems concerning the separation of the photocatalyst from the reaction medium is the application of photocatalytic membrane reactors (PMRs), having other advantages such as the realization of a continuous process and the control of a residence time of molecules in the reactor *(Mozia, 2010)*. In case of polymer membranes, there is a danger for the membrane structure to be destroyed by UV light or hydroxyl radicals. The investigations described *(Chin et al., 2006; Molinari et al., 2000)* show that the lowest resistance exhibit membranes prepared from polypropylene, polyacrylonitrile, cellulose acetate and polyethersulfone, UV light leading to a breakage in the chemical bonds of the methyl group. The least effect of the UV/oxidative environment on the membrane stability was observed in case of polytetrafluoroethylene and polyvinylidene fluoride membranes *(Chin et al., 2006)*.

The self-assembly of TiO2 nanoparticles was established through coordinance bonds with – OH functional groups on the membrane surface, improving reversible deposition, hydrophilicity and flow and diminishing the irreversible fouling (Mansourpanah et al., 2009). TiO$_2$-functionalized membranes may be obtained by several methods, but the sol-gel process is ubiquitous because it has many advantages such as purity, homogeneity, control over the microstructure, ease of processing, low temperature and low cost (Alphonse et al., 2010).

The advanced oxidation processes based on the photo-activity of semiconductor-type materials can be successfully used in wastewater treatment for destroying the persistent organic pollutants, resistant to biological degradation processes. TiO$_2$ is the most attractive semiconductor because of its higher photocatalytic activity and can be used suspended into the reaction medium (slurry reactors) or immobilized as a film on solid material. A very promising method for solving problems concerning the photocatalyst separation from the reaction medium is to use the photocatalytic reactors in which TiO$_2$ is immobilized on support. The immobilization of TiO$_2$ onto various supporting materials has largely been carried out via physical or chemical route.

The application of photocatalysis in water and wastewater treatment has been well established, particularly in the degradation of organic compounds into simple mineral acids, carbon dioxide and water *(Pera-Titus et al., 2004; Cassano and Alfano, 2000)*. Titanium dioxide (TiO$_2$), particularly in the anatase form is a photocatalyst under ultraviolet (UV) light.   A reactor refers to TiO$_2$ powder which is suspended in the water to be treated, while the immobilized catalyst reactor has TiO$_2$ powder attached to a substrate which is immersed in the water to be treated. Immobilised TiO$_2$ has become more popular due to the complications in the TiO$_2$ suspension systems *(Hoffmann et al., 1995)*.

The TiO$_2$ immobilisation procedure developed not long ago can be used in determining a suitable immobilization procedure, particularly if economical and simple equipment is necessary. The overall performance of the TiO$_2$ coating can be affected by various factors depending on the coating methods. In addition, it is also difficult to evaluate the photocatalytic efficiency of the coatings through photocatalytic activity *(Augugliaro et al., 2008)*. Invariably, the TiO$_2$ was immobilized using the sol–gel technique *(Addamo et al., 2008)*. In this case, some problems are noted: decrease of surface area; potential loss of TiO$_2$; decreased adsorption of organic substances on the TiO$_2$ surface; mass transfer limitations.

No polymeric support was considered to immobilise TiO$_2$ since the polymeric material can undergo photo-oxidative degradation by illuminated TiO$_2$ (*Cassano and Alfano, 2000; Augugliaro et al., 2008*).

Researchers have used photocatalytic oxidation to remove and destroy many types of organic pollutants. After photocatalysis was recognized to be a great oxidation mechanism, researchers began testing it on many different compounds, and in many different processes (*Cassano and Alfano, 2000; Hoffmann et al., 1995; Addamo et al., 2008*).

Phenolic compounds, a kind of priority pollutants, often occur in the aqueous environment due to their widespread use in many industrial processes such as the manufacture of plastics, dyes, drugs, antioxidants, and pesticides. Phenols, even at concentrations below 1 lg/L, can affect the taste and odor of the water (*Pera-Titus et al., 2004*). Therefore, the identification and monitoring of these compounds at trace level in drinking water and surface waters are imperative. Chlorophenols represent an important class of very common water pollutants. 4-chlorophenol is a toxic and non-biodegradable organic compound and can often be found in high quantity in the waste waters from various industrial sectors (*Pera-Titus et al., 2004; Augugliaro et al., 2006*). A severe toxicity of 4-chlorophenol requires the development of a simple, sensitive and reliable analytical method.

Among the advanced oxidation processes investigated in thelast decades, photocatalysis in the presence of an irradiated semiconductor has proven to be very effective in the field of environment remediation.The use of irradiation to initiate chemical reactions is the principle on which heterogeneous photocatalysis is based; infact, when a semiconductor oxide is irradiated with suitable light, excited electron–hole pairs result that can be applied in chemical processes to modify specific compounds.The main advantage of heterogeneous photocatalysis, when compared with the chemical methods, is that in most cases it is possible to obtain a complete mineralization of the toxic substrate even in the absence of added reagents.The radical mechanism of photocatalytic reactions, which involve fast attacks of strongly oxidant hydroxyl radicals, determines their un selective features.

Environmental applications of photocatalysis using TiO$_2$ have attracted an enormous amount of research interest over the last three decades (*Hoffmann et al., 1995; Linsebigler et al., 1995; Mills and Le Hunte, 1997; Stafford et al., 1996*). It is well established that slurries of TiO$_2$ illuminated with UV light can degrade to the point of mineralization almost any dissolved organic pollutant. Nevertheless, photocatalysis, particularly in aqueous media, has still not found widespread commercial implementation for environmental remediation. The main hurdle appears to be the cost, which is high enough to prevent the displacement of existing and competing technologies by photocatalysis. TiO$_2$ research has progressed on multiple tiers, whereby the study of fundamental processes promotes material development, allowing no repeated uses that promise to play a larger and larger role in engineering sustainable technologies.

TiO$_2$ is the most used semiconductor because of its higher photocatalytic activity, resistance to the photocorrosion process, absence of toxicity, biological immunity and the relatively low cost *(Nikolaki et al., 2006; Han et al., 2009)*. TiO_2 crystalline powder can be used suspended into a reaction medium (slurry reactors) or immobilized as a film on a carrier material. The anatase form of TiO_2 is reported to give the best combination of photoactivity and photostability. An ideal photocatalyst for oxidation is characterized by the following attributes: photostability; chemically and biologically inert nature; availability and low cost; capability to adsorb reactants under efficient photonic activation *(Bideau et al., 1995, Pozzo et al., 1997, Balasubramanian et al., 2004, Gaya and Abdullah, 2008, Siew-Teng Ong et al., 2009, Hanel et al., 2010).*The support must have the following characteristics: (a) transparent to irradiation; (b) strong surface bonding with the TiO_2 catalyst without negatively affecting the reactivity; (c) high specific surface area; (d) good absorption capability for organic compounds; (e) separability; (f) facilitating mass transfer processes and (g) chemically inert *(Pozzo et al., 1997)*. As solid support, different materials were investigated: natural or synthetic fabrics, polymer membranes, activated carbon, quartz, glass, glass fiber, optical fibers, pumice stone, zeolites, aluminum, stainless steel, titanium metal or alloy, ceramics (including alumina, silica, zirconia, titania), red brick, white cement etc. *(Bideau et al., 1995, Rachel et al., 2002, Balasubramanian et al., 2004, Kemmitt et al. 2004, Gunlazuardi and Lindu, 2005, Hunoh et al. 2005, Chen et al., 2006, Medina-Valtierra et al., 2006, Gaya and Abdullah, 2008, Lim et al., 2009, Siew-Teng Ong et al., 2009, Hanel et al., 2010, Zita et al., 2011)*.

The immobilization of TiO_2 on different supporting materials has largely been carried out via a physical or chemical route: dip coating, porous material impregnation, sol-gel method, reactive thermal deposition, chemical vapor deposition, electron beam evaporation, spray pyrolysis, electrophoresis, electro-deposition *(Rachel et al., 2002, Gunlazuardi and Lindu, 2005, Medina-Valtierra et al., 2006, Hanel et al., 2010, Zita et al., 2011)*.

The methods most commonly used for deposition of TiO_2 on supports are sputtering and sol-gel dip-coating or spin-coating. Often the fixation of TiO_2 on solid supports reduces its efficiency due to various reasons such as reduction of the active surface, a more difficult exchange with solution, introduction of ionic species etc. A degree of 60–70% reduction in performance is reported in aqueous systems for immobilized TiO_2 as compared to the unsupported catalyst *(Gaya and Abdullah, 2008)*, the best results being obtained by the immobilization of TiO_2 on glass fiber *(Rachel et al., 2002, Lim et al., 2009)*. The separation and/or removal technologies based on membrane and photocatalytic processes have a great potential for application in advanced wastewater treatment. Separation membranes have become essential parts of the human life because of their growing industrial applications in high technology such as biotechnology, nanotechnology and membrane based separation and purification processes. Available technologies to deal with chlorophenolic compounds include the advanced oxidation processes (AOPs) based on the formation of hydroxyl radicals with high oxidation potential.

The hybrid method consists in a photocatalytic procedure using a reactor equipped with TiO_2-functionalized membrane (cylindrical shape) and high pressure mercury lamp for UV radiations generation, centrally and coaxially positioned. The TiO_2-functionalized membranes have been obtained by sol-gel method synthesis of TiO_2 (from tetrabutylortotitanate) as nanoparticles, formed directly in porous membrane regenerated cellulose type. Solutions of methyl, ethyl and propyl alcohols have been used as reaction medium. The experiments were performed at 30 ± 2^0C using synthetic solutions of 4-CP (analytical grade reagent). The amount of hydrogen peroxide (30% w/w) used was calculated at 1.5 time H2O2/4-CP stoichiometric ratio. The degradation process was studied by monitoring the organic substrate concentration changes function of reaction time using chemical oxygen demand analysis (COD).

The experimental data show the catalytic role of TiO_2-functionalized membranes in the oxidation process. The oxidation is preceded by an adsorption process and the transfer of 4-chlorophenol from the solution to the photocatalytic reaction zone through the functionalized membrane. Titanium dioxide, deposited on the membrane, acts as a photocatalyst in the presence of UV radiations leading to a higher efficiency of the oxidation process in a short reaction time. The catalytic activity of TiO_2-functionalized membranes is influenced by the nature of the alcohol used in obtaining them. This can be explained by the crystallite size of TiO_2 and their dispersion on membrane. However, at a higher reaction time, the determined solution COD values tend to increase, indicating that the TiO_2-functionalized membranes become unstable. This can be attributed to a partial solubilization process of membrane into reaction medium with a strong oxidizing potential.

The presence of phenol and phenolic derivatives in water induces toxicity, persistence and bioaccumulation in plant and animal organisms and is a risk factor for human health. The technologies of separation and/or removal of phenolic derivatives based on membrane and photocatalytic processes play an important role.

Available technologies to deal with phenolic compounds include the advanced oxidation processes, based on the formation of very active hydroxyl radicals, which react quickly with the organic contaminant. Among the AOPs, the photocatalytic process is one of the most attractive methods because the reagent components are easy to handle and environmentally benign.

3. Wastewater decontamination by processes of absorption

Rapid development of industry and society led to serious environmental problems such as contamination of groundwater and surface chemical treatment with organic compounds coming from agriculture (pesticides, herbicides, other.) or inorganic compounds in industry (pigments, heavy metals, etc.)

A method used for the wastewater decontamination is the contaminants absorption on the catalyst surface. The most known adsorbent substances cleansing practice are: activated carbon, silica gel, discolored soils, molecular sieves, cotton fibers etc.

After the manner in which the contact is realized between wastewater and adsorbent is distinguished static and dynamic adsorption. In the first case finely adsorbent divided is stirred with water and after a time it's separated by decantation or filtration. In the dynamic adsorption case, wastewater passes through a fixed, mobile or a fluidized absorbent layer with a continuous flow.

Another alternative to the wastewater treatment is the use of technologies based on magnetic nanomaterials AB_2O_4 type (A = Co, Ni, Cu, Zn, B = Fe^{3+}) used as catalysts for degradation of organic compounds or absorbents to retain the surface pollutants heavy metals (mercury, arsenic, lead and others). Their importance and complexity led to research programs development on magnetic materials, with new or improved properties in last decades. These properties are dependent on chemical composition and microstructural characteristics, which can be controlled in the fabrication and synthesis processes.

These materials must have a relatively high surface area, a smaller particle size, and porous structure. In particular, the magnetic properties of the powders makes them to be easily recovered by magnetic separation technology after adsorption or regeneration, which overcomes the disadvantage of separation difficulty of common powdered adsorbents (Qu, et.al, 2008).

Challenges in synthesis of nanostructured catalysts are that many reactions employ mixed catalysts consisting of different oxide metals, and that the function of active centers is not only determined by the constituent atoms but also by the surrounding crystal or surface structures; it is thus necessary to accurately control the synthesis of nanostructured catalysts (Rickerby, et.al, 2007).

Magnetite nanoparticles have highest saturation magnetization of 90 emu/g among iron oxides. Therefore, magnetite nanoparticles can be used to adsorb arsenic ions followed by magnetic decantation. Other iron oxides and hydroxides have been reported to have arsenic ability (Hai, et. al, 2009). Oxidation resistance is an important factor for arsenic removal under atmospheric conditions.

By diverse synthesis methods (hydrothermal, ultrasonic hydrothermal, sol-gel, coprecipitation and other), was obtained ferrite nanomaterials derived from magnetite (FeO, Fe_2O_3) substituting the Fe^{2+} ion in different concentrations (0.5, 0.8, 1, 1.2, 1.5) with Co^{2+}, Cu^{2+}, Ni^{2+}, Zn^{2+} ions (Vlazan, 2010; Fannin, et.al, 2011).

4. Membrane oxidative processes in water treatment

This topic gives an overview of the hybrid photocatalysis-membrane processes and their possible applications in water and wastewater treatment. Different configurations of photocatalytic membrane reactors (PMRs) are described and characterized. They include PMRs with photocatalyst immobilized on/in the membrane and reactors with catalyst in suspension. The advantages and disadvantages of the hybrid photocatalysis-membrane processes in terms of permeate flow, membrane fouling and permeate quality are discussed. Moreover, a short introduction to the heterogeneous photocatalysis and membrane processes as unit operations is given.

The detailed mechanism of the photocatalytic oxidation of organic compounds in water has been discussed widely in the literature and will be presented here in brief only. The overall process can be divided into the following steps.

1. Diffusion of reactants from the bulk liquid through a boundary layer to the solution-catalyst interface (external mass transfer).
2. Inter-and/or intra-particle diffusion of reactants to the active surface sites of the catalyst (internal mass transfer).
3. Adsorption of at least one of the reactants.
4. Reactions in the adsorbed phase.
5. Desorption of the product(s).
6. Removal of the products from the interface of the bulk solution.

A photocatalyst should be characterized by: (I) high activity, (II) resistance to poisoning and stability in prolonged at elevated temperatures, (III) mechanical stability and resistance to attrition, (IV) non-selectivity in most cases, and (V) physical and chemical stability under various conditions. Moreover, it is desirable for the photocatalyst to be able to use not only UV, but also visible light and to be inexpensive. Different semiconducting materials, such as oxides (TiO_2, ZnO, CeO_2, ZrO_2, WO_3, V_2O_5, Fe_2O_3, etc.) and sulfides (CdS, ZnS, etc.) have been used as photocatalysts.

Recently, numerous investigations have been focused on different modifications of TiO_2 in order to improve its activity under UV irradiation or to reduce the band gap energy so that it is able to utilize the visible light. The best photocatalytic performances with maximum quantum yields have been always obtained with TiO_2. Anatase is the most active allotropic form of TiO_2 among the various ones available. Unfortunately, due to a wide band gap (about 3.2 eV), TiO_2 is inactive under visible light.

The most important operating parameters which affect the efficiency of the photocatalytic oxidation process can be summarized as follows: reactor design; light wavelength and intensity; loading of the photocatalyst; initial concentration of the reactant; temperature; pH of reaction medium; oxygen content; the presence of inorganic ions *(Mozia, 2005; Gogate and Pandit, 2004; Herrmann, 2006).*The photocatalytic reactors can be divided into two main groups:

i. Reactors with TiO_2 suspended in the reaction mixture: in the case of reactors with TiO_2 suspended in the reaction mixture (I), the photocatalyst particles have to be separated from the treated water after the oxidation process.
ii. Reactors with TiO_2 fixed on a carrier material (glass, quartz, stainless steel, pumice stone, titanium metal, zeolites, pillared clays, membranes etc.).

A very promising method for solving problems concerning separation of the photocatalyst as well as products and by-products of photo-oxidation process from the reaction mixture is application of photocatalytic membrane reactors (PMRs). PMRs are hybrid reactors in which photo-catalysis is coupled with a membrane process. The membrane would play both the role of a simple barrier for the photocatalyst and of a selective barrier for the molecules to be degraded. Membrane processes are separation techniques which are widely applied in various sectors of industry including food, chemical and petrochemical, pharmaceutical, cosmetics and electronic industries, water desalination, water and wastewater treatment and many others. The main advantages of membrane processes are: low energy consumption; low chemicals consumption; production of water of stable quality almost independent on the quality of the treated water; automatic control and steady operation allowing performance of a continuous operation; low maintenance costs; easy scale up by simple connecting of additional membrane modules

Synthetic membranes may be: organic or inorganic materials; homogeneous or heterogeneous; symmetrical or asymmetrical; porous or dense; electrically neutral or charged.

In this sense, the driving forces are: pressure difference; concentration difference; partial pressure difference or electrical potential difference.

Most of the PMRs described in the literature combine photo-catalysis with pressure driven membrane processes such as:

- Micro-Filtration (MF)
- Ultra-Filtration (UF)
- Nano-Filtration (NF)
- Dialysis
- Pervaporation (PV)
- Direct Contact Membrane Distillation (DCMD)

Hybrid photocatalysis-membrane processes are conducted in the installations often called "photocatalytic membrane reactors". However, in the literature, other names for these configuration scan be also found, including "membrane chemical reactor" (MCR), "membrane reactor", "membrane photoreactor", or, more specific, "submerged membrane photocatalysis reactor" and "photocatalysis–ultrafiltration reactor"(PUR). For the hybridization of photocatalysis with membrane process it will be useful to apply a general term of "photocatalytic membrane reactor".

Photocatalytic membrane reactors design show in figure 2 (a, b).

Photocatalytic membranes for the PMRs can be prepared from different materials and indifferent ways. Figure 3 presents two possible types of asymmetric photocatalytic membranes (Bosc et al., 2005). In the first case, photoactive separation layer is deposited on a non-photoactive porous support (Fig. 3a) the photoactive layer, being also the separation layer (skin) is formed on a porous non-photoactive support. In the second case, a non-photoactive separation layer is deposited on a photoactive porous support (Fig. 3b) the separation layer is non-photoactive and is deposited on a porous active support.

The main advantage of PMRs with photocatalytic membranes is that this configuration allows one to minimize the mass transfer resistances between the bulk of the fluid and the semiconductor surface.

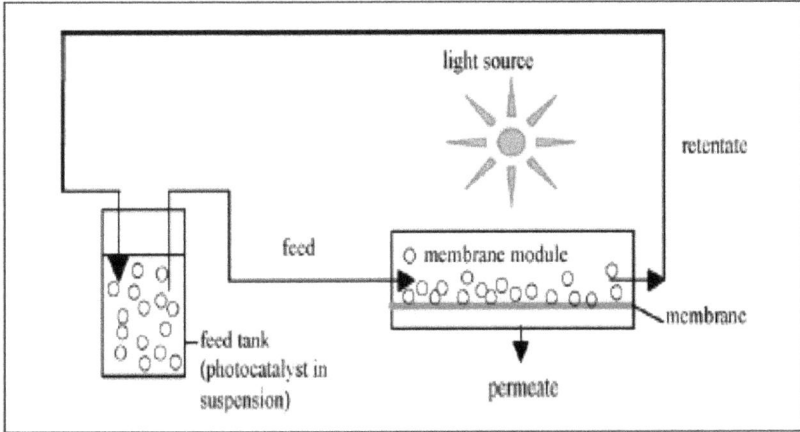

(a)

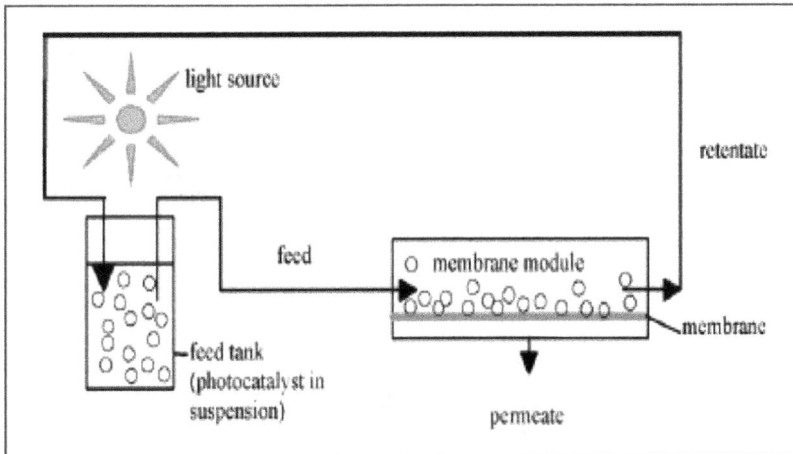

(b)

Figure 2. a. PMR utilizing photo-catalyst in suspension: irradiation of the membrane module *(Mozia S., 2010)*, b. PMR utilizing photo-catalyst in suspension: irradiation of the feed tank (Mozia S., 2010)

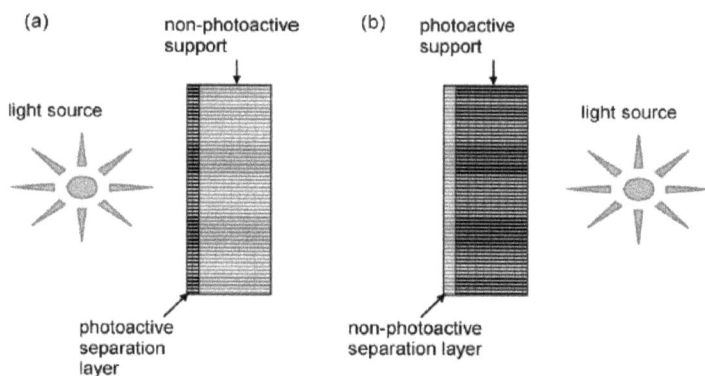

Figure 3. Asymmetric photocatalytic membranes *(Bosc et al., 2005)*

5. Conclusion

Choosing the most suitable method of wastewater treatment studies require both increasing the effectiveness and economic efficiency (operating and investment costs).

Advanced oxidation processes (AOPs) are good alternatives for removal the toxic compunds from wastewater. The AOPs can be successfully used in wastewater treatment to degrade the persistent organic pollutants, the oxidation process being determined by the very high oxidative potential of the HO· radicals generated into the reaction medium by different mechanisms. AOPs can be applied to fully or partially oxidize pollutants, usually using a combination of oxidants. Photo-chemical and photo-catalytic advanced oxidation processes including UV/H_2O_2, UV/O_3, $UV/H_2O_2/O_3$, $UV/H_2O_2/Fe^{2+}(Fe^{3+})$, UV/TiO_2 and $UV/H_2O_2/TiO_2$ can be used for oxidative degradation of organic contaminants. A complete mineralization of the organic pollutants is not necessary, being more worthwhile to transform them into biodegradable aliphatic carboxylic acids followed by a biological process. The efficiency of the various AOPs depend both on the rate of generation of the free radicals and the extent of contact between the radicals and the organic compound.

Photocatalytic oxidation in water treatment has proved its efficiency at many pilot-scale applications. However, wide marketing of commercially available solar detoxification systems is obstructed by the general market situation: a new water treatment procedure has an opportunity to be implemented only when its cost is at least two-fold lower than the cost of a procedure currently in use. Photocatalysis, also called the "green" technology, represents one of the main challenges in the field of treatment and decontamination systems, especially for water and air. Its operating principle is based on the simultaneous action of the light and a catalyst (semi-conductor), which allows for pollutant molecules to be destroyed without damaging the surrounding environment.

In recent years, applications to environmental remediation have been one of the most active subjects in photocatalysis.

Author details

Adina Elena Segneanu, Carmen Lazau, Paula Sfirloaga, Paulina Vlazan,
Cornelia Bandas and Ioan Grozescu*
*National Institute for Research and development in Electrochemistry and Condensed Matter –
INCEMC Timisoara, Romania*

Cristina Orbeci
Politehnica University Bucuresti, Romania

6. References

Addamo, M., Augugliaro, V., Di Paola, A., García-López, E., Loddo, V., Marcì, G., Palmisano, L., (2008), Photocatalytic thin films of TiO₂ formed by a sol–gel process using titaniumtetraisopropoxide as the precursor, *Thin Solid Films* 516, pp. 3802–3807.

Ahmed, S., Rasul, M. G., Martens, W. N., Brown R., Hashib, M. A., (2010), "Heterogeneous photocatalytic degradation of phenols in wastewater: A review on current status and developments", *Desalination*, 261(1-2), pp. 3-18.

Alexy, R., Kümpel, T., Kümmerer, K., (2004), Assessment of degradation of 18 antibiotics in the Closed Bottle Test. *Chemosphere* 57, 505-512.

Alphonse, P., Varghese A., Tendero, C., (2010), "Stable hydrosols for TiO₂ coatings", *Journal of Sol-Gel Science and Technology*, 56(3), 250-263.

Andreozzi R.; Caprio V.; Insola A.; Marotta R. (1999), Advanced oxidation processes (AOP) for water purification and recovery, *Catalysis Today*, 53 (1), 51-59

Anita Rachel, Machiraju Subrahmanyam, Pierre Boule, (2002), Comparison of photocatalytic efficiencies of TiO₂ in suspended and immobilised form for the photocatalytic degradation of nitrobenzenesulfonic acids, *Applied Catalysis B: Environmental* 37(4), 301–308.

Arslan-Alaton, I., Caglayan, A.E., (2006), Toxicity and biodegradability assessment of raw and ozonated procaine penicillin G formulation effluent. *Ecotox. Environ. Safe.* 63, 131-140.

Arslan-Alaton, I., Dogruel, S., Baykal, E., Gerone, G., (2004), Combined chemical and biological oxidation of penicillin formulation effluent. *J. Environ. Manag.* 73, 155-163.

Augugliaro, V., Litter, M., Palmisano, L., Soria, J., (2006), The combination of heterogeneous photocatalysis with chemical and physical operations: A tool for improving the photoprocess performance, Journal of Photochemistry and Photobiology C: *Photochemistry Reviews 7*, 127–144.

Ay, F., Kargi, F., (2010), Advanced oxidation of amoxicillin by Fenton' reagent treatment. *J.Hazard.Mater.* 179, 622-627.

---

* Corresponding Author

Balasubramanian G., Dionysiou D.D., Suidan M.T., Baudin I., Laîné J.M., (2004), Evaluating the activities of immobilized TiO₂ powder films for the photocatalytic degradation of organic contaminants in water. *Applied Catalysis B: Environmental*, 47(2), 73-84.

Bideau, M., Claudel, B., Dubien, C., Faure, L., Kazouan H., (1995), On the "immobilization" of titanium dioxide in the photocatalytic oxidation of spent waters, *Journal of Photochemistry & Photobiology A: Chemistry*, 91(2), 137-144.

Bosc, F., Ayral, A., Guizard, C., (2005), Mesoporous Anatase Coatings for Coupling Membrane Separation and Photocatalyzed Reactions" *J. Membr. Sci.* 265, 13-19.

Bound, J.P., Voulvoulis, N., (2004), Pharmaceuticals in the aquatic environment – a comparison of risk assessment strategies. *Chemosphere* 56, 1143-1155.

Byrappa, K., Rai, K.M.L., Yoshimura, M. (2002), "Hydrothermal preparation of TiO₂ and photocatalytic degradation of hexachloro cyclohexane and dichlorodiphenyl trichloromethane", *Journal of Environmental Science and Technology*, 21, pp 1085–1090;

Cassano, A. E., Alfano, O. M. (2000) "Reaction Engineering of Suspended Solid Heterogeneous Photocatalytic Reactors", Catalysis Today, 58(2-3), pp. 167-197.

Carp, O., Huisman, C.L., Reller, A., (2004), Photoinduced reactivity of titanium dioxide, Prog. *Solid State Chem.* 32(1-2), 33-177.

Chakinala, A.G., Gogate, P.R., Burgess, A.E., Bremner, D.H., (2009), Industrial wastewater treatment using hydrodynamic cavitation and heterogeneous advanced Fenton processing. *Chemical Engineering Journal*, 152, 498-502.

Chen S.Z, Zhang P.Y, Zhu W.P., Chen L, Sheng-Ming Xu, (2006), Deactivation of TiO₂ photocatalytic films loaded on aluminium: XPS and AFM analyses, *Appl. Surf. Sci.* 252 (20), 7532-7538.

Chin, S. S., Chiang, K., Fane, A. G., (2006), "The stability of polymeric membranes in a TiO₂ photocatalysis process", *Journal of Membrane Science*, 275(1-2), pp. 202-211.

Daneshvar N., Salari D., Khataee A.R., (2003), Photocatalytic degradation of azo dye acid red 14 in water: investigation of the effect of operational parameters, *Journal of Photochemistry and Photobiology A: Chemistry*, 157 111–116;

Daneshvar N., Aber S., Seyed Dorraji M. S., Khataee A. R., Rasoulifard M. H., (2007), Preparation and Investigation of Photocatalytic Properties of ZnO Nanocrystals: Effect of Operational Parameters and Kinetic Study, *World Academy of Science, Engineering and Technology* 29;

Daneshvar N., D. Salari, A.R. Khataee, (2003), Photocatalytic degradation of azo dye acid red 14 in water: investigation of the effect of operational parameters, *Journal of Photochemistry and Photobiology A: Chemistry* 157, 111–116.

De Laat J., Truong Le G., Legube B., (2004), A comparative study of the effects of chloride, sulfate and nitrate ions on the rates of decomposition of H₂O₂ and organic compounds by Fe(II)/H₂O₂ and Fe(III)/H₂O₂. *Chemosphere*, 55(5):715-23.

Du Y., Zhou M., Lei L., (2006), Role of the intermediates in the degradation of phenolic compounds by Fenton-like process, *Journal of Hazardous Materials*, 136, 859-865.

Duarte F., Madeira L.M., (2010), Fenton and Photo-Fenton-Like Degradation of a Textile Dye by Heterogeneous Process with Fe/ZSM-5 Zeolite, *Separation Science and Technology*, 45, 1512-1520.

Elmolla, E., Chaudhuri, M., (2009), Degradation of the antibiotics amoxicillin, ampicillin and cloxacillin in aqueous solution by photo-Fenton process. *J. Hazard. Mater.* 172, 1476-1481.

Esplugas, S., Gimenez, J., Contreras, S., Pascual, E., Rodriguez M., (2002), Comparison of different advanced oxidation processes for phenol degradation, *Water Res.*, 36, p. 1034-1042

Fang Han, Venkata Subba Rao Kambala, Madapusi Srinivasan, Dharmarajan Rajarathnam, Ravi Naidu, (2009), Tailored titanium dioxide photocatalysts for the degradation of organic dyes in wastewater treatment: A review. *Applied Catalysis A, General*, 359 (1-2), 2009, 25-40.

Fannin P.C., Marin C.N., Malaescu I., Stefu N., Vlăzan P., Novaconi S., Popescu S., (2011), Effect of the concentration of precursors on the microwave absorbent properties of Zn/Fe oxide nanopowders; *Journal of Nanoparticle Research*; 13, 311-319.

Feng J., Hu X., Yue P. L., (2005), Discoloration and mineralization of Orange II by using a bentonite clay-based Fe nanocomposite film as a heterogeneous photo-Fenton catalyst, *Water Research*, 39, 89–96.

Gao, Y., Liu, H., (2005), Preparation and catalytic property study of a novel kind of suspended photocatalyst of TiO_2-activated carbon immobilized on silicone rubber film, *Mater. Chem. Phys.* 92(2-3), 604-608.

Gayaa U. I., Abdullaha A. H., (2008), Heterogeneous photocatalytic degradation of organic contaminants over titanium dioxide: A review of fundamentals, progress and problems, *Journal of Photochemistry and Photobiology C: Photochemistry Reviews* 9, 1–12.

Getoff N., (2001), Comparison of radiation and photoinduced degradation of pollutants in water: synergistic effect of O2, O3 and TiO2. A Short Review, *Research on Chemical Intermediates*, Vol. 27, Numbers 4-5, 343-358.

Li Puma G., Bonob A., Krishnaiah D., Collin J.G.,(2008) Preparation of titanium dioxide photocatalyst loaded onto activated carbon support using chemical vapor deposition: A review paper, *Journal of Hazardous Materials* 157. 209–219.

Gogate P. R., Pandit A. B., (2004), A review of imperative technologies for wastewater treatment II: hybrid methods, *Advances in Environmental Research*, 8, 553–597.

Guillard C.; Disdier J.; Herrmann J.-M.; Lehaut C.; Chopin T.; Malato S.;Blanco J., 1999, Comparison of various titania samples of industrial origin in the solar photocatalytic detoxification of water containing 4-chlorophenol, *Catalysis Today*, 54, (2,-3), 217-228.

Hänel, A., Moren, P., Zaleska, A., Hupka, J. (2010), Photocatalytic activity of TiO_2immobilized on glass beads, *Physicochem. Probl. Miner. Process.* 45, 49-56

Hansen H. K., Nunez P., Raboz D., Schippacasse I., Grandon R., (2007), Electrocoagulation in wastewater containing arsenic: Comparing different process designs, *ScienceDirect*, *Electrochimica Acta* 52, 3464-3470.

Hai N. H., Phu N. D., 25 (2009), Arsenic removal from water by magnetic $Fe_{1-x}Co_xFe_2O_4$ and $Fe_{1-y}Ni_yFe_2O_4$ nanoparticles; *VNU Journal of Science, Mathematics - Physics* 15-19.

He J., Ma W., Song W., Zhao J., Qian X., Zhang S., Yu J. C., (2005), Photoreaction of aromatic compounds at a-$FeOOH/H_2O$ interface in the presence of H_2O_2: evidence for organic-goethite surface complex formation, *Water Research*, 39, 119–128.

Hepel M., Hazelton S., (2005), Photoelectrocatalytic degradation of diazo dyes on nanostructured WO_3 electrodes, *Electrochimica Acta* 50. 5278–5291.

Hoffmann, M. R., Martin, S. T., Choi, W., Bahnemann, D. W., (1995) „Environmental Applications of Semiconductor Photocatalysis", Chemical Reviews, 95, pp. 69-96.

Hu, K.H., Yong Kui Cai Y.K., Li S., (2011), Photocatalytic Degradation of Methylene Blue on MoS2/TiO2, Nanocomposite, Advanced Materials Research, Volumes 197 – 198.

Hua, Zi-Le, Shi, Jian-Lin, Zhang, Wen-Hua, Huang, Wei-Min, (2002), Direct synthesis and characterization of Ti-containing mesoporous silica thin films, Materials Letters 53(4-5), 299-304.

Kaniou, S., Pitarakis, K., Barlagianni, I., Poulios, I., (2005), Photocatalytic oxidation of sulfamethazine. *Chemosphere* 60, 372-380.

Jarnuzi Gunlazuardi, Winarti Andayani Lindu, (2005), Photocatalytic degradation of pentachlorophenol in aqueous solution employing immobilized TiO_2 supported on titanium metal, *Journal of Photochemistry and Photobiology A: Chemistry*, 173(1), 51-55.

Jiří Zita, Josef Krýsa, Urh Černigoj, Urška Lavrenčič-Štangar, Jaromír Jirkovský, Jiří Rathouský, (2011), Photocatalytic properties of different TiO_2 thin films of various porosity and titania loading, *Catalysis Today*, 161 (1), 29-34.

João Rocha, Artur Ferreira, Zhi Lin, Michael W. Anderson, (1998), Synthesis of microporous titanosilicate ETS-10 from $TiCl_3$ and TiO_2: a comprehensive study, *Micropor. Mesopor. Mater*, 23(5-6), 253-263.

Jorge Medina-Valtierra, Manuel Sánchez-Cárdenas, Claudio Frausto-Reyes, Sergio Calixto, (2006), Formation of smooth and rough TiO_2 thin films on fiberglass by sol-gel method, *J. Mex. Chem. Soc.*, 50(1), 8-13.

Kavitha V., Palanivelu K., (2004), The role of ferrous ion in Fenton and photo-Fenton processes for the degradation of phenol, *Chemosphere*, 55, 1235-1243.

Kakuta S., Abe T., (2009), Photocatalysis for water oxidation by Fe_2O_3 nanoparticles embedded in clay compound: correlation between its polymorphs and their photocatalytic activities, *Journal of Materials Science*, Vol. 44, Number 11 2890-2898.

Khelifa A., Moulay S., Naceur A.W., (2005), Treatment of metal finishing effluents by the electroflotation technique, Desalination 181, 27-33.

Kondo Y., Fujihara S., (2011), Solvothermal Synthesis of WO_3 Photocatalysts and their Enhanced Activity, Key Engineering Materials (Volume 485) 283-286.

Hu K.H., Yong Kui Cai , Sai Li, (2011), Photocatalytic Degradation of Methylene Blue on MoS_2/TiO_2 Nanocomposite, Advanced Materials Research , vol 197 – 198, 996-999;

Kümmerer, K., (2009), Antibiotics in the aquatic environment - A review - Part I. *Chemosphere* 75, 417-434.

Kwon, B. G., Lee, D. S., Kang, N., Yoon, J., (1999), Characteristics of p-chlorophenol oxidation by Fenton's reagent.*Water Res*. 33 (9), 2110-2118.

Lazau C., Ratiu C., Orha C., R. Pode, F. Manea, (2011), Photocatalytic activity of undoped and Ag-doped TiO_2-supported zeolite for humic acid degradation and mineralization, *Materials Research Bulletin* 46, 11, Pages 1916–1921.

Leland J. K., Bard A. J., (1987), Photochemistry of colloidal semiconducting iron oxide polymorphs, Journal of Physical Chemistry, 91, 5076–5083.

Li Puma G., Bonob A, Krishnaiah D., Collin J.G., (2008), Preparation of titanium dioxide photocatalyst loaded onto activated carbon support using chemical vapor deposition: A review paper, *Journal of Hazardous Materials*, 157 209–219.

Li, J.-H., Xu, Y.-Y., Zhu, L.-P., Wang, J.-H., Du, and C.-H., (2009), Fabrication and characterization of a novel TiO_2 nanoparticle self-assembly membrane with improved fouling resistance, *Journal of Membrane Science*, 326(2), pp. 659-666.

Li, X.Y., Cui, Y.H., Feng, Y.J., Xie, Z.M., Gu, J.D., (2005), Reaction pathways and mechanisms of the electrochemical degradation of phenol on different electrodes. Water Res. 39, 1972-1981.

Lim, L. L. P., Lynch, R. J., In, S. I. (2009), Comparison of simple and economical photocatalyst immobilisation procedures, Applied Catalysis A: General, 365(2): 214–221.

Liu Xinsheng. Kerry Thomas. J., (1996), Synthesis of microporous titanosilicates ETS-10 and ETS-4 using solid TiO_2 as the source of titanium, *Chem. Commun.*, 12, 1435-1436.

Lopez T., M. Alvarez, F. Tzompantzi, M. Picquart, Photocatalytic degradation of 2,4-dichlorophenoxiacetic acid and 2,4,6-trichlorophenol with ZrO2 and Mn/ZrO2 sol-gel materials, *J Sol-Gel Sci Techn* (2006) 37: 207–211.

Lu M. C., Chen J. N., Chang C. P., (1997), Effect of inorganic ions on the oxidation of dichlorvos insecticide with Fenton's reagent. *Chemosphere*, 35(10), 2285–2293.

Mansourpanah, Y., Madaeni, S. S., Rahimpour, A., Farhadian, A., Taheri, A. H. (2009), Formation of appropiate sites on nanofiltration membrane surface for binding TiO_2 photo-catalyst: Performance, characterization and fouling-resistant capability", *Journal of Membrane Science,* 330(1-2), pp. 297-306.

Molinari, R. , Mungari, M., Drioli, E., Di Paola, A., Loddo, V., Palmisano L., Schiavello,M., (2000), Study on a photocatalytic membrane reactor for water purification", *Catalysis Today*, 55(1-2), pp. 71-78.

Moura F. C. C., Araujo M. H., Costa R. C. C., Fabris J. D., Ardisson J. D., Macedo W. A. A., Lago R. M., (2005), Efficient use of Fe metal as an electron transfer agent in a heterogeneous Fenton system based on Fe^0/Fe_3O_4 composites, *Chemosphere*, 60, 1118-1123.

Moura F. C. C., Oliveira G. C., Araujo M. H., Ardisson J. D., Macedo W. A. A., Lago R. M., Highly reactive species formed by interface reaction between Fe–iron oxides particles: An efficient electron transfer system for environmental applications, *Applied Catalysis A:* General, 307, (2006), 195-204.

Mozia, S., (2010), "Photocatalytic membrane reactors (PMRs) in water and wastewater treatment. A review", *Separation and Purification Technology*, 73(2), pp. 71-91.

Munoz I., Rieradevall J., Torrades F., Peral J., Domenech X., (2005), Environmental Assessment of Different Solar Driven Advanced Oxidation Processes", *Sol. Energy*, 79, p. 369.

Munter R., (2001), Advanced Oxidation Processes – Current Status and Prospects, *Proc. Estonian Acad. Sci. Chem.* 50(2) p. 59.

Nie Y., Hu C., Zhou L., Qu J., (2008), An efficient electron transfer at the Fe^0/iron oxide interface for the photoassisted degradation of pollutants with H_2O_2, *Applied Catalysis B; Environmental*, 82, 151-156.

Nikolaki, M. D., Malamis, D., Poulopoulos, S. G., Philippopoulos, C. J. (2006), Photocatalytical degradation of 1,3-dichloro-2-propanol aqueous solutions by using an immobilized TiO_2 photoreactor, *Journal of Hazardous Materials* 2006, 137(2), 1189–1196.

Oh S.H., Kim J.S., Chung J.S., Kim E.J., Hahn S. H., (2005), Crystallization and Photoactivity of TiO2 Films Formed on Soda Lime Glass by a Sol-Gel Dip-Coating Process, Chem. Eng. Comm. 192 (3), 327-335.

Ong S.T., Lee C.K., Zainal Z., Keng P.S. Ha S.T., (2009), Photocatalytic degradation of basic and reactive dyes in both single and binary systems using immobilized TiO_2, *Journal of Fundamental Sciences*, 5(2), 88-93.

Orbeci, C., Untea, I., Kopsiaftis, G., (2008), The influence of inorganic species on oxidative degradation of 4-chlorphenol by photo-Fenton type process, *Revista de Chimie* 59(9), 952-955.

Orbeci, C., Untea, I., Dancila, M., Stefan, D.S., (2010), Kinetics considerations concerning the oxidative degradation by photo-Fenton process of some antibiotics. *Environ. Eng. Manage.* J., 9, 1-5.

Ortiz de la Plata G. B., Alfano O. M., Cassano A. E., (2010), Decomposition of 2-chlorophenol employing goethite as Fenton catalyst. I. Proposal of a feasible, combined reaction scheme of heterogeneous and homogeneous reactions, *Applied Catalysis B: Environmental*, 95, 1-13.

Papp J., Shen H.S., Kershaw R, Dwight K., Wold A., (1993), Titanium(IV) Oxide Photocatalysts with Palladium, *Chem. Mater.* 5, 284;

Pera-Titus M., Garcia-Molina V., Banos M. A., Gimenez J., Esplugas S., (2004), Degradation of chlorophenols by means of advanced oxidation processes: a general review, *Applied Catalysis B: Environmental*, 47, 219-256.

Pignatello, J.J., (1992), Dark and photoassisted Fe^{3+} catalyzed degradation of chlorophenoxy herbicides by hydrogen peroxide, *Environ.Sci.Technol.* 26, 944–951.

Qu J., (2008), Research progress of novel adsorption processes in water purification: A review; *Journal of Environmental Sciences* 201–13.

Ratiu C., Manea , F. Lazau, C., Grozescu, I. Radovan C., Schoonman, J., Electrochemical oxidation of p-aminophenol from water with boron-doped diamond anodes and assisted photocatalytically by TiO_2-supported zeolite, *Desalination* 260 (2010) 51–56.

Ratiu C., Manea , F. Lazau, C., Orha, C., Burtica, G., Grozescu, I., Schoonman, J., Photocatalytically-assisted electrochemical degradation of p-aminophenol in aqueous solutions using zeolite-supported TiO_2 catalyst, *Chemical Papers* 65 (3) 289–298 (2011).

Riga, A., Soutsas, K., Ntampegliotis, K., Karayannis, V., Papapolymerou G., (2007), Effect of System Parameters and of Inorganic Salts on the Degradation Kinetics of Procion Hexl Dyes. Comparison of H_2O_2/uv, Fenton, and photo-Fenton, TiO_2/UV and TiO_2/UV / H_2O_2 processes, *Desalination*, 211, 72-86.

Pozzo, R., Baltanás, M., Cassano, A., Supported titanium oxide as photocatalyst in water decontamination: *State of the art, Catalysis Today*, 39 (3), 1997, 219-231.

Rozas, O., Contreras, D., Mondaca, M.A., Pérez-Moya, M., Mansilla, H.D., (2010), Experimental design of Fenton and photo-Fenton reactions for the treatment of ampicillin solutions. *J. Hazard. Mater.* 177, 1025-1030.

Shimizu N., Ogino C., Farshbaf Dadjour M., Murata T., (2007), Sonocatalytic degradation of methylene blue with TiO2 pellets in water, *Ultrasonics Sonochemistry*, 14, 184–190.

Siedlecka, E. M., Wieckowska, A., Stepnowski, P. (2007), Influence of inorganic ions on MTBE degradation by Fenton's reagent, *Journal of Hazardous Materials*, 147, 497–502.

Spacek, W., Bauer, R., Heisler, G., (1995), Heterogeneous and homogeneous wastewater treatment Comparison between photodegradation with TiO_2 and the photo-Fenton `reaction, *Chemosphere* 30, 477–484.

Trovó, A.G., Nogueira, R.F.P., Agüera, A., Fernandez-Alba, A.R., Malato, S., (2011), Degradation of the antibiotic amoxicillin by photo-Fenton process – Chemical and toxicological assessment. *Water Res.* 45, 1394-1402.

Venkata K., Rao S., Subrahmanyam M., Boule P., (2004), Immobilized TiO_2 photocatalyst during long-term use: decrease of its activity, *Applied Catalysis B: Environmental* 49 239–249.

Vinita M., Dorathi R. P. J., Palanivelu K., (2010), Degradation of 2,4,6-trichlorophenol by photo Fenton's like method using nano heterogeneous catalytic ferric ion, *Solar Energy*, 84, 1613-1618.

Vlazan P., Vasile M., (2010), Synthesis and characterization $CoFe_2O_4$ nanoparticles prepared by the hydrothermal method; *Optoelectronics and Advanced Materials-Rapid Communications*; 4; 1307-1309;

Xu, Hongwu, Zhang, Yiping, Navrotsky, Alexandra, (2001), Enthalpies of formation of microporous titanosilicates ETS-4 and ETS-10, *Microporous and Mesoporous Materials*, 47, 2-3, 285-291.

Zhou H., Smith, D.W., (2002), Advanced technologies in water and wastewater treatment, *J. Environ. Eng. Sci.* 1: 247–264.

Zazo J. A., Casas J. A., Mohedano A. F., Gilarranz M. A., Rodríguez J. J., (2005), Chemical Pathway and Kinetics of Phenol Oxidation by Fenton's Reagent, *Environmental Science and Technology*, 39, 9295-9302.

Wang H., Wang J., (2007), Electrochemical degradation of 4-chlorophenol using a novel Pd/C gas-diffusion electrode, *Applied Catalysis B; Environmental*, 77, 58-65.

Yu Y., Ding Y., Zuo S., Liu J., (2011) Photocatalytic Activity of Nanosized Cadmium Sulfides Synthesized by Complex Compound Thermolysis, *International Journal of Photoenergy*, 1-5;

Chapter 8

Groundwater Chemistry and Treatment: Application to Danish Waterworks

Erik Gydesen Søgaard and Henrik Tækker Madsen

Additional information is available at the end of the chapter

1. Introduction

Groundwater plays a pivotal role in Denmark. It is used for both agricultural and industrial purposes, but most importantly all Danish drinking water is produced from groundwater. To comprehend and discuss the processes and issues involved in the production of drinking water in Denmark, an understanding of the composition and the formation of groundwater is highly important.

2. Groundwater formation

Groundwater is formed by rain infiltrating the soil and subsurface, and as a result, the final composition of the water depends on both the specific geological formations and the residence time of the water in these. With respect to groundwater, the subsurface may be divided into two zones: the unsaturated zone and the saturated zone. In the unsaturated zone, the voids between particles are a mixture of water and air, while in the saturated zone all the voids have been filled with water. The transition from the unsaturated to the saturated zone marks the beginning of the water bearing layers; the groundwater. This is also called the water table.

As water infiltrates the subsurface, it moves from the highest hydraulic head to the lowest. Since land is generally higher elevated than water bodies such as rivers, lakes and the sea, these will usually be the final destination for the water. For a given hydrological area, this is called the discharge area, while the area in which the water infiltrates is called the recharge area. If the recharge area is far from the discharge area, the water will move almost vertically downward until it reaches a confining layer, see Figure 1. From here the water moves horizontally towards the discharge area, until it meets an opposing force that forces it upwards. If the discharge area is a river or a lake, this opposing force may be water coming from other directions, and if the discharge area is the sea, it will be the seawater, which will force the groundwater upwards due the difference in density. If the recharge area is closer to the discharge area, the path of the water will be more curved. It will not reach the same depth and will have a considerably shorter residence time. Also, as the water nears the discharge area, the flow of water will increase due to the incompressibility of water.

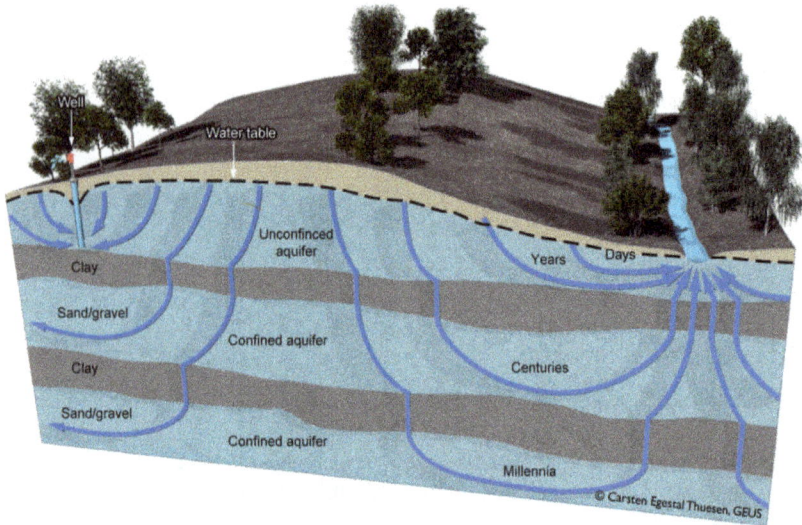

Figure 1. Illustration of groundwater flow and retention times as a function of distance between recharge and discharge area and depth of aquifer. Modified from *Viden om Grundvand* (15).

The movement of groundwater depends on the permeability of the water bearing layers. Layers such as clay have a low permeability and tend to inhibit water flow, whereas sand or chalk layers have a high permeability and promote water flow. In the subsurface, low permeable layers will act as confining beds, while high permeable layers will be water bearing layers. A geological unit from which groundwater may be extracted is called an aquifer, and there may be distinguished between two types: unconfined and confined aquifers. A confined aquifer is a water bearing layer completely enclosed by confining layers. These aquifers will only slowly recharge, but are also well protected against anthropogenic pollution from the surface. An unconfined aquifer is in direct contact with the surface, and will as such rapidly recharge depending on the amount of rainfall, but will also be more exposed to activities on the surface.

In Denmark, the majority of cities are situated near the ocean, and the available aquifers will most often be unconfined and placed close to the surface. If water is extracted from greater depths, it will be salty because the dense seawater forces the fresh groundwater further inland. The groundwater directly underneath the cities will as such be heavily influenced by the activity on the surface, and in recent years it has been found to be polluted with compounds such as chlorinated organic solvents and pesticides originating from industries and park maintenance (1). Furthermore, because water from the entire recharge area flows past the city on its way to the discharge area, pollution of the water in the recharge area may end up affecting the water quality in the city. Because Denmark is heavily populated and cultivated relative to its size, most recharge areas are farming land. This has led to increasing problems with pesticides and fertilizers used by the farmers, even though these farming chemicals are applied far from the city.

3. Groundwater composition

The most important factors affecting the composition of groundwater is the composition of the water after the immediate infiltration of the top layers of the surface, the geology of the subsurface, and the flow rate of the water through the subsurface. As the water moves through the subsurface, it is constantly approaching equilibrium with the surrounding geological layers. The type of equilibrium reactions are determined by the initial composition of the water and the specific geology of the subsurface, which therefore becomes very important for the final composition of the groundwater. The flow rate of the water controls the time available for the water to reach equilibrium with the surroundings, which is important since the equilibrium reactions vary in rate of reaction. In the upper part of the subsurface, the composition of the water is mainly determined by pH and redox conditions, and because of differences in rate of reaction, specific zones and fronts will be formed. These fronts and zones are general, and will be found in most places. At deeper levels, the retention time for the water is greatly increased, and slower reactions become influential. Here the specific geological conditions determine the composition of the groundwater, and this may result in very different types of groundwater. The type of groundwater is defined based on a division of its constituents into a number of groups as seen in Table 1. To understand the presence of these constituents, a more detailed description of the before mentioned processes is necessary.

Group		Constituents
Main components	Cations	Ca^{2+}, Na^+, NH_4^+, K^+, Mg^{2+}, Fe^{2+}, Mn^{2+}
	Anions	HCO_3^-, NO_3^-, SO_4^{2-}, PO_4^{3-}
Uncharged species		H_4SiO_4
Trace components		Al^{3+}, Ni^{2+}, Zn^{2+}, F^-, H_3AsO_3 and others
Gases		CO_2, H_2S, CH_4, O_2
Organic compounds		Humus
Anthropogenic compounds		Pesticides, chlorinated solvents, and others

Table 1. Division of groundwater constituents into component groups.

The first factor to influence the groundwater composition is the type of precipitation, which depends on its place of origin. Denmark has a coastal climate, and the rain will as such have a relatively high content of salts compared to rain formed from water evaporated inland. The atmospheric conditions also affect the composition of the rain. Combustion of fossil fuels may result in formation of SO_x and NO_x gases, which will dissolve in the rain drops and form sulfuric and nitric acid. Ammonia evaporation from farming industry may also lead to the formation of nitric acid, see equation 5.

On the surface and in the upper layers of the soil, substances like pesticides, fertilizers and organic solvents may be present due human activity (anthropogenic compounds) together with naturally occurring compounds, and these may dissolve in the water. Which compounds that are present, and how they dissolve in the water depends on the type of land. Different types of land such as forest, farming, or meadow, affects the degree evaporation and biological activity. A high evaporation will result in an increasing concentration of the dissolved compounds, and places with high biological activity may have a large uptake of these compounds and hereby change the composition of the water. The degree of biological activity will also influence the acidity of the water. Besides the anthropogenic acidifiers in the atmosphere, the natural content of CO_2 in the air will equilibrate with the rain drops, but in places with high biological activity this CO_2 contribution only plays a minor role. The microbial degradation of organic matter produces concentrations of CO_2 in the air trapped in the pores, which may be between 10-100 times higher than the concentration in atmospheric air (2). It will as such determine the acidity of the rain to a larger degree than the atmospheric CO_2. The acidity of CO_2 stems from its equilibrium with water, in which it dissolves and forms carbonic acid.

$$CO_{2(g)} + H_2O_{(l)} \rightleftharpoons H_2CO_{3(aq)} \tag{1}$$

Carbonic acid is a diprotic acid and may convert to either bicarbonate or carbonate depending on pH ($pK_{a1} = 6.351$, $pK_{a1} = 10.329$; T = 25 °C, zero ionic strength (3)).

$$H_2CO_{3(aq)} + H_2O_{(l)} \rightleftharpoons HCO_{3(aq)}^- + H_3O_{(aq)}^+ \tag{2}$$

$$HCO_{3(aq)}^- + H_2O_{(l)} \rightleftharpoons CO_{3(aq)}^{2-} + H_3O_{(aq)}^+ \tag{3}$$

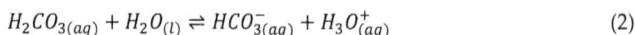

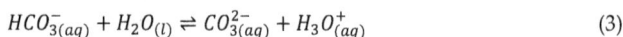

As the water infiltrates the soil, it will initially contain O_2 from equilibrium with the atmospheric air, and this promotes redox processes, which may also affect the acidity. In places with pyrite minerals, bisulfate can be formed during the oxidation of the sulfide minerals.

$$2FeS_{2(s)} + 7O_{2(aq)} + 2H_2O_{(l)} \rightarrow 4HSO_{4(aq)}^- + 2Fe_{(aq)}^{2+} \tag{4}$$

Also, when ammonium is present under oxidizing conditions, it may be oxidized to nitric acid.

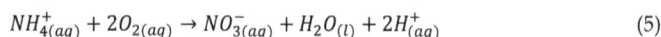

$$NH_{4(aq)}^+ + 2O_{2(aq)} \rightarrow NO_{3(aq)}^- + H_2O_{(l)} + 2H_{(aq)}^+ \tag{5}$$

As mentioned previously, the content of the water will change progressively as the water infiltrates deeper into the subsurface. For the pH driven processes, the order of reaction is determined by the pK_a values of the minerals in the subsurface. For the redox processes, the redox potential is the driving force. In reality pH and redox will often both affect the solubility of minerals, but for the sake of the overview, a distinction is made between important pH and redox driven processes.

pH driven processes

Some of the most important pH driven processes are:

- Dissolution of $CaCO_3$
- Dissolution and conversion of silicates
- Ion exchange with H^+
- Dissolution of hydroxide minerals

The processes are listed in the order they become influential as the pH is lowered.

Dissolution of CaCO₃ and hardness

Because of its basic nature, calcium carbonate is highly affected by pH, and its low solubility makes it one of the main issues in the use of groundwater for drinking water. Calcium carbonate will be in equilibrium with the Ca^{2+} and $CO_3{}^{2-}$ ions in the water, and as pH is lowered, carbonate ions will convert to bicarbonate and more calcium carbonate dissolves.

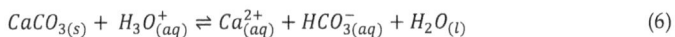

$$CaCO_{3(s)} + H_3O^+_{(aq)} \rightleftharpoons Ca^{2+}_{(aq)} + HCO^-_{3(aq)} + H_2O_{(l)} \tag{6}$$

Over time, the result is a removal of calcium carbonate, and the longer the surface has been exposed to rainfall the further down into the subsurface, calcium carbonate will have been dissolved. This creates a front, which is known as the acidic front. Below the front, calcium carbonate acts as a buffer, and the pH will be between 7 and 8. Above the acidic front there is no calcium carbonate to neutralize the CO_2, and if the water table is above the acidic front, the groundwater here will have a higher content of CO_2. Because this may lead to corrosion in drinking water equipment, this type of groundwater is said to contain aggressive CO_2.

High concentrations of dissolved calcium carbonate can also be an issue. When groundwater is exposed to the atmosphere or is heated, CO_2 will diffuse out of the solution, which then becomes supersaturated with calcium carbonate. Groundwater's content of calcium carbonate is commonly given in units of hardness, which is actually a measure of the content of Ca^{2+} and Mg^{2+} ions in the water. Magnesium is included since it is often found along with calcium minerals and has similar characteristics. Hardness is divided into three types: Total, transient and permanent. The total hardness is the total sum of Ca^{2+} and Mg^{2+} ions. The transient hardness is the amount of Ca^{2+} and Mg^{2+} ions dissolved as a result of CO_2.

Hardness classification	Total hardness °dH	$Ca^{2+} + Mg^{2+}$ meq/L	Ca^{2+} mg/L *
Very soft	0-4	0 - 1.4	0 - 2.0
Soft	4-8	1.4 - 2.8	28 – 56
Medium hard	8-18	2.8 - 6.4	56 – 128
Hard	18-30	6.4 - 11	128 – 220
Very hard	>30	>11	> 200

*The conversion from meq/L depends on the molar mass, and the calcium concentration in the last column has been calculated by assuming that only calcium is present.

Table 2. Hardness classification as a function of total hardness (2).

When CO_2 is driven from the water, this is the amount of Ca^{2+} and Mg^{2+} ions that will precipitate out of solution. The permanent hardness is the difference between total and transient hardness, and is caused by Ca^{2+} and Mg^{2+} dissolved by other acids than carbonic acid or CO_2. In cases where the concentration of HCO_3^- is greater than the content of Ca^{2+} and Mg^{2+} ions, the permanent hardness is zero. Hardness, in German units °dH, is calculated by converting the concentration of Ca^{2+} and Mg^{2+} ions from mg/L to meq/L and multiplying by 2.8. The typical classification of hardness is given in Table 2.

Dissolution and conversion of silicates

As pH drops below 6.7, silicate minerals become unstable and start to convert to clay minerals or hydroxides under loss of cations such as K^+, Ca^{2+} and Mg^{2+} and silicic acid. The reactions are complex and slower than the dissolution of calcium carbonate. Examples are given below (4).

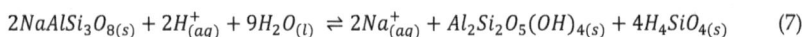

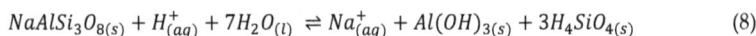

$$2NaAlSi_3O_{8(s)} + 2H_{(aq)}^+ + 9H_2O_{(l)} \rightleftharpoons 2Na_{(aq)}^+ + Al_2Si_2O_5(OH)_{4(s)} + 4H_4SiO_{4(s)} \qquad (7)$$

$$NaAlSi_3O_{8(s)} + H_{(aq)}^+ + 7H_2O_{(l)} \rightleftharpoons Na_{(aq)}^+ + Al(OH)_{3(s)} + 3H_4SiO_{4(s)} \qquad (8)$$

Ion exchange of H^+

As pH decreases, the adsorption equilibrium of the of H^+ ions/cations will result in especially Na^+, K^+ and Ca^{2+} ions being released as H^+ adsorbs to the minerals. The process becomes more pronounced as the concentration of oxonium increases.

Dissolution of hydroxide minerals

As the pH drops below 5, hydroxide minerals, most commonly iron and aluminum hydroxides, become partly unstable and start to dissolve.

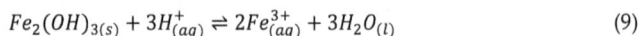

$$Fe_2(OH)_{3(s)} + 3H_{(aq)}^+ \rightleftharpoons 2Fe_{(aq)}^{3+} + 3H_2O_{(l)} \qquad (9)$$

Because the solubility of iron and aluminum minerals is highly dependent on the oxidation state as well, these processes will also be affected by the redox conditions.

Redox driven processes

The most important processes that are influenced by the redox conditions are:

- Nitrification
- Denitrification/sulfide oxidation
- Sulfate reduction
- Methane formation.

All of these are caused by microbes seeking to extract energy from the environment, and they will use the oxidizing agent that produces the largest gain in energy. The available oxidizing agents in groundwater are oxygen, nitrate and sulfate, listed in order of falling redox potential. As a result, the redox processes occurs in different zones. Near the surface, oxygen from the atmosphere is present in the water, and the redox conditions are oxidizing.

As the oxygen becomes depleted, the microorganisms start using nitrate if present, and redox conditions become weakly reducing. The transition from the oxidizing O_2/NO_3^- environment to the reducing NO_3^- environment is often called the nitrate front and the anoxic zone.

Nitrification

In the nitrification process, ammonium is oxidized to nitrite, which is further oxidized to nitrate.

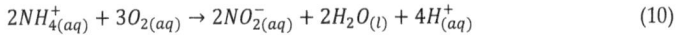

$$2NH_{4(aq)}^+ + 3O_{2(aq)} \rightarrow 2NO_{2(aq)}^- + 2H_2O_{(l)} + 4H_{(aq)}^+ \tag{10}$$

$$2NO_{2(aq)}^- + O_{2(aq)} \rightarrow NO_{3(aq)}^- \tag{11}$$

Denitrification and sulfide oxidation

Under anaerobic conditions, nitrate may be used as the oxidizing agent in denitrification processes. The reducing agent in these reactions may be organic matter, or pyrite, which are oxidized to CO_2 and sulfate. The ferrous ions (Fe^{2+}), released during pyrite oxidation, may contribute further to the denitrification under oxidation to ferric ions (Fe^{3+}), which may then precipitate as ferrihydrite compounds. The following three reaction schemes can be used to describe the denitrification processes.

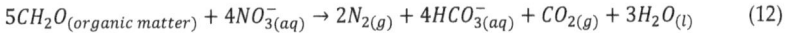

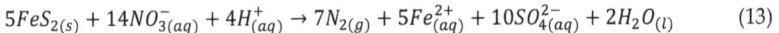

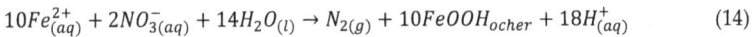

$$5CH_2O_{(organic\ matter)} + 4NO_{3(aq)}^- \rightarrow 2N_{2(g)} + 4HCO_{3(aq)}^- + CO_{2(g)} + 3H_2O_{(l)} \tag{12}$$

$$5FeS_{2(s)} + 14NO_{3(aq)}^- + 4H_{(aq)}^+ \rightarrow 7N_{2(g)} + 5Fe_{(aq)}^{2+} + 10SO_{4(aq)}^{2-} + 2H_2O_{(l)} \tag{13}$$

$$10Fe_{(aq)}^{2+} + 2NO_{3(aq)}^- + 14H_2O_{(l)} \rightarrow N_{2(g)} + 10FeOOH_{ocher} + 18H_{(aq)}^+ \tag{14}$$

Sulfate reduction

When nitrate is depleted sulfate may be used as the oxidizing agent. This occurs under its reduction to hydrogen sulfide gas.

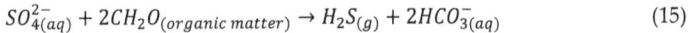

$$SO_{4(aq)}^{2-} + 2CH_2O_{(organic\ matter)} \rightarrow H_2S_{(g)} + 2HCO_{3(aq)}^- \tag{15}$$

In the presence of iron, hydrogen sulfide may precipitate as pyrite, but often some hydrogen sulfide will be left in the groundwater. Hydrogen sulfide has to be removed since it has a very pungent smell, which will ruin the quality of the water, and make it corrosive.

Methane formation

Below the sulfate zone the environment will be highly reducing, and organic matter may be reduced to methane.

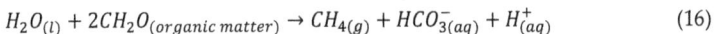

$$H_2O_{(l)} + 2CH_2O_{(organic\ matter)} \rightarrow CH_{4(g)} + HCO_{3(aq)}^- + H_{(aq)}^+ \tag{16}$$

Even though this zone is usually found deep underground, the groundwater from shallower aquifers may still contain methane since it will diffuse upward after formation.

4. Drinking water production - Simple water treatment

In Table 3, the usual composition of Danish groundwater is compared to the drinking water standards.

Parameter	Unit	Groundwater	Threshold limit after WW
Ca^{2+}	mg/L	10-200	< 200
Mg^{2+}	mg/L	2-30	50
Hardness	°dH		5-30
Na^+	mg/L	10-100	175
NH_4^+	mg/L	0.08-6	0.05
Fe	mg/L	0.02-40	0.05
Mn	mg/L	0.001-3	0.02
HCO_3^-	mg/L	10-400	>100
Cl^-	mg/L	30-70	250
NO_3^-	mg/L	0.5-110	50
NO_2^-	mg/L	-	0.01
SO_4^{2-}	mg/L	20-100	250
H_2S	mg/L	-	0.05
Agg. CO_2	mg/L	-	2
CH_4	mg/L	-	0.01
O_2	mg/L	0	10/5 *

* The concentration of oxygen at the tap must be 5 mg/L, and the waterworks therefore strive to saturate the water with oxygen, which will often result in a concentration around 10 mg/L after the waterworks.

Table 3. Groundwater and drinking water composition as specified in Danish law by executive order no. 1024 (2,5).

As seen, iron, manganese and ammonium often exceed the limits regarding concentration of the ionic species. Also, nitrate is sometimes found in concentrations above the threshold limit, which is commonly due to anthropogenic pollution. For the non-ionic species, it is mainly hydrogen sulfide and methane that must be removed.

Because of the natural filtration taking place during the formation of groundwater, usually only a simple treatment consisting of aeration and sand filtration is necessary to produce drinking water. In standard Danish drinking water production, the water is pumped to the waterworks, where it is aerated to remove dissolved gasses, and then led through a sand filter where solids are retained, before it is finally stored in a holding tank. Earlier, the treated water was stored in water towers or elevated containers, but this practice has been abandoned due to hygienic considerations. Instead, buried storage tanks and pumping systems are used today to create the necessary pressure in the water system. In Figure 2, addition of $Ca(OH)_2$ is also shown. The purpose of $Ca(OH)_2$ is to remove aggressive CO_2, and may be necessary in areas with soft water. It is however not a part of the standard simple treatment, but belongs to the category of extended treatment techniques, which is covered after the simple treatment. As a final general remark, it should be noted that one of the unique features in the Danish drinking water treatment process is the missing disinfection step. No chlorine or other disinfectants are added to the water at any time during the drinking water process, resulting in high quality water.

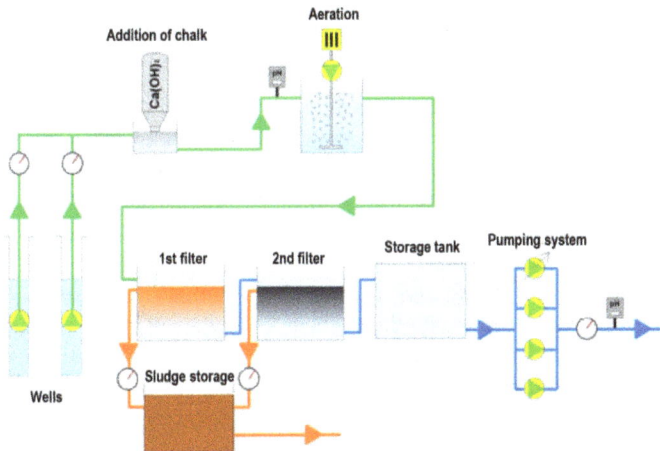

Figure 2. General process diagram for a Danish waterworks. The addition of $Ca(OH)_2$ is not part of the standard simple treatment. Reprinted with permission from Esbjerg Water Supply (Forsyningen Esbjerg).

Aeration

The aeration process serves two purposes:

1. A physical venting of the gasses H_2S and CH_4 (and CO_2) from the water
2. An increase of the oxygen content in the water to facilitate chemical/biological oxidation.

There exist several aerations methods. Earlier fountains or aeration stairs were widely used in the Danish drinking water sector due to their simplicity. In these methods, the water is forced to run down a number of steps where the water will be mixed with air during the drops and the turbulence created by the impact on the following step (6).

Cascade systems and trickling filter trays have also been used for aeration. These are variations of the fountain/stair concept, and show a somewhat improved aeration level. In the tray method, the water is led out onto perforated stainless steel plates with ø3-5 mm holes. This creates a large number of falling water jets that increases the total surface area, and hereby making the aeration more efficient. A drawback of the method is that the small holes tend to get clogged. The cascade system is similar to the fountain/stair system. Here the water runs down through a series of closed boxes with a small drop from box to box. This creates a negative pressure that sucks air down through the system and increases its efficiency compared to the open fountain/stair systems (6).

In cases where a large degree of venting is necessary, an INKA system can be applied. Here the water flows across a perforated plate, while air is blow up from below. This creates a large air to liquid ratio of up to 50-200, and ensures a very effective venting. Usually, this degree of venting is not necessary, and it might also affect the calcium carbonate balance negatively, since it will vent off a large degree of CO_2, leading to precipitation of calcium carbonate as pH increases. Also, the process is energy demanding and as such expensive (6).

Today, the most wide spread aeration method is bottom diffusers. The air inlet can be placed in a highly porous polyurethane sponge, which will ensure formation of small bubbles to give an efficient transfer of oxygen from air to water. The diffuser system has a better ability to saturate the water with oxygen, as well as to vent unwanted gasses compared to the previously used techniques. One of the main advantages of the diffusor system is that it can be modified to handle variations in the water flow and type of groundwater. An increase in the oxygen demand can be met by increasing the air flow; something, that cannot be done with the methods using the fall of the water for aeration. These must be designed to meet the specific oxygen demand for each waterworks. The oxygen demand is determined both by the flow and the groundwater's content of oxygen consuming species. The oxygen consuming species and their oxygen demand are shown in Table 4 along with the respective oxidation reactions:

- Oxidation of Fe^{2+} and Mn^{2+} to insoluble solids
- Oxidation of NH_4^+ to NO_3^-
- Oxidation of H_2S and CH_4 to SO_4^{2-} and CO_2

The relatively high oxygen demand of the methane and hydrogen sulfide oxidation reactions makes it important to vent these gases. Otherwise, the residual oxygen concentration might not meet the limit of 5 mg/L. Low oxygen concentration in the drinking water may result in anaerobic conditions in the piping system, leading to unwanted microbial growth. In case of microbial growth, nitrate may be reduced to nitrite, which may then increase to a level above the threshold limit. After aeration, the water may be lead to a reaction tank to allow for sufficient reaction time for the chemical oxidation reactions, but often the water is led straight to the sand filter(s).

Substance	Oxidation	Oxygen demand per mg
Fe^{2+}	$4H^+_{(aq)} + O_{2(aq)} + 4Fe^{2+}_{(aq)} \rightarrow 4Fe^{3+}_{(aq)} + 2H_2O_{(l)}$	0.14
Mn^{2+}	$4H^+_{(aq)} + O_{2(aq)} + 2Mn^{2+}_{(aq)} \rightarrow 2Mn^{4+}_{(aq)} + 2H_2O_{(l)}$	0.29
NH_4^+	$2NH^+_{4(aq)} + 3O_{2(aq)}$ $\rightarrow 2NO^-_{2(aq)} + 2H_2O_{(l)} + 4H^+_{(aq)}$	3.6
	$2NO^-_{2(aq)} + O_{2(aq)} \rightarrow 2NO^-_{3(aq)}$	
CH_4	$CH_{4(aq)} + 2O_{2(aq)} \rightarrow CO_{2(aq)} + 2H_2O_{(l)}$	4.0
H_2S	$2H_2S_{(aq)} + O_{2(aq)} \rightarrow 2S_{(aq)} + 2H_2O_{(l)}$	0.51
	$2S_{(aq)} + 3O_{2(aq)} + 2H_2O_{(l)} \rightarrow H_2SO_{4(aq)}$	0.79

Table 4. Oxygen demand for main oxidizable components in groundwater (5).

Sand filtration

In the sand filtration, solid precipitates are filtrated from the water, and it is here that the largest part of the oxidation reactions takes place.

The oxidation of ferrous to ferric ions results in iron precipitating from the water as ferrihydrite because of its low solubility (pK$_{sp}$ = 38.8, T = 25 °C zero ionic strength, (3)).

$$Fe^{3+}_{(aq)} + 3OH^-_{(aq)} \rightarrow Fe(OH)_{3(s)} \tag{17}$$

The ferrihydrite coats the sand grains, where it leads to autocatalysis of the oxidation reaction (7, 8). The autocatalytic reaction makes the iron oxidation very efficient, and removes the need for a reaction tank before the sand filter (6). Because of the redox potential for the oxidation of ferrous to ferric iron, the use of oxygen as the oxidizing agent is sufficient, see Figure 3. In some cases the chemical catalytic oxidation can be supported by iron oxidizing bacteria, which can increase the rate of oxidation/precipitation even further. An iron oxidizing bacteria often found in sand filters of Danish waterworks is *Gallionella ferruginea* that has been found to enhance the oxidation and precipitation velocity due to their production of exopolymers. The exopolymers give a denser structure of the iron precipitate, and allows for more iron to be removed by the sand filter before a backwash of the sand filter for its cleaning is necessary (9).

The redox potential for the oxidation of manganese(II) to manganese(IV) is higher compared to the iron(II) to iron (III) oxidation, see Figure 6, and with oxygen as the oxidizing agent, the process is relatively slow. At neutral pH, the reaction between Mn(II) and O$_2$ is around 10^6 times slower than the reaction between Fe(II) and O$_2$ (7). However, two processes in the sand filter aid the oxidation of manganese: Surface catalyzed oxidation and co-precipitation with ferrihydrate. Both processes can be illustrated by the two step reaction scheme in Figure 4.

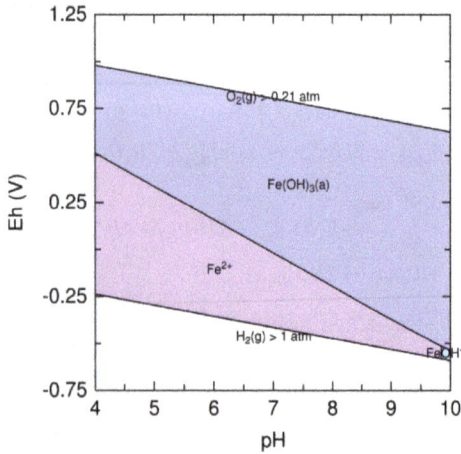

Figure 3. Pourbaix diagram for iron, showing the most thermodynamical stable form as a function of pH and redox potential.

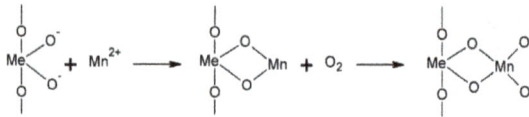

Figure 4. Reaction scheme for the oxidation of manganese in sand filters, where Me symbolizes a transition metal ion.

In the first step, Mn(II)-ions is oxidized to MnO_2. As with iron, the oxidation of manganese is catalyzed by metal oxide surfaces on the sand grains. The hydroxyl groups on the metal oxide surface (Me-OH) attracts the Mn(II) ions and promote the oxidation, as illustrated in step two. If the metal surface is manganese oxides, the process is autocatalytic, but because of the ratio of iron and manganese in groundwater, the metal surface is most likely iron oxides in the first filtration step where manganese is then said to co-precipitate with iron (7). In the second filtration step precoating of the sand grains with manganese oxide can help oxidizing the adsorbed Mn(II) ions, see step three. The result is that it is not necessary to use stronger oxidizing agents than the oxygen found in atmospheric air to remove iron and manganese.

Also present in the sand filter are nitrification bacteria. These will oxidize ammonium first to nitrite and afterwards to nitrate (6).

To ensure an efficient oxidation, a double filtration system is commonly employed. The sand filters may be open or closed, with variations from waterworks to waterworks. The filters are back washed at regular time intervals, in a process where first air followed by water are sent backwards through the filter system. The air will remove and lift the colloids adsorbed to the sand grains producing a floating sludge on top of the filter. Later it will be washed away by help of the back wash water. At some drinking water treatment plants, the backwash water is returned to the plant where it is treated with UV-light, filtrated, oxidized again, and brought to the drinking water container. When a sand filter is changed, some of the old sand is mixed with the new to preserve the microbiological environment and to increase the rate of re-population.

To demonstrate the effect of the processes included in a simple treatment at Danish waterworks, data has been collected at a specific waterworks at different points along the treatment process.

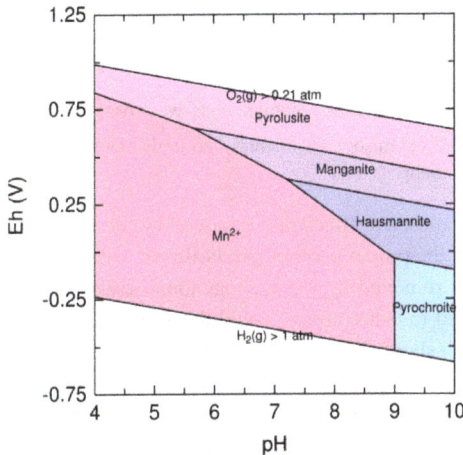

Figure 5. Pourbaix diagram for manganese, showing the most thermodynamical stable form as a function of pH and redox potential.

Simple drinking water treatment – Case Spangsbjerg waterworks, Esbjerg

Spangsbjerg waterworks is one of four waterworks in the city of Esbjerg. It is one of the old waterworks placed in the city, but today 75 % of the groundwater is produced from wells 40 km away in central Jutland, from two deep aquifers: Boegeskov and Sekaer. The waterworks is equipped with a diffusor system for aeration and two open sand filters in series. After treatment the water is stored in a buried drinking water storage tank outside the waterworks.

To evaluate the effect of the treatment, the concentration of Fe^{2+}, Mn^{2+}, Ca^{2+}, Mg^{2+} and NO_3^- was measured with ICP-AES at five places during the treatment.

1. Raw groundwater
2. After aeration
3. After 1st filter
4. After 2nd filter
5. Storage tank

The results are plotted in Figure 6 with the x-axis representing the transport through the waterworks.

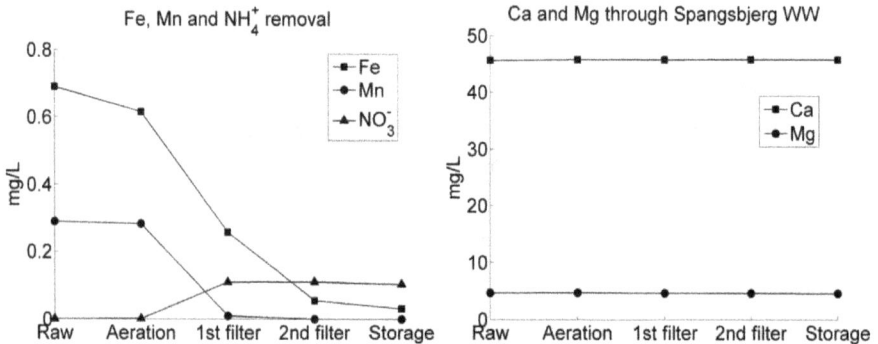

Figure 6. Concentration levels of major constituents of groundwater through Spangsbjerg Waterworks simple drinking water treatment.

As seen, iron and manganese are effectively removed by the sand filters. Manganese are not affected significantly by the pure oxidation, but is almost completely removed already in the first filter. Some iron is removed by the homogenous oxidation, and needs to go through both filters to be reduced to below the accepted threshold limit. Ammonium is not directly measured, but the graph shows that the ammonium in the water is effectively converted to nitrate in the first filtration step. This is important since it shows that the conversion is complete in the first sand filter, meaning that no nitrite is left in the treated drinking water. Calcium and magnesium passed unaffected through the waterworks together with other macro ions as K^+, Na^+, Cl^-, SO_4^{2-} and the main part of HCO_3^-. The unchanged hardness is also to be expected since Spangsbjerg Waterworks only applies aeration and sand filtration, which do not affect the solubility of calcium and magnesium minerals significantly.

5. Extended water treatment in the Danish drinking water sector

It requires a special permit if a Danish waterworks is to apply more advanced water treatment techniques than those already described in simple water treatment[1]. To achieve a permit, the waterworks must submit a technical, economic and environmental report for the choice of treatment technology according to § 14, part 2 of executive order no. 1451. Furthermore, according to § 9 of executive order no. 1451 it is necessary to obtain a statement from the National Board of Health, which is represented by the medical health inspectors (10). The need for a special permit is based on the principle that if techniques capable of removing pollution are allowed, the incentive towards avoiding pollution of the groundwater in the first place will be smaller. Therefore, if a well is found to be polluted, common practice is to close the well or use it for purposes other than drinking water, instead of applying more sophisticated water treatment techniques. With the structural reform of the Danish municipalities in 2007, the power to give permissions for extended

---

[1] From here on techniques other than what constitutes the simple water treatment will be classified as extended water treatment techniques.

water treatment was moved from the counties to the municipalities, which are expected to be more willingly to allow for further treatment (11). However, the conservative approach to the use of extended water treatment techniques is still widespread throughout the Danish drinking water sector, and the use is very limited.

Before the municipality reform was enforced, a compilation of the applications for use of extended water treatment techniques was made, and it was found that the sources for need of further treatment of the water was distributed into four categories (11). Treatment for:

1. Main constituents of groundwater: Aggressive carbon dioxide, Ca^{2+} and hardness, SO_4^{2-}, Cl^-, NO_3^-, Fe/humus-complexes, humus and color.
2. Inorganic trace compounds: As and Ni.
3. Problems caused during simple treatment: $Fe/Mn/NH_4^+$ removal and increased bacterial count.
4. Organic micropollutants: Pesticides and chlorinated organic solvents

An overview of the application is given in Table 5.

Problem causing compound	Number of applications	Geographical placement
Main constituents	**28**	
Agg. CO_2	9	Bornholm, west- and eastern Jutland
Ca and hardness	6	Funen, west- and southern Jutland, Zealand
SO_4^{2-}	1	Copenhagen
Cl^-	1	Southern Zealand
NO_3^-	4	Northern Jutland
Humic-bound Fe	5	Jutland
Humus	1	Western Jutland
Colour	1	Southern Jutland
Inorganic trace compounds	**11**	
As	10	Funen, Eastern Jutland, Southern Zealand
Ni	1	Copenhagen
Operational problems	**3**	
Mn/NH_4^+	1	Western Jutland
CFU	2	Western Jutland
Organic micropollutants	**19**	
Pesticides	13	Funen, Copenhagen, north, east and central Jutland, Zealand
Chlorinated solvents	6	Funen Copenhagen
Total	**61**	

Table 5. Overview of compounds causing need for extended water treatment in the Danish drinking water sector (11).

As a result of these applications, there were 29 plants operating with extended water treatment in 2006. 14 of these were treating for issues with main constituents, nine for problems with inorganic trace compounds and five treating for pesticides and chlorinated solvents. The total amount of produced water from these plants was 2.5 million m^3 annually (11). The main problems may be divided into two groups: One correlated with the Danish geology, and a second with the anthropogenic activity.

Problems with calcium carbonate scaling and arsenic contamination belong to the first category, and are most prominent from Eastern Jutland and eastwards. This part of the country was covered by ice during the last ice age, while the western part of Jutland was left uncovered. As a result, carbon dioxide in the rain has dissolved much of the calcium carbonate in the underground in this part of the country. The soil is also more sandy in the western part of Denmark, while the soil in East Denmark has a high content of clay; a fact which also influences the vulnerability of the groundwater aquifers (1, 12).

The second category, pollution caused by anthropogenic activity, is distributed over the entire country, although it is also influenced by the geology, mainly the type of soil. The two main threats to groundwater quality are nitrate and pesticides. Chlorinated organic solvents are also a concern, but they are often found together with pesticide pollution (1).

In Table 6 an overview of the techniques applied as extended treatment in the Danish drinking water sector is given. It is seen that for some of the problems only one type of technique has been investigated, as with the use of active carbon filtration for removal of pesticides, whereas for other problems, a wider range of techniques have been applied. The use of different techniques is also correlated to the number of times the problem has been encountered.

Problem	Treatment technique
Aggressive CO_2	NaOH, $Ca(OH)_2$
Calcium and hardness	CO_2, fluid-bed softener, magnetic treatment
Chloride	Reverse osmosis
Nitrate	Nitrate-redox method
Humic-Fe and humus	$Al_2(SO_4)_3$, $KMnO_4$, $AlCl_3$
Arsenic	$FeSO_4$, $FeCl_2$
Chlorinated solvents	Active carbon + UV, extended aeration (Microdrop)
Pesticides	Active carbon + UV

Table 6. Overview of extended water treatment techniques applied for the different problems encountered in the Danish drinking water production (11).

To investigate the use of some of these techniques in greater detail from here on, a case study approach will be used.

6. Galten waterworks – removal of arsenic

In 2003 the threshold limit for arsenic in drinking water was lowered from 50 to 5 µg/L. As a result, many waterworks situated in places with marine clay sediments got issues with removal of arsenic (12). In Figure 7, it can be seen where in Denmark arsenic has been found in the drinking water, and it is clear that the clay rich eastern part of Denmark from east Jutland and eastward is the most affected. Arsenic is often bound to iron minerals and released when ferrihydrite (Fe(OH)$_3$) is reduced or pyrit (FeS) is oxidized. However, it has been found that by applying reduced iron, arsenic can be made to co-precipitate (13). At Galten waterworks near Aarhus, the concentration of arsenic was found to be 21 µg/L, and experiments with addition of FeSO$_4$ were made at the smaller waterworks Galten Vestermarks. The method has been found to be effective. Based on these results, in 2004 Galten waterworks applied for permission to use FeSO$_4$, in concentrations of 5 mg/L, for removal of arsenic, to be able to meet the threshold limit. The permission was given, but only for a two year period based on the recommendation of the health inspector (11). In 2008 Galten waterworks got permission to use FeCl$_2$ to remove arsenic (14).

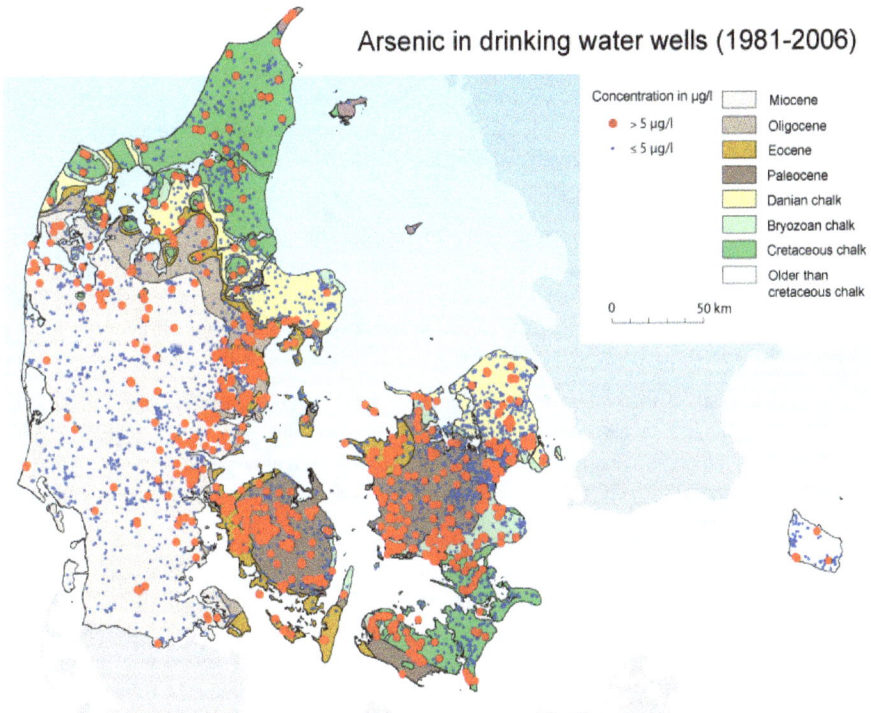

Figure 7. Map of drinking water wells in which arsenic has been found in the period from 1981 to 2006. Modified from *Viden om Grundvand* (15).

Areas with arsenic concentrations above the threshold limit in more than 100 water wells are shown by red colour (15).

Frederiksberg waterworks – removal of chlorinated solvents

In many places in Denmark, the groundwater is polluted with organic micropollutants such as chlorinated solvents. These originate from varying sources, including landfill leachate, colouring and varnish industry, pesticide production industry, gas stations and dry cleaning industry. Because the pollution is industrial related, it is often found close to population centers, where also many drinking water wells have been placed (1).

Frederiksberg waterworks is found close to central Copenhagen and produces 2,500,000 m³ drinking water annually, which is 45 % of the total consumption for Frederiksberg. The remaining water is purchased by Copenhagen Energy. In 1997 the waterworks filed for permission to establish an active carbon filtration system because the wells had been found to be polluted with a wide range of mainly chlorinated solvents. In the wells 1,1,2-trichloro-ethene, cis/trans-1,2-dichloro-ethene, 1,1-dichloro-ethene, tetrachloro-ethene, vinylchloride, 1,2-dichloro-ethane, MTBE, toluene and benzene have been found in concentrations of 0.02-5 μg/L (15). In the application the waterworks assessed that it was not possible to find new wells without pollution. Also, although it was possible to purchase water from Copenhagen Energy, it was stressed that this would result in increased pressure on the environment of Zealand from where the water would be drawn. Already, Zealand is relatively poor in received rainfalls, and it was viewed as important to use the water resources as efficiently as possible (11).

In June 1997 permission was given to use active carbon for a five year period. Originally, Frederiksberg waterworks had applied for a permanent use of carbon filtration, but the health inspector would only agree to the five year permission. Later the permission has been extended on several occasions, latest in 2009, on the conditions that the filtration system is regularly checked for efficiency in removal of the chlorinated solvents.

The long term goal for the waterworks is to supply water that has only gone through simple water treatment, and to lower the content of chlorinated solvents in the water. A new extraction strategy was constructed in 2003, in which two new wells were established. However, this has not been sufficient to reduce the concentrations, and carbon filtration continues to be necessary.

At Frederiksberg waterworks, the water first undergoes a simple treatment with aeration and sand filtration before it is stored in the water tanks. The carbon filtration system is placed after the storage tanks, and consists of two closed filters in parallel, followed by UV disinfection. Parallel filters have larger treatment capacity and are cheap with regards to installation costs, but also have higher risk for a breakthrough compared to serial setups, where the breakthrough can be measured on the first filter. To compensate for the higher risk of breakthrough, sampling points have been installed on the carbon columns to measure the saturation front. The filtration system has a capacity of 500 m³/h, contains 16 tons coal per filter and each filtration tank has a volume of 40 m³.

7. Hvidovre waterworks – removal of pesticides

One of the biggest issues in the Danish drinking water sector is contamination of the groundwater with pesticides, and it is estimated that between 1993 and 2009 around 130 wells all over the country have been closed due to pesticide pollution (16). Since 1993 the degree of pesticide pollution of the Danish groundwater has been monitored by the Danish geological service (GEUS), and during the years, an ever increasing amount of the aquifers has been found to be contaminated, as seen in Figure 9. This is not so much a result of an increasing actual pollution, as it is a result of more and more pesticides being included into the monitoring program. In the latest report, it was found that between 1990 and 2010, 50.7 % of the monitored aquifers had been polluted with pesticides, and that 24.5 % of the wells used by the water works contained pesticides (1). In Figure 8 it can be seen that all parts of Denmark, bot rural and cities, are affected by the pollution. However, the monitoring program does only cover parts of the aquifers in Denmark, and the pesticides currently in the monitoring program only constitute 29 % of the total sale of pesticides in Denmark from 1988 to 2010. The degree of pesticide pollution may as such be expected to increase over the coming years, forcing more waterworks to initiate extended treatment.

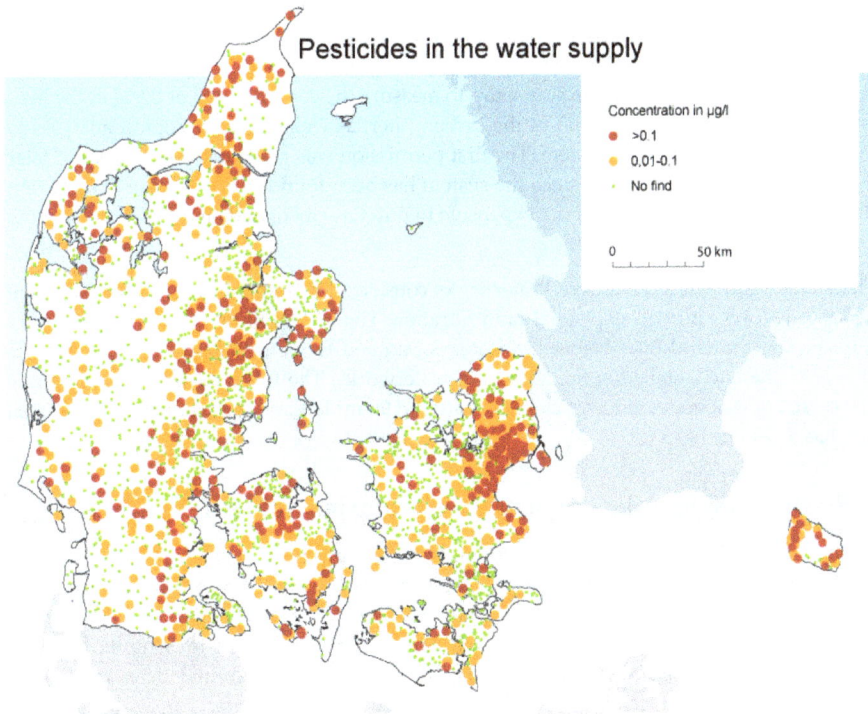

Figure 8. Map of groundwater wells (1993-2004) and drinking water wells (1993-2010) where pesticides have been found (1,15). Modified from *Viden om Grundvand* (15).

Of the pesticides currently in the program, 2,6-dichlororbenzamide (BAM), is the biggest problem, being found in 20 % of the analyses which have been found to contain pesticides (1). In all cases where extended treatment has been used to remove pesticides, it has been BAM that has been the polluting pesticide (11). BAM is a degradation product of the pesticide dichlobenil, which was earlier a widely used pesticide for removal of weed on uncultivated areas like farmers gravel covered courtyards, public parks and along railways.

Hvidovre Waterworks is situated in South East Copenhagen and produces around 800,000 m^3 of drinking water annually. Besides this, 2,400,000 m^3 is purchased annually from Copenhagen Energy to supply the waterworks own production. In October 1995, when BAM was taken into the analysis program of the waterworks, it was found in all the wells of Hvidovre waterworks, and in several of these wells BAM was found in concentrations above the threshold limit of 0.1 µg/L. In 1996, Hvidovre waterworks applied for permission to use active carbon filters followed by UV disinfection. In the application, the waterworks listed two scenarios: 1) Increase the amount of purchased water from Copenhagen Energy to lower the concentration below the threshold limit through dilution, or 2) Apply active carbon filtration to remove the pesticides. The use of active carbon was the cheapest solution, and also the ability to produce its own water was important to Hvidovre Municipality. On the 27th of June 1996, a temporary permission was given on the grounds that it was important to maintain a local drinking water production (health inspector), and because the county assessed the technology to be relatively simple. As a condition for the permission, an analysis program was setup to measure the concentration of BAM in the inlet and outlet of the waterworks and in the carbon filter, as well as bacteriological analyses to monitor the effect of the UV system. The first permission was given for three years and later extended on several occasions, since the system has been found to be effective at removing BAM and because it would be more expensive to purchase increased amounts of water from Copenhagen Energy (11).

The water treatment at Hvidovre Waterworks consists of a simple treatment with aeration on cascade trays and two step serial sand filtration. The carbon filtration unit is installed at the outlet of the sand filter before the water is pumped to the drinking water tank, and is similar to the one installed at Frederiksberg waterworks. The full capacity of the system is 150 m^3/h, but it operates at lower capacities around 90 m^3/h. Each filter contains 6.3 tons coal and has a volume of 15 m^3.

8. Sjaelsoe waterworks – removal of mycotoxins from surface water

Today all drinking water in Denmark originates from groundwater. However, because of the low rainfall on Zealand, it has been necessary on occasions to use surface water. One such waterworks, capable of treating surface water is Sjaelsoe Waterworks in Rudersdal Municipality, north of Copenhagen, where it supplies drinking water to Gentofte, Lyngby-Taarbaek, Hoersholm and Karlebo municipalities. It produces around 5.5-6 million m^3 annually, but has a capacity of 11 million m^3 (17). Because the water sources have a high variation in the composition of the water, the waterworks consists of three main facilities handling each type of water:

- Plant I: Is a traditional waterworks with aeration on a cascade tray and subsequent sand filtration. It handles water from one water source.
- Plant II: Is equipped with a more efficient INKA aeration system, followed by sand filtration. The plant receives water from six water sources.
- Plant II: Is equipped to handle surface water from Sjaelsoe, a local lake. The plant consists of a sand filtration unit followed by flocculation, sedimentation, ozone treatment and a final scrubbing with biological active carbon filtration.

Due to the use of fertilizers, many of the inland waters in Denmark have suffered from eutrophication, which has also been the case in Sjaelsoe. Toxin analyses have shown the blue green alga from Sjaelsoe to contain microcystins in amounts from 1-59.1 µg per gram dry matter. To evaluate the plant's efficiency in removing these compounds, experiments were made with microcystins extracted from dried algae on a pilot plant. The results showed that sand filtration and sedimentation did not affect the concentration of the microcystins, but that ozone was very effective for removing these compounds. By using an ozone concentration of 2 mg/L, the concentration of microcystins was reduced below the detection limit, see Figure 10 (18).

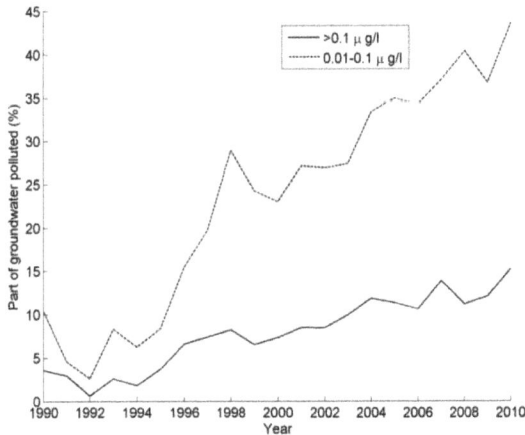

Figure 9. Development in the percentage of the Danish groundwater reserve where pesticides have been found from 1990 to 2010 (1).

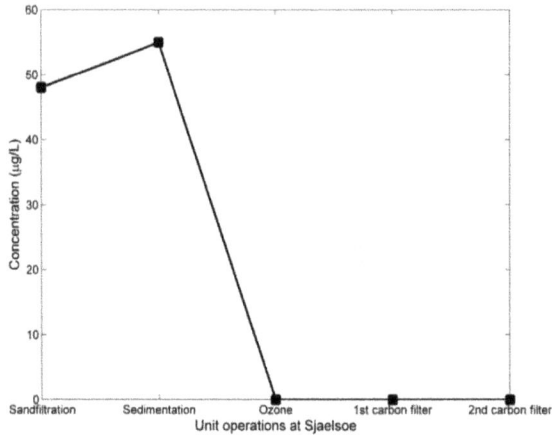

Figure 10. Removal of microcystin in a pilot plant at Sjaelsoe Waterworks. Modified from *Blue green algae in bathing and drinking water* (18).

The plant at Sjaelsoe has as such been found to be effective in ensuring safe drinking water. Even so, due to the goal of using groundwater in the drinking water production, the plant has not been in operation since 1998.

Author details

Erik Gydesen Søgaard and Henrik Tækker Madsen
Section of Chemical Engineering, Aalborg University Esbjerg, Denmark

9. References

[1] Thorling L., Hansen B., Larsen C. L., Brüsch W., Møller R. R., Mielby S. and Højberg A. L. (2011) Grundvand - Status og udvikling 1989-2010 (Groundwater – Status and development 1989-2010). Technical report, GEUS, Denmark.

[2] Elberling B. (2002). Grundvandskemi (Groundwater chemistry). In: *Vandforsyning (Water supply)*, Karlby H. and Sørensen I., 2nd edition, Ingeniøren I bøger, Copenhagen, Denmark, pp. 83-102.

[3] Harris D. C. (2007). *Quantitative Chemical Analysis*, 7th edition, W. H. Freeman and Company, New York.

[4] Appelo C. A. J. and Postma D. (2005). *Geochemistry, Groundwater and Pollution*, 2nd edition, Taylor and Francis.

[5] Danish Ministry of the Environment, Executive order no. 1024, Executive order for water quality and inspection of waterworks, 2011.

[6] Jensen E. D. (2002). Normalbehandling af grundvand (Simpel treatment of groundwater). In: *Vandforsyning (Water supply)*, Karlby H. and Sørensen I., 2nd edition, Ingeniøren I bøger, Copenhagen, Denmark, pp. 321-383.

[7] Martin S. T. (2005). Precipitation and Dissolution of Iron and Manganese Oxides. In: *Environmental Catalysis*, Grassian V. H., Taylor and Francis Group, Boca Raton, Florida, United States of America, pp. 61-83.

[8] Chaturvedi S. and Dave P. N. (2012). Removal of iron for safe drinking water. *Desalination*.

[9] Søgaard E. G., Aruna R., Abraham-Peskir J. and Koch C. B. (2001). Conditions for biological precipitation of iron by *Gallionella ferruginea* in a slightly polluted ground water. *Applied Geochemistry*. 16, 1129-1137.

[10] Danish Ministry of the Environment, Executive order no. 1451, Executive order for water extraction and waterworks, 2007.

[11] Juhl M. M., Jensen T. O. and Jensen K. M. (2007). Erfaringsopsamling inden for tilladelser til videregående vandbehandling (Compilation of experiences with permits for extended water treatment). Technical report, Danish Environmental Agency, Denmark.

[12] Larsen C. L. and Larsen F. (2003). Arsen i danske sedimenter og grundvand (Arsenic in Danish sediments and groundwater). *Vand & Jord*. 10, 147-151.

[13] Søgaard E. G. and Sønderby C. (2007). Elimination of Arsenic Substances from Groundwater for Production of Drinkable Water using Microdrop Reactor System. *Chemical Water and Wastewater Treatment: The International Gothenburg Symposium on Chemical Treatment of Water and Wastewater*. IWA Publishing Company, 9, 389-400.

[14] Skanderborg Municipality, Vandforsyningsplan 2008-2016 (Water distribution plan 2008-2016).

[15] Rosenstand U., Marvil C., Thomsen H. H., Warna-Moors P. and Thuesen C. E. (2012). Viden om Grundvand (Knowledge about Groundwater). Geological Survey of Denmark and Greenland (GEUS), http://www.geus.dk.

[16] Thorling L., Hansen B., Larsen C. L., Brüsch W., Møller R. R., Mielby S. and Højberg A. L. (2010) Grundvand - Status og udvikling 1989-2009 (Groundwater – Status and development 1989-2009). Technical report, Geological Survey of Denmark and Greenland, Denmark.

[17] Hoersholm Municipality, Vandforsyningsplan 2007-2016 (Water distribution plan 2007-2016).

[18] Danish Environmental Agency (1998). Blågrønalgetoksiner i bade- og drikkevand (Blue green algae in bathing and drinking water). Technical report, project no. 435, Denmark.

Treating of Waste Water Applying Bubble Flotation

Ramiro Escudero, Francisco J. Tavera and Eunice Espinoza

Additional information is available at the end of the chapter

1. Introduction

Recently, it might be exaggerated the impact of man activities as the solely cause of the tremendous changes in the global climate resulting from its activity effects on the environment nature; it has been suggested that it has started a new geological era, the "antroposoic age;" however, there are some clues, together with man influence in climate changes, that sun activity has a determining effect on the subject. Nevertheless, from the beginning of man species presence in the geological record it is observed an interaction of these species with the surroundings that increases with their degree of social and technological development.

The association between man and its environment presents a continuous growth because the development of its intellect and the complex nature of its thought; these features allowed man to adapt to a variety of geographical situations, spreading towards the entire planet, building technologies as a result of its social development needs.

In the historic record of civilization, in modern times, between the second half of the 1700's and the middle of the 1800's, there is an exponential increase in natural resources exploitation because the advance in the scientific knowledge and the technological development. The expansion trend of these features, the science and technology, has been maintained with no interruption, as well as their effect on the surroundings producing a deleterious consequence on the environs quality.

It is well understood that the equilibrium in the planet nature is very sensitive to man activities, therefore, it is necessary to intensify the carefulness to preserve the ecosystem quality in order to prevent catastrophic environmental disasters.

A critic facet of natural environment is related to water pollution. Depletion of fresh water reserves makes it important to propose effective technologies to clean and recover water from polluted bionetwork. Therefore, advances in possible water treating improvements are issued in this chapter.

2. Treating of waste water applying bubble flotation

From the time when natural resources started to exploit in a massive way, water is used as a vehicle to transport and process materials. Water is the most important solvent in nature, and the contact of water with a variety of substances makes it the origin of water pollution through the formation of dissolved solutes solutions which decrease the quality of the liquid.

Water is often encountered with high concentration of pollutants that could be in solution or dispersed as insoluble phases frequently forming emulsions and colloids. Among pollutants encountered in fresh water are heavy metals and organics. These materials regularly are difficult to process because to the enormous amounts of water that is produced by urban and industrial activities; therefore, the ideal process to clean polluted water should be operating in a continuous way to minimise fixed and operating costs.

2.1. Lead carbonate colloid in residual water: continuous ion flotation

The ion flotation term was first introduced by Sebba [1], defined it as a technique used to collect a material which is in solution in an aqueous phase, and it may be even in a colloid structure. By adding modifiers of the chemical conditions of the aqueous media, and with the addition of a collector reagent, adequately electrically charged, the dissolved substance, or the colloid, is transformed in a product with hydrophobic sites which promotes its adsorption on the liquid/vapour "surface" of a gas bubble and, therefore, to float in the aqueous media up to its surface producing a foam that contains a concentrate of the original substance which can be collected.

By looking into the previous statement, it can be thought that there is a concentration relationship between the amount of dissolved substance and the required amount of collector reagent to create hydrophobic sites. Therefore, ion flotation technique is restricted to treat dilute aqueous solutions of such ion or colloid material, otherwise, the required amount of collector reagent may exceed the micelle critic concentration and the desired process would not be attained.

The goal idea in the application of ion or colloid flotation techniques may be to treat residual industrial or urban water, which means that it has to deal with enormous amounts of water and, therefore, a continuous process should be available to separate the solute impurity.

In the case of flotation as applied to mineral particles, it requires the gas bubbles to be large enough to produce a solid-bubble aggregate with a lower relative density than the mineral pulp density which is processed and, in this case, bubble shear must be controlled to avoid the detachment of solids from the bubble.

However, in ion flotation the material to be separated from the liquid presents a similar density as that of the liquid bulk, and the inertia of this material in front of the moving gas bubbles is such that it can be easily drawn by the liquid that flows surroundings the bubble. Consequently, it may be expected that the contact between the hydrophobic material and the gas bubble will be likely to occur by decreasing bubble size; in addition, when the formation of small bubbles takes place, it creates a larger gas surface area which improves the flotation process [2, 3].

Most of the experimental work reported on ion flotation has been carried out in flotation columns may be under the influence of a better performance of column flotation in mineral processing. In spite of this, the ascending bubbles in a liquid column produce axial mixing and turbulences that prevent an effective contact among the floatable material and the gas bubble.

At the present time, the ion flotation technique has not reached any commercial appliance; nevertheless, there are good expectations for different engineering applications. A possibility in commercial scale might be the treating of aqueous media containing heavy metals.

The presence of heavy metals in residual water is commonly observed as a product of mining and metallurgical works; if these waters are discharged directly this will cause a major environmental problem; therefore, it is necessary to treat these waters to remove impurities to acceptable levels to avoid any damage to the environment. However, still today, in some locations, the exigencies to keep a close control on the application of environmental regulations are poor and serious pollution problems exist [4].

To correct the problem, engineering work must be dedicated to produce a realistic procedure for its application in large scale as it is required in mining and metallurgical locations.

It has been proposed and validated in laboratory, and pilot plant tests, the use of froth flotation to treat lead polluted water in a continuous scheme operation. In that case, lead is present in the aqueous phase as a lead colloidal carbonate, with a lead concentration of 20 ppm. A series of spargered flotation cells are used to perform the separation of the lead specie [5]; Figure 1 shows a representation of the series of flotation cells.

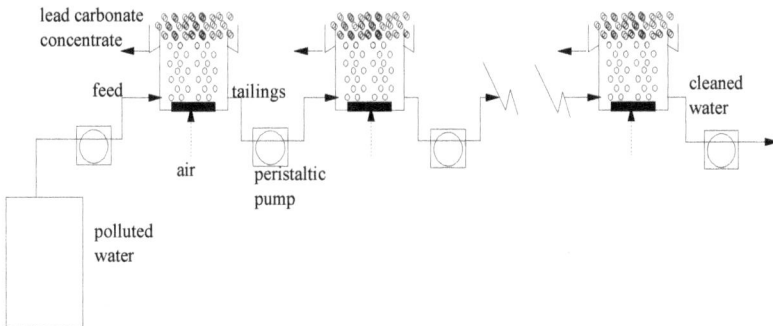

Figure 1. Illustration of a series of spargered flotation cells where porous gas spargers are installed to inject-disperse air in the form of small gas bubbles (< 1.5 mm).

In this flotation process, the water containing the lead colloidal carbonate is adequately conditioned by the addition of chemical reagent to fix an adequate pH to make possible to complex the lead specie with a suitable flotation reagent. Gas hold-up was measured below the froth region in the flotation cells by pressure measurements [3, 6], and locally at different points measuring electrical conductivity [7]; these measurements were related to the efficiency of producing large gas surface areas to perform the collection process of hydrophobic species.

In designing the chemical characteristics of the process, for example the pH of the aqueous phase, and the type of collector reagent, lead carbonate particles were separated and analysed in their Zeta potential (ζ) behaviour. Figure 2 shows the ζ performance of the lead carbonate.

The polluted water presented a pH of 7.4, therefore a cationic collector was used to produce a hydrophobic behaviour to the lead carbonate particles. The series of flotation cells were operated until the steady state conditions were achieved. At this point an apparent residence time was estimated ($\tau = [h_{c\alpha}/J_l] \times [1 - \epsilon_g]$; where $h_{c\alpha}$ is the collection zone height measured from the bottom to the froth interface, J_l is the superficial liquid velocity, and ϵ_g is the gas hold-up); as a first approximation this representation of residence time may be used to express the nature of mixing in the series of flotation cells.

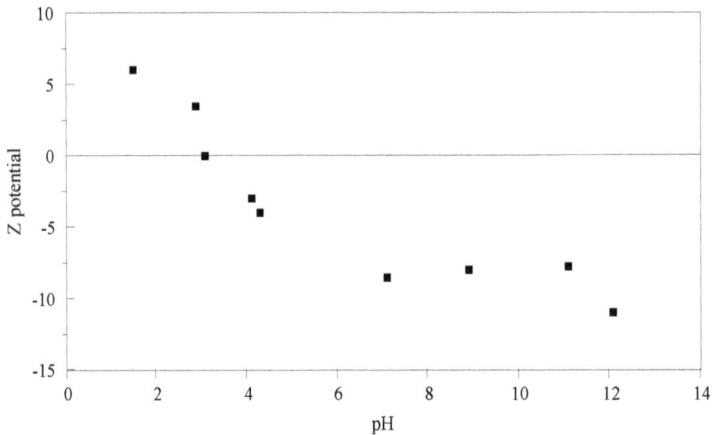

Figure 2. Measurements on the behaviour of Zeta potential (ζ) lead carbonate; measurements through the pH range between 1.4 and 12.

The different streams (concentrate, tailings, and feed) were sampled to analyse their concentration of lead, and to determine the recovery of this material through the separation process. The recovery of lead may be related to the characteristics of the process in terms of the variables ϵ_g, bubble size, the superficial bubble surface flux, and the kinetics of the process.

During the operation of the series of flotation cells, gas hold-up was monitored continuously in each flotation cell using measurements of pressure and electrical conductivity [6, 9].

Figure 3 presents the lead recovery in concentrate as a function of gas hold-up change in the first flotation cell.

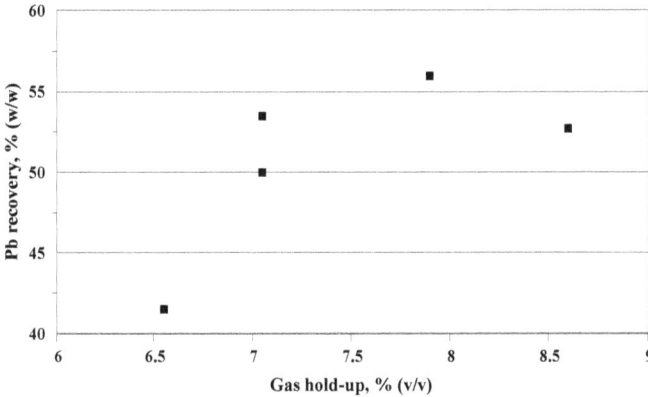

Figure 3. Typical graphic of experimental results on the flotation of lead represented as the lead recovery % and the corresponding gas hold-up.

The lead reported to the concentrate shows that its recovery increases as the amount of gas in the gas-water dispersion rises, as a result of the increase in the availability of bubble surface area to perform the collection process of hydrophobic species. The flotation of the lead specie presents a steady increase with increasing until the gas hold-up reaches 7.7% (v/v) above which there is a sudden reduction on that operation variable due to the formation of excessive turbulence produced by a considerable flow of gas in the cell.

Nevertheless, the magnitude of gas hold-up is related to bubble size, beside to the amount of gas which is fed into the flotation system. This may imply that the production of large gas hold-ups is related to formation of small gas bubbles. In this way, the recovery of lead from the aqueous media should be upgraded with the bubble size reduction. This effect is shown in Figure 4.

The bubble dimension was estimated by two methods: first a direct method using image analysis [10]; and the indirect method of drift flux analysis [3].

The experimental practice indicates that the decrease in the diameter of gas bubbles produces larger bubble surface area which "flows" through the flotation system collecting the hydrophobic species as they have a collision.

Production of large bubbles presents a hydrodynamic effect in the collection process, in addition to a smaller bubble surface area, which decreases the collection of lead carbonate colloid by minimizing the possible collision between the gas bubble and the hydrophobic lead carbonate surface. The contribution of these two effects of the bubble surface area is shown in fig. 4.

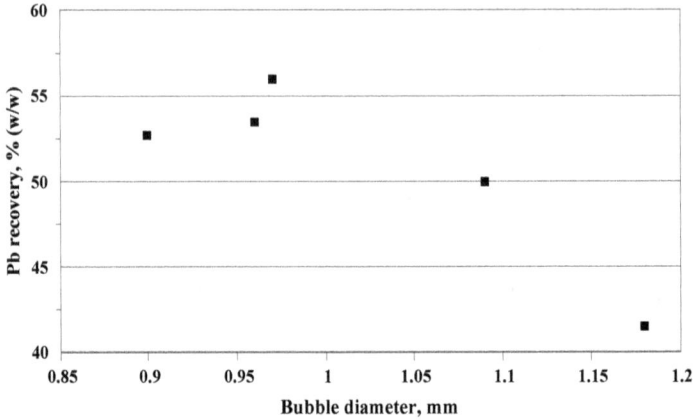

Figure 4. Effect of the bubble size on the separation of lead from the aqueous media.

In terms of the bubble surface area, the concept of "superficial bubble surface flux," Sb, has been suggested as a controlling ultimate flotation variable. Sb contains the chemical, hydrodynamics, and mechanical characteristics of the flotation system; Sb can be thought as the amount of bubble surfaces that flow through a unit of a cross sectional area normal to the bubbles motion, by unit of time, therefore with units of time^{-1}; therefore from geometry considerations, Sb = 6 (J_g/d_b), where J_g is the superficial gas velocity, and d_b is the bubble diameter [11]. It has been demonstrated that Sb is directly related to the flotation – separation kinetics [12]. Figure 5 shows the relationship between the recovery of floated lead and Sb.

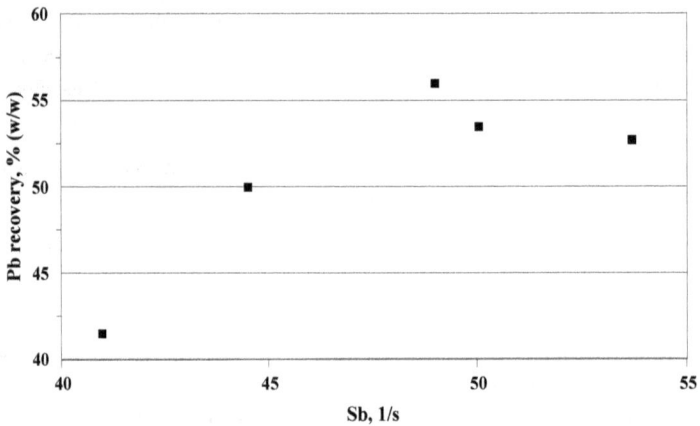

Figure 5. Relationship between the separation of lead carbonate from the aqueous phase by flotation, and the superficial bubble surface flux.

It can be noticed that the separation of lead from the water increases as the available bubble surface rises. Nevertheless, there is a maximum value of the lead recovery which corresponds to a given Sb value.

Above that Sb value, the relationship between Pb recovery and Sb changes due to an increase of mixing and circulation in the flotation system. This information indicates that the flotation of the lead colloid carbonate is very sensitive to the bubble size change, and therefore, to the superficial gas velocity and the superficial liquid velocity.

The behaviour of the flotation cells follow a model of the perfect mixer, therefore, flotation constant rate may be expressed as $\kappa = R/[\tau(1-R)]$, where R is the lead recovery expressed as the weight fraction.[13]

The apparent rate constant for the lead colloid carbonate flotation is presented in Figure 6 as a function of the gas hold-up in the collection zone.

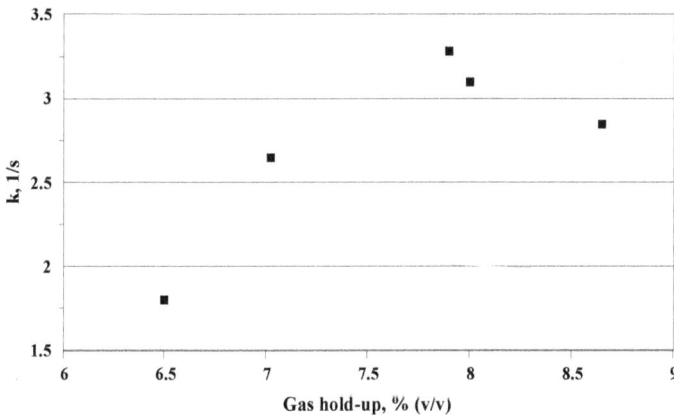

Figure 6. Behaviour of the apparent rate constant, as a function of the change on gas hold-up in the flotation of lead colloid carbonate.

It is observed that the flotation rate constant follows a linear relationship with respect to the change in the gas hold-up if the last is kept below 8%. Above this gas hold-up value the correlation between the flotation kinetics and the gas hold-up changes. This information suggests that the lead colloid carbonate flotation rate is strongly dependent from the hydrodynamic conditions of the system.

The previous statement indicates that the flotation rate constant must be related to the bubble size. Figure 7 presents the relationship between the flotation rate constant for the separation of the lead colloid carbonate, and the estimated bubble size.

This information describes that when the bubble size is between 0.97 mm and 1.2 mm presents a linear correlation with the apparent flotation rate constant. In this range of bubble size it can be seen that the kinetics of the flotation process is favoured with the decrease in the bubble diameter. But with the increase in bubble size above 1.2 mm the possibility to have a collision between the lead colloid carbonate and the bubble decreases, because the hydrodynamics of the bubble that moves faster in the liquid bulk.

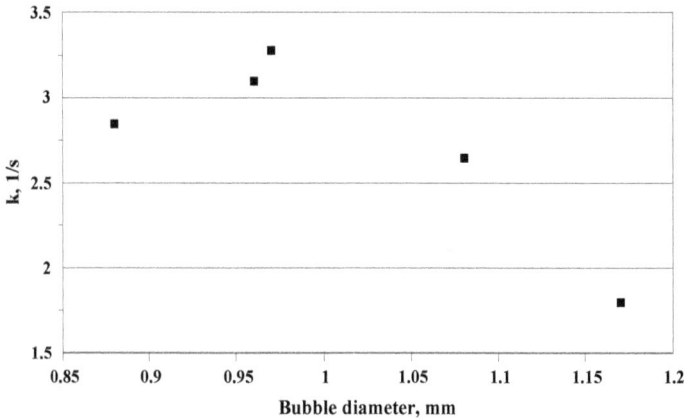

Figure 7. Relationship between the apparent rate constant and bubble size in the lead carbonate colloid separation.

When bubble size is smaller than 0.97 mm the linear trend between the apparent rate constant and bubble size is distorted because the effect of axial mixing. This axial mixing is formed by radial differences of gas distribution, becoming more evident as the gas hold-up increases.

In the series of flotation cells, the superficial gas velocity was maintained at 0.8 cm/s in each flotation cell, and the superficial liquid velocity was kept at 0.19 cm/s.

The lead carbonate colloid cumulative recovery, in the series of flotation cells, was estimated from the mass assaying in each cell.

Figure 8 represents the cumulative recovery of lead as a function of the number of flotation cells in the series of five flotation cells.

It is observed in fig. 8 that the cumulative recovery of lead through the series of 5 flotation cells is 93% (w/w). The total residence time of water in the flotation circuit is 13 minutes. This water processing time gives the impression that is short, as compared with the reported processing times in laboratory flotation columns testing (between 30 to several hundreds of minutes) [2].

This information indicates that the flotation cells instrumented with porous gas bubble generators can produce satisfactory bubble sizes to perform colloid flotation; it seems that differences in bubble size distribution are low enough to reduce axial mixing as compared with the natural circulation presented in flotation columns [13]. In this case lead content in water is reduced from 20 ppm to about 1 ppm.

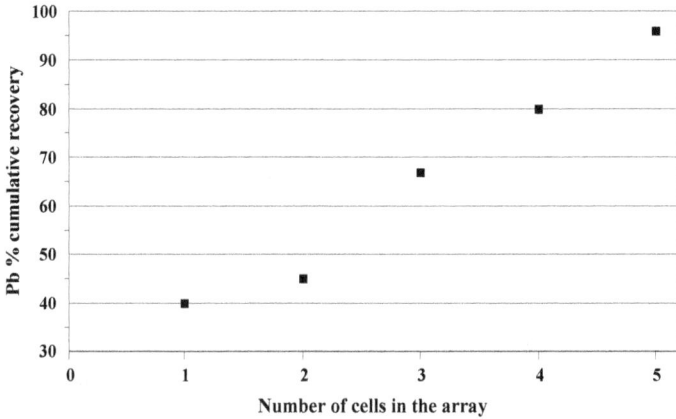

Figure 8. Cumulative recovery of floated lead as a function of the number of flotation cells series.

2.2. Water de-oiling using column flotation

Insoluble organic compounds in water are often encountered as pollutants, particularly in locations near urban centres or in regions associated with industrial and agricultural activities. Among these pollutants oily materials are frequently present in discharged water and should be removed from the aqueous phase before its emission.

In the production of food products such as vegetable cooking oil, residual water contains vestiges of oil, soap, and suspended solid particles from the processed oily seeds. Commonly, pollutants and water forms stable emulsions where the different phases are not easily coagulated and separated by differences of their densities.

In treating such emulsion products, different several approaches are used to eliminate the problem. Bacterial processing is often used to decompose the organic materials generating some non pollutant products; in these biological processes special care must be taken in providing adequate aeration and pH conditions of the polluted water in treating reactors where materials must stay for a while to be processed by living microorganisms. Some other techniques include water evaporation – condensation, but the cost tends to be high by the energy required to evaporate the water from the emulsion.

In searching for a short residence time process that allows to treat continuously large amounts of water, it was studied the possible separation of the organics content in water using flotation techniques.

The use of flotation columns is reported as a possible solution to the problem [14], which implies a simple operation at a very low cost [3].

It was processed industrial residual water from a vegetable oil producer. The polluted water contained 40% v/v of organics consisted of residual oil, soap and solid particles, originated from the oil extraction and refining operations, in processing oily vegetable seeds.

Laboratory flotation column

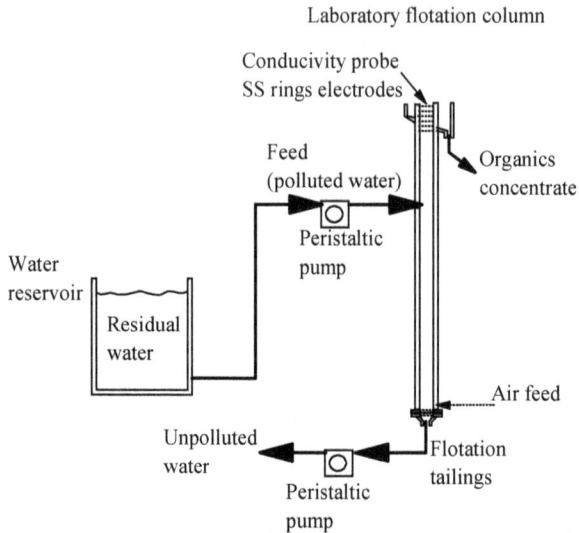

Figure 9. Representation of the laboratory flotation column assembly used to remove organics (oil, soap, solids) from water by froth flotation.

The flotation column was 0.1 m in diameter and 4 m height, and was fully instrumented to be operated automatically. The raw residual polluted water derived from the outlet water circuit at the industrial plant was placed in a 3 m³ reservoir from where the flotation column water feed was continuously taken. The flotation column feed and tailings flow rates were controlled to their desired values using automated peristaltic pumps. The gas hold-up was estimated by measurements of hydrostatic pressure at two heights in the column by means of electronic pressure transmitters. The froth depth was maintained at 10 cm (! 5 cm) from the column lip by varying the feed pump operation to rise or drop the froth zone-collection zone interface; the position of the froth level was monitored using an electrical conductivity probe which consisted of 6 steel ring electrodes placed flushed at the internal wall of the upper part of the column, within 5 cm separately.

The superficial liquid velocity in the column was controlled by controlling to desired values the liquid flow rate controlling the tailings pump. The air was feed into the flotation column through a porous gas sparger made with filter cloth. The air flow rate in the column was measured and controlled in desired values by means of a mass air flow rate controller. The bubble diameter was estimated using drift flux analysis and validated against direct photographic image analysis. The flotation column was made of transparent acrylic, in such manner is possible to visually observe the presence of bubble coalescence, circulation and mixing, under different operating conditions.

In the first approximation a single laboratory flotation column was operated varying the superficial gas velocity under predetermined superficial liquid velocity conditions. The flotation measurements indicate that the flotation of organics is highly affected by the relative velocities between the liquid flow rate and the gas flow rate. The performance in the flotation column is reported to increase circulation and mixing as the liquid and gas flow rates increase in the column, this in turns will increase the bubble coalescence producing a rise in the collection zone turbulences, consequently decreasing the collision possibility between the bubbles and the oily globules. Nevertheless, when the superficial liquid velocity is in a small value range (0.1 cm/s – 0.15 cm/s) the organics recovery has an increase with the increase in the liquid velocity, perhaps due to an increase of the retention of the air bubbles by the liquid flow in the collection zone. This effect can be noticed in fig. 11 where the organics recovery is plotted as a function of the superficial liquid velocity.

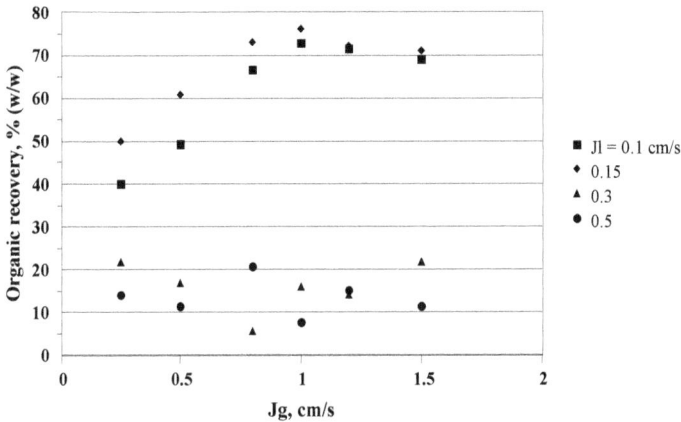

Figure 10. Organic recovery from water in the flotation column as the superficial gas velocity changes; the system is under different superficial liquid velocity.

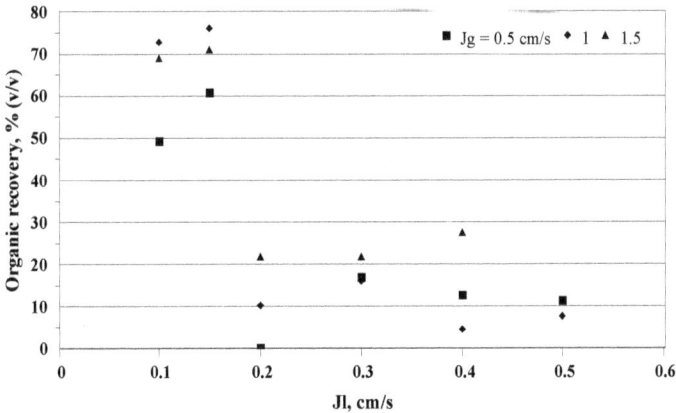

Figure 11. Organic recovery from water, reported in the flotation column concentrate, as a function of the superficial liquid velocity.

It can be observed that the de-oiling process efficiency is high when the superficial liquid velocity is kept below 0.2 cm/s, but above that value there is a drastic decline on the separation efficiency, this behaviour is observed in whole range of superficial gas velocities reported here. Also, it is thought that as the superficial liquid velocity increases in the collection zone of the flotation column, the possibility that small gas bubbles charged with hydrophobic organic materials are gone with the tailings flow out of the column, and for this reason, decreasing the organics recovery.

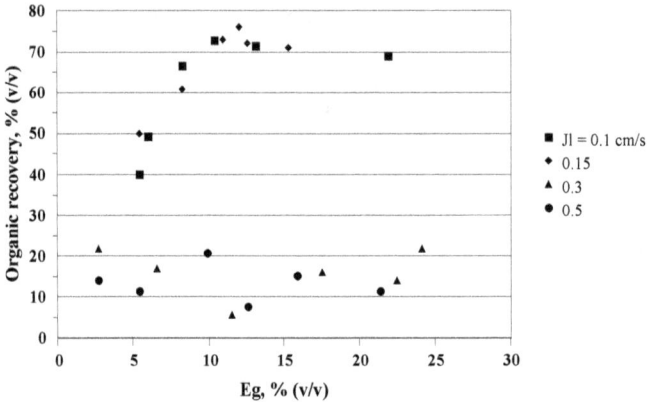

Figure 12. The effect of the superficial liquid velocity on the organic recovery/gas hold-up behaviour.

The experimental results plotted in fig. 12 suggest that large gas hold-up values are associated to small bubbles formation which will go down to the bottom of the flotation column because the effect of the downwards liquid stream; these experimental results support the impression that as the liquid flow rate increases the bubbles entrapment in the tailings stream also intensifies, producing low organic recoveries reported to the concentrate run. It can be noticed that under superficial liquid velocities above 0.15 cm/s the organic recovery in the concentrate ranges between 5% and 20%, which may be associated to large bubbles covered with the hydrophobic materials (which yields to such 5 to 20% organic recovery) and, therefore, most of the organic material collected by the gas bubbles is associated with the interactions with small bubble sizes. This idea is consistent with the hydrodynamic behaviour of gas bubbles moving in a liquid phase, since the oily globules present a lower density than the aqueous phase.

This explanation is supported by the gas bubble size estimates, from both image analysis and drift flux analysis, which are shown in Figure 13 in terms of the organic recovery in the concentrate.

The experimental bubble size estimates show the range of bubble dimensions is more constrained as the superficial liquid velocity is increased in the collection zone of the flotation column. This suggests that bubble-bubble coalescence is induced as the slip velocity among the phases (gas water counter-current flow) is increased.

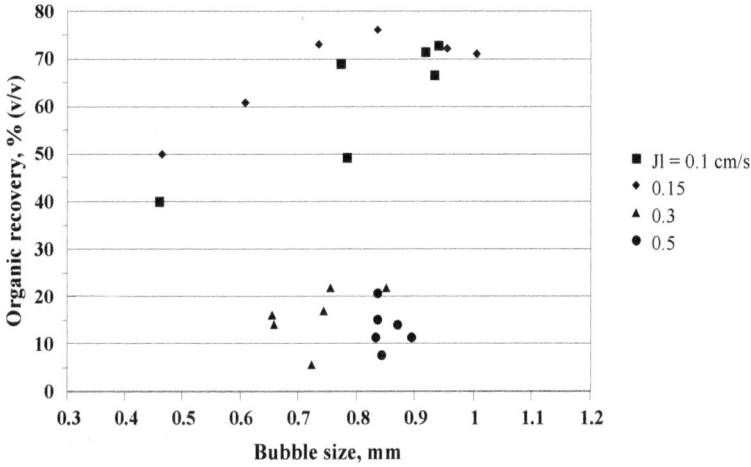

Figure 13. Effect of the bubble dimension on the recovery of organic floated.

Figure 14 presents the column flotation results on the water de-oiling process in terms of the organic recovery – superficial bubble surface flux behaviour. Under the lower values of the superficial liquid velocity in the collection zone of the column, it is very clear that the hydrophobic organic flotation increases when the gas surface area which flows in the flotation column increases.

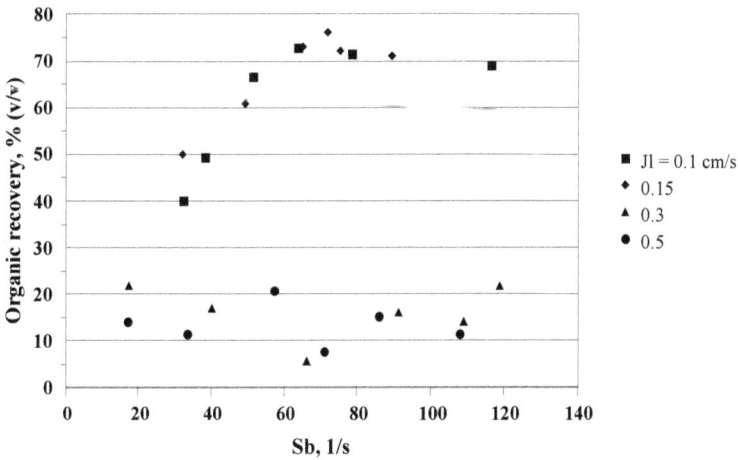

Figure 14. Behaviour of the organic recovery in concentrate with the change in the superficial bubble surface flux (Sb). The water de-oiling process in series stages of flotation columns. The water de-oiling process by column flotation was performed using a series of flotation stages. In this experience arrays of 2, 3 and 4 flotation column sequence were operated.

The experimental results and the estimates of the superficial bubble surface flux show that the hydrodynamic behaviour of the flotation column determines the separation efficiency and affects the relationship between the separation process and the available bubble surface area to perform the collection process. These experiences present that when the superficial liquid velocity is kept under the range between 0.1 and 0.15 cm/s, the change among the organic recovery and the superficial bubble surface flux follows a smooth pattern; however, when the superficial liquid velocity goes above 0.15 cm/s, the relationship between these process characteristics becomes erratic.

The experimental flotation columns array is schematically presented in fig. 15; this presentation shows that the residual water is fed into the first flotation column and the tailings from this column is used as the second column feed, and this arrangement is repeated up to four flotation columns.

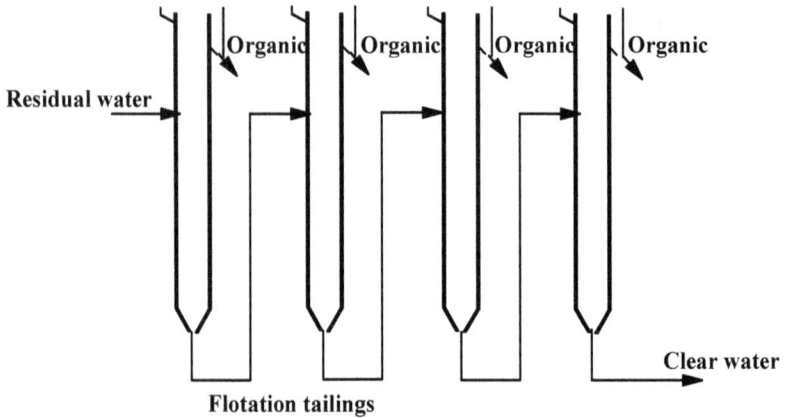

Figure 15. Experimental set up of flotation column series.

The experimental information shows that the de-oiling process can be performed effectively using a series of flotation columns. In this experience about 99% organic recovery is achieved using a series of four columns. The minimum residence time reported in order to achieve 99% organic recovery from the very stable water-oil (~40% oil) suspension is in the order of 100 minutes in a continuous four flotation columns operation [14].

2.3. Waste water treatment by crystallization – Cleaning of water contaminated with copper, lead, and nickel

It is well known the fact that although three quarters of this planet is occupied by water, not all is available for human consumption. 97% of this water is in oceans and seas whereas only 3% is called fresh water or sweet water. Of that 3%, the 77.6% is frozen polar ice, the 21.8% is groundwater, and only the 0.6% is available surface water for human consumption [15].

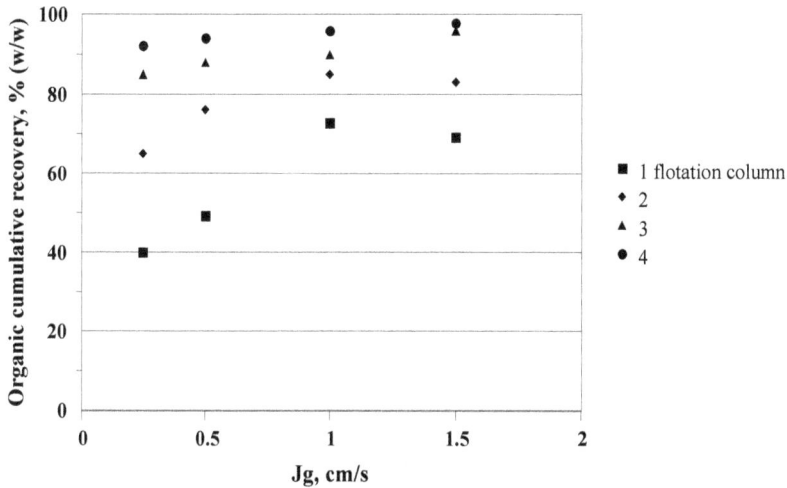

Figure 16. Experimental results of water de-oiling in one flotation column and a series of 2, 3 and 4 flotation columns.

According to the United Nations Industrial Development Organization (UNIDO), it is likely that in the year 2025 the industrial activity will consume twice as much water than at present; on the other hand, pollution may be multiplied by four due as well to this industrial activity [16]. Researchers at the Swiss Institute of Environmental Science and Technology recently reported that the rainwater that fall on the European Continent contains so many toxic pesticides dangerous for human consumption. In addition to acid rain, many other pollutants are deposited in water sources or streams, as in the third world, where 90% of the waste water is ending up in local rivers and streams [17]. The consumption of this contaminated water by the population leads to serious health problems such as gastrointestinal and cardiovascular disorders, malformations in newborns, among others. The World Health Organization (WHO) establishes the recommended levels of trace elements in drinking water (set for an average daily consumption), for both that help the body's metabolism as representing toxicity to humans. For instance, in the case of lead the recommended values are from 0.01 to 0.05 mg/l, whereas for copper is among 0.5 and 2.0 mg/l, depending of the government of each country [18, 19].

To solve the problem of contaminated effluents, are applied some processes in order to leave the water clear, colorless, odorless, and disinfected. The suspended solids, organic compounds, and pathogenic organisms, in most of the waste water treatment plants are removed; nevertheless, these processes do not eliminate the heavy metals dissolved in the effluent [20].

If the heavy metals are in solid state they may be extracted through a primary treatment (i.e., sedimentation or flotation) [21], or secondary (i.e., activated sludge) [22]. On the other hand, if heavy metals are in solution, the use of lagoons, hydroponic bed, activated zeolite and membranes, phytoremediation, bird feathers, nanostructures, bioremediation, chemical cristallyzation, among others, have been proposed [23, 24, 25, 26, 27, 28].

Most of the processes mentioned above observe a certain limit as regards efficiency since the physicochemical aspects affecting such processes are not well understood, and the pollution problem is partially solved since the new "aggregate" with the removed contaminant remains as such.

This part of the chapter establishes from a thermodynamic point of view (pH, electrochemical potential, ionic strength, activity coefficient), the conditions to predict the formation of certain species (precipitated or dissolved) in distilled water contaminated with lead, copper, and nickel, and open to the atmosphere. The knowledge of the mechanisms of crystallization or chemical precipitation of the ionic species contribute to selectively agglomerate and separate the copper, nickel, and lead species in order to clean water contaminated with these heavy metals.

The role of heavy metals in reactions involving liquid-liquid and liquid-solid relationships is not enough studied yet. In the case of waste water treatment several research has been done although the mechanisms of precipitation of certain species and the selectivity of their capture in not clear [26, 27, 28]. For liquid-liquid, and solid-liquid systems, the interaction between species is ruled by the following expressions:

$$I = \frac{1}{2} \sum m_i z_i^2 \tag{1}$$

Where I is the ionic strength, which is a measure of the intensity of the electric field in the system [29, 30], m is the molality of i; and z is the charge of the corresponding ion i (i represents every specie involved in a given reaction). In this work the involved species are the salts: $Pb(NO_3)_2$, $CuSO_4$, and $NiSO_4 \cdot 6H_2O$, the water and the pH modifiers, H_2SO_4, and KOH. The chemical activity is a corrected concentration [31], and physically is the actual amount of reagent that takes part during the reaction; in this case is the concentration of metallic ions in the media that affectively react. The average activity coefficient is calculated as follows:

$$\gamma \pm = 10^{\left(A|z+z-|\sqrt{I} \right)} \tag{2}$$

Being A the constant value from Debye –Hückel equation for liquid media and pressure of 1 atmosphere, $|z + z -|$ is the absolute value of the sum of the electric charge of the dissolved ions. The activity a of given i species can be calculated according to:

$$a[i] = \gamma_i m_i \tag{3}$$

Where m is the molarity of the i species. In order to calculate the electrochemical potential, Eh, the equation proposed by Garrels [29] was applied:

$$Eh = E° - \left(\frac{0.05916}{Z} \right) \log Q \tag{4}$$

Being $E°$ the standard potential, Z the number of electrons participating during the reaction, and Q is the reaction quotient. $E°$ can be calculated through the following expression:

$$E° = \frac{-\Delta G°}{ZF} \tag{5}$$

Where $\Delta G°$ is the Gibbs free energy of the corresponding reaction, and F is the Faraday constant (96487 C/mol = 23 060.9 Cal/Vol•mol). By applying the former equations there is possible to build $Eh - pH$ diagrams and to use them as tools to understand the conditions under which given ionic or precipitated species are chemically stable.

In this part of the chapter lead, nickel, and copper salts were dissolved in distilled water and physicochemical parameters such as ionic strength, activity coefficient, activity, and electrochemical potential were calculated, in a pH ranging from 3 to 13. Lead, nickel, and copper precipitates were identified and the corresponding formation reactions were established in order to build a Pourbaix diagram.

The obtained information makes possible at first to design a procedure to clean water contaminated with the heavy metals mentioned here through the route sedimentation-flotation or filtering-flotation. The experimental results also provide information regards deposited species on mineral surfaces during milling which affect the behavior of collectors during flotation decreasing its metallurgical performance.

2.3.1. Preparation of diluted solutions of Cu, Ni, and Pb in distilled water

Lead nitrate ($Pb(NO_3)_2$), copper sulphate ($CuSO_4$), and hexahydrated nickel ($NiSO_4 \cdot 6H_2O$) were dissolved separately and simultaneously in distilled water. The pH of the media was varied in 3,5,7,9,11, and 13. After 24 hours the precipitated solids were separated and analyzed through X-ray diffraction (XRD), and scanning electron microscopy (SEM) techniques. The remaining lead, copper, and nickel in every solution were quantified by atomic absorption spectroscopy (AAS) analysis. The pH was modified by adding sulfuric acid (H_2SO_4), and potassium hydroxide (KOH). The initial metal concentration in each solution was 40 ppm.

The chemical analysis of precipitates was carried out by X-ray diffraction (XRD), and scanning electron microscopy (SEM). On the other hand, the quantitative chemical analysis from liquids, were carried out by atomic absorption spectroscopy (AAS).

With the quantitative and qualitative chemical analysis data, the values of activity, activity coefficient, ionic strength, and electrochemical potential were calculated. The former information was used to calculate the corresponding transformation lines as function of the pH. The resulting equilibrium diagrams are shown below.

2.3.2. Precipitation of lead species

Visually the formation of lead precipitates starts at pH 3, although these solids are re-dissolved at pH 5. Lead crystals are formed again at pH 7, and finally the precipitates are dissolved once more at pH 11. From XRD analysis at pH 3 the detected species are the lead sulfate ($PbSO_4$), and hydrated lead nitrite ($Pb(NO_2)_2 \bullet (H_2O)$), which indicates the decomposition and hydration of the salt originally dissolved. The precipitated solids at pH from 7 to 11 correspond to an hydroxicarbonate $Pb_3(CO_3)_2 \bullet (OH)_2$, also known as hydrocerusite.

Taking into consideration that hydrocerusite forms under alkaline conditions, and in absence of ionic sulfate, the following reaction is suggested:

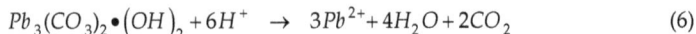

$$Pb_3(CO_3)_2 \bullet (OH)_2 + 6H^+ \rightarrow 3Pb^{2+} + 4H_2O + 2CO_2 \qquad (6)$$

The re-dissolution of hydrocerusite is carried out according to the following reaction proposed by Pankow [31], and thermodynamically it occurs at pH 11.2:

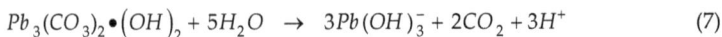

$$Pb_3(CO_3)_2 \bullet (OH)_2 + 5H_2O \rightarrow 3Pb(OH)_3^- + 2CO_2 + 3H^+ \qquad (7)$$

On the other hand, the formation of carbonated species is explained by considering the replacement of sulfates or sulfites to carbonates in an open system to air, according the following reaction proposed by Azareño [32]:

$$PbSO_4 + CO_3^- \rightarrow PbCO_3 + SO_4^- \qquad (8)$$

In the case of the decomposition of lead sulfate:

$$PbSO_4 + 2H^+ \rightarrow Pb^{2+} + H_2SO_4 \qquad (9)$$

From the above reactions it is possible to observe that the dissolution or precipitation of lead species just depends on pH. The Pourbaix diagram built according the calculated variables is shown in Figure 17. As known, in the following equilibrium diagrams the dashed lines represent the zone where aqueous species are stable; within these lines both the aqueous and precipitated species co-exist. For the copper case, the vertical lines correspond to the transformations shown in reactions (6), (7), and (9).

In light of the above, from pH up to 3.9, both ($PbSO_4$), and Pb^{2+} co-exist; whereas from pH 3.9 to 6 the all lead is dissolved. In the pH range from 6.0 to 11.2 the steady species are the Pb^{2+} and the $Pb_3(CO_3)_2 \bullet (OH)_2$. For pH larger than 11.2 the hydrocerusite is dissolved again and a lead hydroxide is formed.

According to the results from quantitative chemical analysis, the concentration of Pb^{2+} in the liquid media decreases because of the presence of lead precipitates at pH higher than 5, and the ionic lead increases again for pH higher than 11.

Table 1 shows the calculated values of activity for equations (6), and (9), as well as their Gibbs free energy, and the equilibrium pH. Thermodynamically the lead precipitation starts at pH 3.9 ($PbSO_4$), although visually it is noticed at pH 3. In the case of the $Pb_3(CO_3)_2 \bullet (OH)_2$, this visually starts at pH 7, whereas according to thermodynamics the precipitation of such specie would initiate at pH 6. Differences between observations and calculations are due to human errors and it suggests the use of another technique (i.e., conductivity measurements) to detect accurately the moment at which the precipitation phenomena take place [33].

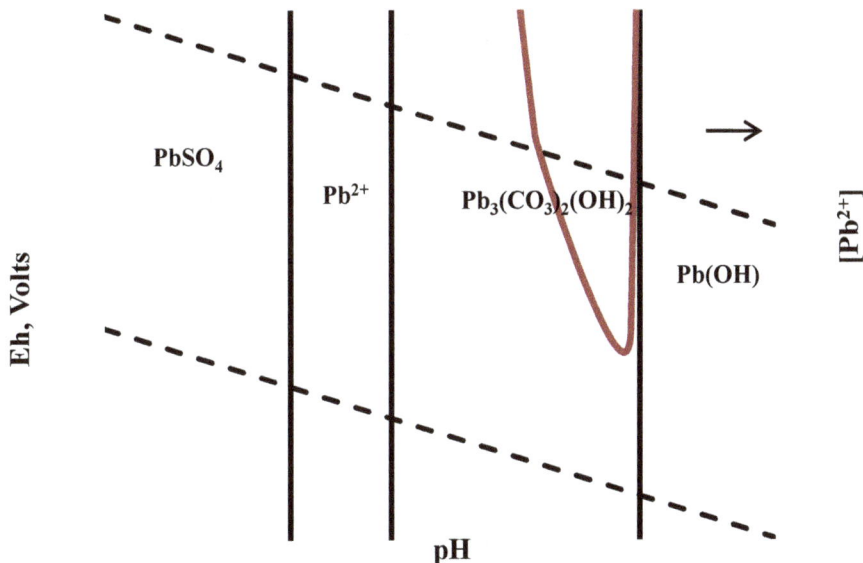

Figure 17. Transformation lines with changes on the metal concentration in the bulk solution. Pb – S – H_2O system.

Reaction	aPb^{2+}	$\Delta G°$ Reaction (Kcal/mol)	Equilibrium pH
$PbSO_4 + 2H^+ \rightarrow Pb^{2+} + H_2SO_4$	3.84E-04	-5975.14	3.9
$Pb_3(CO_3)_2(OH)_2 + 6H^+ \rightarrow 3Pb^{2+} + 4H_2O + 2C$	2.14E-04	-24617.8	6.0

Table 1. Calculated values of activity of Pb^{2+}, Gibbs free energy, and the equilibrium pH for equations (6), and (9).

2.3.3. Precipitation of copper species

From qualitative DRX analysis the detected crystalline species are the hydrated copper hydroxisulphate ($Cu_4SO_4(OH)_6 \cdot H_2O$), cupric hydroxide ($Cu(OH)_2$), and cupric oxide (CuO). In the pH range from 3 to 5.5 there is not precipitation of any copper specie, whereas among pH 7.5 and 10.5 the three former species co-exist. From pH 5.5 to 7.5, and

10.5 to 13, the precipitated species are the hydrated copper hydroxisulphate, and the cupric oxide, respectively.

The solid, $Cu_4(SO_4)(OH)_6 \bullet H_2O$, in a given pH transforms itself and co-exist with both the cupric hydroxide and the cupric oxide, and finally the cupric hydroxide transforms to cupric oxide at pH 10.5.

In the case of formation of the hexahydrated copper hydroxisulphate :

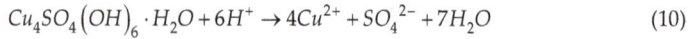

$$Cu_4SO_4(OH)_6 \cdot H_2O + 6H^+ \rightarrow 4Cu^{2+} + SO_4^{2-} + 7H_2O \tag{10}$$

For the precipitation of the cupric hydroxide:

$$4Cu(OH)_2 + 2H^+ + CuSO_4 \rightarrow CuSO_4(OH)_6 \cdot H_2O + Cu^{2+} + H_2O \tag{11}$$

Being the formation of the cupric oxide defined by the following reaction:

$$4CuO + CuSO_4 + 3H_2O + 2H^+ \rightarrow Cu_4SO_4(OH)_6 \cdot H_2O + Cu^{2+} \tag{12}$$

The Pourvaix diagram for the Cu – S – H₂O system is shown in Figure 18, and Table 2: it shows the Gibbs free energy values for reactions (10), (11), and (12).

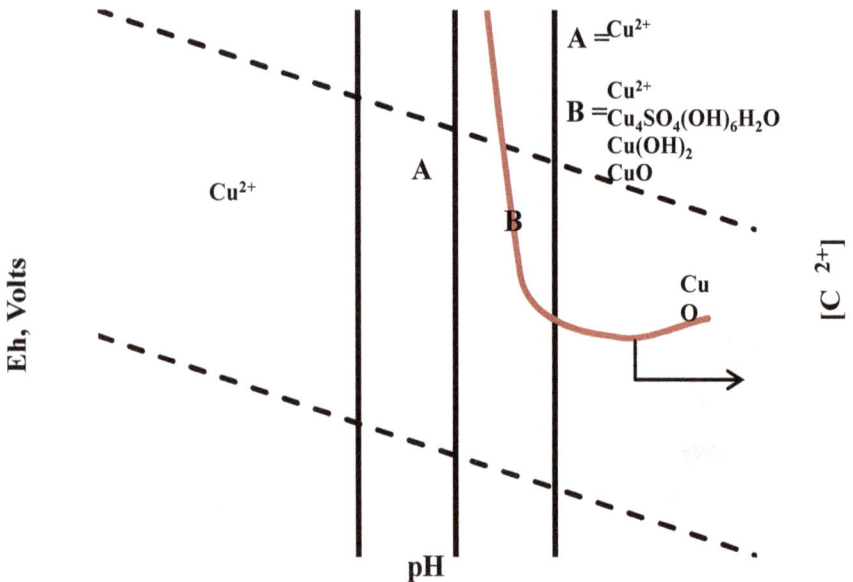

Figure 18. Pourvaix diagram for the Cu-SO₄-H₂O system. A and B means the liquid and solid species that co-exist in a respective region.

Reaction	aCu²⁺	ΔG° Reaction (Kcal/mol)	Equilibrium pH
$Cu_4SO_4(OH)_6 \cdot H_2O + 6H^+ \rightarrow 4Cu^{2+} + SO_4^{2-} + 7H_2O$	8.00E-04	193.6	5.6
$4Cu(OH)_2 + 2H^+ + CuSO_4 \rightarrow CuSO_4(OH)_6 \cdot H_2O + Cu^{2+} + H_2O$	4.08E-05	176.37	7.6
$4CuO + CuSO_4 + 3H_2O + 2H^+ \rightarrow Cu_4SO_4(OH)_6 \cdot H_2O + Cu^{2+}$	4.08E-05	143.4	9.7

Table 2. Activity, Gibbs free energy, and equilibrium pH values for reactions (10), 11), and (12).

2.3.4. Precipitation of nickel species

In this case and according the DRX analysis the only detected specie was the nickel hydroxide (Ni(OH)₂). Visually the precipitation of such specie is detected at pH 9; although, thermodynamically the nickel hydroxide starts forming at pH 7.6. Figure 19 shows the Eh-pH diagram for the Ni-SO₄-H₂O system.

From the above information the proposed reaction is as follows:

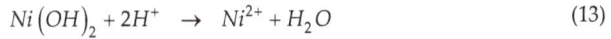

$$Ni(OH)_2 + 2H^+ \rightarrow Ni^{2+} + H_2O \qquad (13)$$

The corresponding thermodynamic values are shown in Table 3. Sean et al., [20], Chanturiya et. al., [21], Liu [22], Shigehito et. al., [23], and Guo-riu Fu [24], reported the precipitation of two nickel phases named α – nickel hydroxide, and β – nickel hydroxide, within the pH range established in this work. These species were detected by using both: XRD and IR analysis techniques.

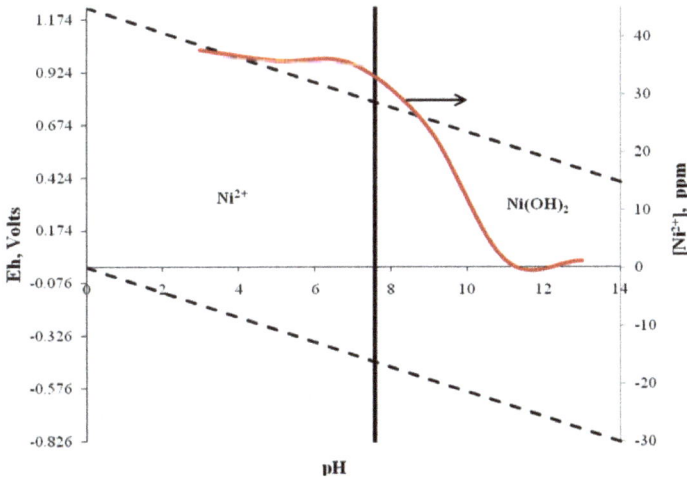

Figure 19. Pourbaix diagram for the Ni – SO₄ – H₂O system. Changes on concentration of the ionic nickel are included.

Reaction	aCu²⁺	ΔG° Reaction (Kcal/mol)	Equilibrium pH
$Ni(OH)_2 + 2H^+ \rightarrow Ni^{2+} + H_2O$	4.47E-03	-17447.1	7.57

Table 3. Activity coefficient, Gibbs free energy, and equilibrium pH for reaction (13).

2.4. Selective precipitation of copper, nickel, and lead

With all the above information it is possible to design a cleaning route for water contaminated with lead, copper and nickel. By adjusting the pH of the media selectively any of the metals can be precipitated and then separated by filtration or sedimentation.

Lead, copper, and nickel salts were simultaneously dissolved in distilled water. The amount of every salt was adjusted to fix the initial concentration of every metal at 160 g/l.

According to the experimental data and the thermodynamic calculations, each metal can be selectively precipitated and removed from the contaminated effluent. After each pH adjustment of the liquid, the solids were separated by centrifugation. The proposed procedure is as follows:

a.   At pH values up to 4 the only precipitated specie is the lead sulfate (PbSO₄), remaining in the liquid media both the nickel and the copper. Figure 20 shows the XRD analysis of the solids crystallized under pH 3.

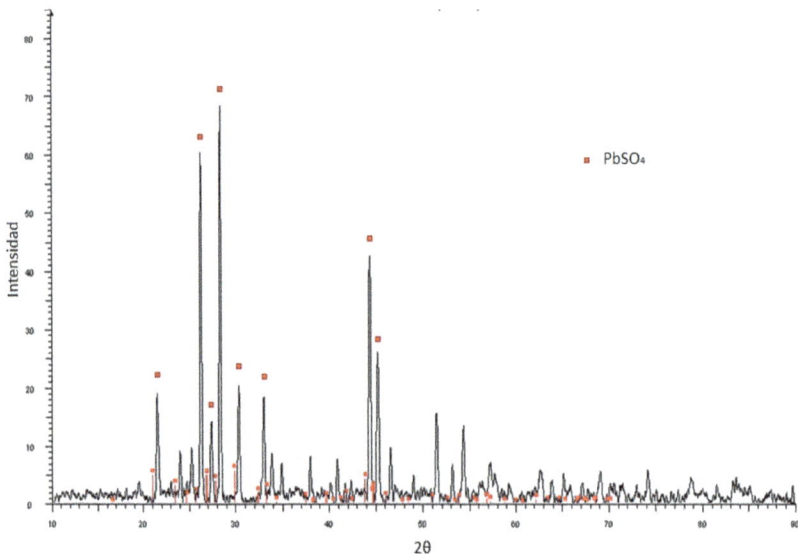

Figure 20. XRD analysis for solids precipitated at pH 3. Experimental results for the case of lead, copper, and nickel salts simultaneously dissolved in distilled water.

b. By keeping the pH at 7.5, the copper species will precipitate as hydroxysulphate, and only the nickel is expected to remain in solution. Figure 21 shows the XRD analysis of the solids precipitated at this stage. Considering that the pH of equilibrium for the Ni-H$_2$O system is 7.6, it seems normal to observe the onset of precipitation of some nickel species.

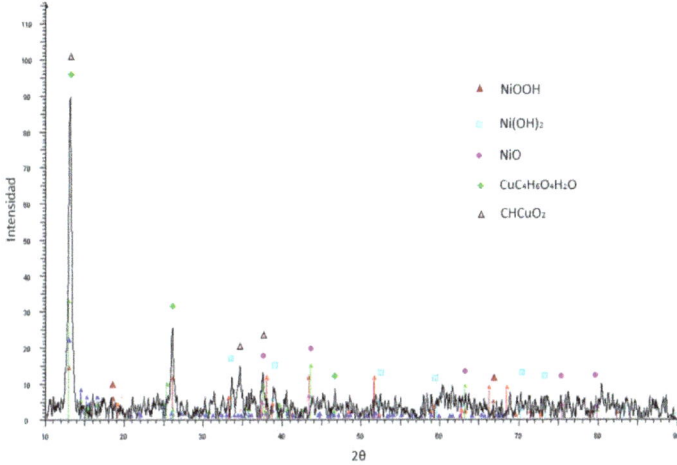

Figure 21. XRD analysis for solids precipitated at pH 7.5. Experimental results for the case of lead, copper, and nickel salts simultaneously dissolved in distilled water.

c. Finally, if the pH of the liquid is fixed at values larger than 10, the nickel will precipitate as hydroxide. Figure 22 shows the XRD result after fixing the pH at 11.

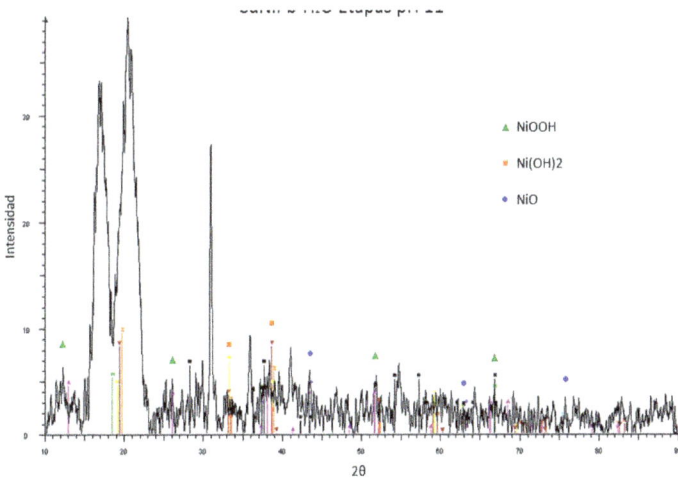

Figure 22. XRD analysis for solids precipitated at pH 11. Experimental results for the case of lead, copper, and nickel salts simultaneously dissolved in distilled water.

2.5. Waste water treatment de-oiling by using microorganisms

The biological treatments are procedures for cleaning waste water effluents; they have in common the use of microorganisms (i.e., bacteria), to carry out the removal of undesirable components of the water, using the metabolic activity of such microorganisms. The traditional application is the removal of biodegradable organic matter, either soluble or colloidal, and the elimination of compounds containing nutrients (nitrogen and phosphorous). The following describes an experimental procedure, carried out at laboratory level, to eliminate vegetable oils, first from a synthetic emulsion and then, from an industrial effluent. The microorganisms tested here, are both aerobic and nonpathogenic. The bioreactor used was a flotation cell made of transparent acrylic material.

2.5.1. Determining the lipolytic capacity of microorganisms

With the appropriate conditions for the growth of the bacteria colony (pH, temperature, dissolved oxygen, surface tension, and nutrients), it was possible to quantify the lipolytic capacity of the microorganisms: *Candida Kefir, Candida Parapsilosis, Pseudomonas Fluorescens, Bacillus Cereus,* and *Bacillus Coagulans,* individually and in bulk. Since the purpose of this stage of the work was to establish the lipolytic character of the above mentioned microorganisms, the measurements of residual oil in water were carried out at the beginning and at the end of the experimental digestion. The experimental results are shown in Table 4.

SAMPLES	RESIDUAL OIL (g/l)	RESIDUAL OIL (%)
Undigested sample	0.241	100
Candida Kefir	0.233	96.5
Candida parapsilosis	0.023	9.78
Bacillus cereus	0.059	24.5
Pseudomonas fluorescens	0.005	2.3
Bacillus coagulans	0.019	7.9
Bulk	0.023	94.7

Table 4. Residual oil in the synthetic emulsion after microbial action.

From table 1 the microorganism with higher lipolytic capacity is the *Pseudomonas fluorescence,* leaving only 2.3 % of residual oil in the emulsion. The same experimental results expressed in terms of the biomass developed (growth of colony) is shown in Table 5. As can be observed, the amount of biomass increases with the oil removed from the emulsion, due to the amount of nutrients (lipids) ingested by the microorganisms.

The microorganism with more biomass developed during the ingestion of lipids was the *Pseudomonas fluorescens* (253.5%), followed by the *Bacillus coagulans* (197.9%), and at the end appears the *Candida parapsilosis* (178%). Table 6 shows the variations in pH, temperature, and dissolved oxygen, monitored during this first stage of experimentation and corresponding to the particular case of the *Pseudomonas fluorescens.* The entire experiment finished after 132 hours.

SAMPLES	BIOMASS (g/l)	BIOMASS (%)
Undigested sample	0.247	100
Candida Kefir	0.328	132.7
Candida parapsilosis	0.344	139.2
Bacillus cereus	0.440	178.0
Pseudomonas fluorescens	0.626	253.5
Bacillus coagulans	0.489	197.9
Bulk	0.329	158.5

Table 5. Development of biomass in the degradation of a synthetic emulsion.

Sampling (hrs)	pH	Temperature (°C)	Surface Tension (dyn/cm)	Dissolved Oxygen (mg/l)	Residual Oil (%)	Biomass (%)
0	5.9	17	36.8	6.4	100	100
12	5.7	21	38.1	6.6	95	104
24	5.1	18	36.6	7	87	100
36	4.7	20	36.5	6.8	81	102
48	4.6	18	36.6	7	71	105
60	4.4	20	36.9	6	67	108
72	4.3	17	38.9	6.5	58	101
84	4.3	19	36.9	6.1	55	111
96	4.1	17	41.3	5.9	47	112
108	4.1	20	41.2	6	41	116
120	4.1	18	41.5	6	33	127
132	4.1	21	42.1	6	23	146

Table 6. Experimental results after digestion of the synthetic emulsion by the microorganism *Pseudomonas fluorescens*.

From table 6 the pH of the system decreased during the first 100 hours of the experiment. After this time, the pH of the emulsion remained constant at pH 4.1. This acidification is caused by the rupture of triglycerides, releasing fatty acids and glycerol, together with the carbonic acid formed when water combines with the carbon dioxide from the bacterial metabolism. Regarding the surface tension of the emulsion, this increases with the amount of lipids ingested by the microorganisms, because of the ionic species eliminated from the emulsion, increasing the cohesive forces of the fluid [39]. This increase in the surface tension value is more evident after 96 hours of starting the cleaning process of the synthetic emulsion.

The increase in the surface tension coincides with the increase in the biomass after the 96 hours above referred. As shown in table 6, inoculation lasted approximately 80 hours, and the bacterial colony growth remained. Then after 52 hours, the exponential growth phase of the colony is observed, increasing 45% compared with the initial biomass. It is not observed a decrease in the biomass because there are still nutrients (oil) to degrade by the microorganisms.

The behavior of the dissolved oxygen (DO) in the emulsion also has a direct relationship with the colony growth; during the inoculation phase, the DO practically remains constant (~ 6.6 mg/l), then there is a decrease in the amount of dissolved oxygen by the growth of the colony of microorganisms.

2.5.2. De-oiling of an industrial effluent

A sample of the effluent from an edible vegetable oils factory was collected. The amount of oil, quantified according the Soxhlet technique [40], was of 41.45 mg/l, which is above the maximum permissible value for the environmental standards. The initial pH of the sample was 5.6 and this value does not affect the growth of the *Pseudomonas fluorescens* bacteria, since this microorganism is naturally acidophilus. Table 7 shows the measured variables to the original sample.

Characteristic	Result
pH	5.8
Surface Tension (dyn/cm)	37.9
Dissolved Oxygen (mg/l)	0.35
Oil content (mg/l)	41.5

Table 7. Characteristics of an effluent from an edible vegetable oils factory.

A set of experiments were run by using a transparent acrylic plexiglass cell as biological reactor (30 cm by side and 40 cm in height), with 4 taps properly located to adjust 4 sensors for measuring continuously: pH, temperature, dissolved oxygen, and surface tension of the system. A bubble generator (made of filter cloth) was installed at the bottom of the reactor. The air feed to the cell was measured and controlled through a mass flow controller. The system was operated under stationary regimen, and open to the atmosphere. Table 4.5 shows the experimental variables applied in this particular research.

Variable	Value
Volume of emulsion fed to the cell	7.0 liters
pH	5.6
Air fed to the cell	25 l/min
Microorganism	*Pseudomonas fluorescense*
Sampling time	Every 48 hours
Volume of each sample	20 ml
Test duration	8 days

Table 8. Experimental conditions for de-oiling of an industrial effluent using a laboratory biological reactor.

Figure 23 shows changes in pH during the bio-treatment of the industrial sample. As can be observed the pH of the liquid media increases with the time (from 5.6 to 8.5). This is associated with the stressful conditions in the system: the special nutrients are depleted, waste products accumulate, bacterial overpopulation occurs (causing high demand of oxygen for aerobic metabolism), conditions that limit the growth colony and cause cell death.

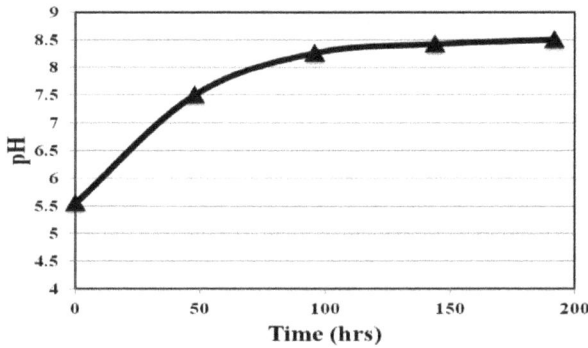

Figure 23. Changes in pH during the bio-treatment of the industrial effluent. Pseudomonas fluorescens bacteria were used to degrade oil.

Even the pH of the media begins to be unsuitable for cell growth. At this time there are more dead than alive bacteria, then the scarcity of nutrient sources (sugars and lipids), the protein degradation process of the bacteria begins, and generates ammonia as degradation product. This process provides organic nitrogen sources as nutrient which transforms the system into alkaline, increasing the pH value [41].

On the other hand, as in the case of the synthetic emulsion, the surface tension of the effluent increases with the time, because of the amount of lipids ingested by the microorganisms. As mentioned above, the cohesive forces of the fluid increase because of the ionic species eliminated from the emulsion (see Figure 24).

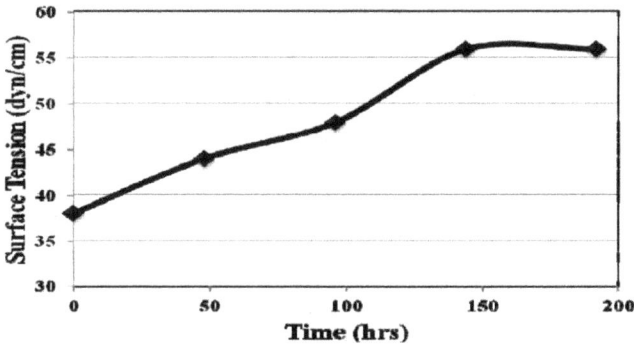

Figure 24. Changes in the surface tension of the effluent with the time, due to the increment in the amount of lipids digested by the microorganisms.

The lipolytic activity of the microorganisms is presented in Figure 25. The initial concentration of fat in the effluent was equal to 41.5 mg/l (taken as 100%); at 48 hours after initiation of the experiment the oil content in the emulsion decreased to 34.2 mg/l (82.4%); at 96 hours of incubation, was reached a concentration below 10 mg/l (24%), and finally, after 144 hours of the vegetable oil content in the effluent was 4.6 mg/l (11%) and remained constant from this point.

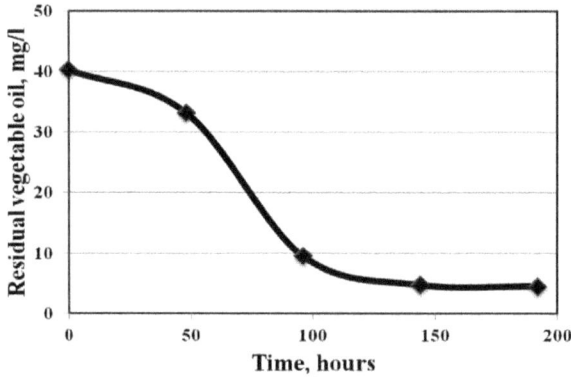

Figure 25. Variations of the vegetable oil content in the industrial effluent treated by the microorganism *Pseudomonas fluorescens*.

The constant trend in the last two experimental points, shown in figure 3, coincides with the trends of both: pH and surface tension, shown in figures 1 and 2, respectively, which also agrees with the behavior of the biomass in the last two points of sampling, as can be seen in Figure 26.

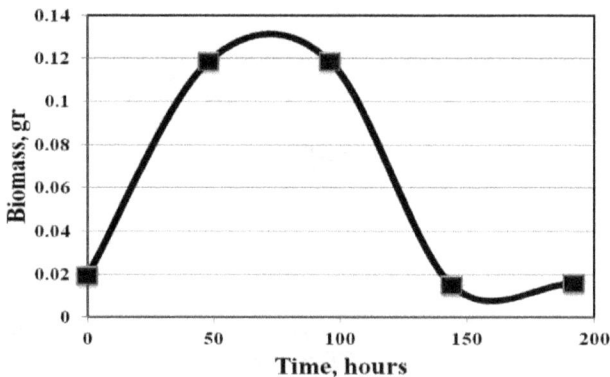

Figure 26. Biomass development (Pseudomonas fluorescens) during the degradation of an industrial effluent.

Is observed in figure 26 that biomass increased five times their initial concentration in a period of 48 hours of incubation, this development continued until approximately 100 hours of testing; this period matches with the maximum lipolytic degradation in the effluent. At this time the fat concentration of the effluent reached the lowest value (4.6 mg/l); the bacterial growth was limited to the shortage of nutrients and metabolic waste buildup in the effluent.

2.6. Conclusions

The information derived from the set of experiments conducted to establish the feasibility of using flotation devices in non-mineral systems, such as waste water treatment, can be used to set the following conclusions:

The experimental information shows that the de-oiling process can be performed effectively using a series of flotation columns. In this experience about 99% organic recovery is achieved using a series of four columns. The minimum residence time reported in order to achieve 99% organic recovery from the very stable water-oil (~40% oil) suspension is in the order of 100 minutes in a continuous four flotation columns operation.

Aereated flotation cells can be used as oxygen suppliers to aerobic nonpathogenic microorganisms (i.e., *pseudomonas fluorescens* bacteria), used during the biological treatments for cleaning waste water effluents; these microorganisms degrade the fat components of the water (either ionic or colloidal), using their metabolic activity. The experimental results shows that through these bacteria, around the 90% of the vegetable oil content in the effluent can be removed.

On the other hand, by precipitating heavy metal ions, it is possible to design a process for cleaning waste water contaminated with heavy metals (i.e., lead, copper, and nickel), through a sedimentation-flotation route.

The thermodynamic al analysis of the metal-water systems (activity, activity coefficient, ionic strenght, electrochemical potential), allows accurately predict the crystallization of the ionic species under controlled conditions of temperature and pH.

The experimental results shows that these data converge well enough with the thermodynamic calculations for dissolved or precipitated lead species, being possible the selective precipitation of the three metals tested here, and leaving the water without any kind of metal, ionic or solid.

Author details

Ramiro Escudero*, Francisco J. Tavera and Eunice Espinoza
Instituto de Investigaciones Metalúrgicas, Universidad Michoacana de San Nicolás de Hidalgo, Morelia, Michoacán, México

---

* Corresponding Author

Acknowledgement

The authors deeply appreciate the support given to this investigation to the Universidad Michoacana de San Nicolas de Hidalgo, and the Mexican Government through the National Council for Science and Technology.

3. References

[1] Sebba, F. 1962. Ion Flotation. Elsevier Publishing Co., New York.

[2] Doyle, F. M. 2003. Ion flotation – its potential for hydrometallurgical operations. Int. J. Miner. Process., 72, pp. 387 – 399.

[3] Finch, J. A., Dobby, G. S. 1990. Column Flotation. Pergamon Press, Oxford.

[4] Lara, R. H. C., Cruz, R. G., Monroy, M. G. F. 2005. Mecanismos de la alteración de la galena asociados al efecto ambiental en medios calcáreos. Artículo presentado en el XL Congreso Mexicano de Química, Sociedad Química de México, 25 – 29 de Septiembre, Morelia, México.

[5] Tavera, F. J., Reyes, M., Escudero, R., Patiño, F. 2010. On the treating of lead polluted wáter: flotation of lead coloidal carbonate. J. Mex. Chem. Soc., vol. 54 (4), pp. 204 – 208.

[6] Tavera, F. J., Escudero, R., Uribe, A., Finch, J. A. 2000. Ni – DETA flotation in aqueous media: application of flotation columns. AFINIDAD, LVII (190), pp. 415 – 423.

[7] Tavera, F. J., Gomez, C. O., Finch, J. A. 1996. A novel gas hold-up probe and application in flotation columns. Trans. Instn. Min. Met., 105 (C), pp. 99 – 104.

[8] Maxwell, J. C. 1892. A Treatise of Electricity and Magnetism. 3rd Edition, 1 (II), Chapter IX, Oxford University Press, London, pp. 435 – 449.

[9] Tavera, F. J., Gomez, C. O., Finch, J. A. 1998. Conductivity flow cells for measurements on dispersions. Can. Met. Quart., 37 (1), pp. 19 – 25.

[10] Escudero, R. 2000. Characterisation of rigid porous spargers by permeability and its relevance to scaling-up. Ph. D. thesis, McGill University, Montreal, Canada.

[11] Tavera, F. J., Escudero, R., Finch, J. A. 2001. Gas hold-up in a flotation column: laboratory measurements. Int. J. Miner. Process., 61, pp. 23 – 40.

[12] Gorain, B. K. 1997. The effect of bubble surface area flux on the kinetics of flotation and its relevance to scale – up. Ph. D. thesis, University of Queensland, Australia.

[13] Levenspiel, O. 1972. Chemical reaction engineering. 2nd ed., McGraw Hill, Inc., New York.

[14] Tavera, F. J., Escudero, R., Sánchez, A. 2004. Treating of water polluted with organics in flotation columns. Proceedings of the International Symposium on Environment; ISBN 970-703-293-6, Morelia, Mexico, pp. 95 – 105

[15] Grover V. I. (2006). *Water: global common and global problems*. Science Publishers.

[16] [www.unido.org]

[17] Barlow M., & Clarke, T. (2004). *Oro azul: las multinacionales y el robo organizado de agua en el mundo*. Ed. Paidós.

[18] Alvez, Z. C. (1995). Qualidade da água. Meio Ambiente por inteiro. *Jornal da Secretaria Municipal do Meio Ambiente* (1), 7.

[19] www.who.int/

[20] Crites, R., and Tchbanoglous, G. (2004). *Tratamiento de Aguas Residuales en Pequeñas Poblaciones*. (M. Camargo, L. P. Pardo, & G. Mejía, Trads.) Mc Graw Hill.

[21] Metcalf, and Eddy, INC. (1994). *Ingeniería Sanitaria, tratamiento, evacuación y reutilización de aguas residuales* (Tercera Edición ed.). (J. d. Montsoriu, Trad.) Ed. Labor.

[22] Celis Hidalgo, J., Junod Montano, J.,and Sandoval Estrada, M. (2005). *Recientes Aplicaciones de Depuración de Aguas Residuales con Plantas Acuáticas, Teoría* (Vol. 14).

[23] Alba Elias, F., Vergara Gonzalez, E. P., Ordieres, J. B., Martinez de Piston, F., González Marcos, A., and Ortiz Marcos, I. (s.f.). Tratamiento de aguas contaminadas con metales pesados empleando compost usado de champiñon como agente bioregenerador. 16.

[24] López Martinez, A., Cuapio Ortiz, L., Cárdenas Puebla, S., Balcazar, M., Jauregui, V., and Bonilla Petriciolet, A. (2007). Determinación de Concentración de Cr y Cd Adsorbido por Plumas de Pollo. *Tesis de Maestría* . (I. T. Aguascalientes, Ed.)

[25] Soto, E., Miranda, R. D., Sosa, C. A., and Loredo, J. A. (2006). Optimización del Proceso de Remoción de Metales Pesados de Agua Residual de la Industria Galvánica por Precipitación Química. *Información Tecnológica* , 17 (2), 33-42.

[26] Aksu S. and Doyle F. M., Potential-pH Diagrams for Copper in Aqueous solutions of various organic complexing agents, Electrochemical Society Proceedings Vol. 2000(14), 258-269 (2000).

[27] Reyes Pérez M. 2005, Tratamiento continuo, de aguas contaminadas con Cu y Pb, por flotación iónica en celdas con dispersores porosos; efecto de las propiedades de la dispersión aire-líquido en la separación, Tesis de maestría, IIM, UMSNH, (2005).

[28] Barakat M. A., Removal of Cu (II), Ni (III) and Cr(III) Ions from Wastewater Using Complexation - Ultrafiltration Technique, Journal of Environmental Science and Technology, Vol. 1 (3): 151-156, (2008).

[29] Garrels, R. M. and Christ, C. L., Minerals, Solutions, and Equilibria, Harper & Rowe, N. Y. 450pp, (1965).

[30] Cisternas L. A., Diagramas de fases y su aplicación, Reverte, (2009).

[31] Pankow J. F., Aquatic chemistry concepts, CRC Press, (1991).

[32] Azareño O.A., Núñez J. P., Figueroa L. A., León D. E., Fernández S. S., Orihuela S. R., Caballero R. M., Bazán R. R., and Yi Choy A. S. Flotación de Minerales Oxidados de Plomo. Revista del Instituto de Investigación de la Facultad de Ingeniería Geológica, Minera, Metalúrgica y Geográfica. Vol. 5 (10), 34-43, (2002).

[33] Tavera F. J., Colwell D., Escudero R., and Finch J., (2000). Estimation of Gas Holdup in Froths by Electrical Conductivity: Aplication of the Standard Addition Method. Revista de Química Teórica y Aplicada AFINIDAD, Barcelona Vol. 57(486), 139-142, España, Abril,.

[34] Sean R. S., and Thomas S.D. (2002). Raman Spectroscopy of Co(OH)2 at High Pressures: Implications for Amorphization and Hidrogen Repulsion. Physical Review. Vol. 66 B, pp 134301-1 – 134301-8.

[35] Chanturiya, V.A., Matveeva T.N., and Lantsoba L.B. (2003). Investigation into Products of Dimethyl Dithiocarbamate and Xantate Sorption of Sulfide Minerals of Copper-Nickel Ores. Journal of Mining Science. Vol. 39 (3), pp 281-286.

[36] Liu, Z., & Doyle, F. M. Modeling Metal Ion Removal in Alkylsulfate Ion Flotation Systems. Minerals and Metallurgical Processing , Vol. 18 (3), 167-171, (2001).

[37] Shigehito D., Akinobu H., Bienvenu B.A., and Mizuhata M. (2009). α - Ni(OH)$_2$ Films Fabricated by Liquid Phase Deposition Method. Thin Solids Films, Vol. 517 (5), pp 1546-1554.

[38] Guo-riu F., Zhong-ai H., Li-jing X., Xiao-qing J., Yu-long X., Yao-xian W., Zi-yu Z., Yu-ying Y., and Hong-ying W. (2009). Electrodeposition of Nickel Hydroxide films on Nickel Foil and its Electrochemical Performances for Supercapacitor. International Journal of Electrochemical Science. Vol. 4, pp 1052-1062.

[39] Atkins, Peter. Jones, Loretta. "Principios de Química: los caminos del descubrimiento". 3ra edición, 1ra reimpresión 2006, Ed. Médica-Panamericana. B.A. Arg. pp 17.

[40] Marta S.M., and Francisca S.C., (2000). The Chemical Composition of "Multimistura" as a Food Supplement. Food Chemistry. Vol. 68, No. 1, pp 41-44.

[41] www.biologia.edu.ar

Chapter 10

State of the Art Treatment of Produced Water

Rangarajan T. Duraisamy, Ali Heydari Beni and Amr Henni

Additional information is available at the end of the chapter

1. Introduction

Produced water is water trapped during subsurface formations which is brought to the surface along with oil or gas. It contributes the largest volume of waste stream associated with oil and gas production. Globally, 77 billion bbl of water are produced per annum. The conventional methods to handle waste stream are reinjection into the well, direct discharge or reuse in case of thermal loop. Out of these, the most efficient way of handling produced water is to re-inject it into disposal wells. The disposal cost, which includes transportation cost, capital cost and infrastructure maintenance cost, may be as much as $4.00/bbl. On the other hand, many oil producing regions (West Texas, Middle East and the Central Asian Republics) have scarcity of potable water. An affordable water treatment process could convert produced water into an asset. The harmful effects of produced water and the depletion of usable water resources act as a driving force for the treatment of produced water.

Produced water contains soluble and insoluble organic compounds, dissolved solids, production chemicals (corrosion inhibitors, surfactants etc.) and solid particles due to leaching of rocks and corrosion of pipelines. The methods available for treating produced water are physical, chemical, biological and membrane treatment processes.

Stringent water quality parameters can be achieved efficiently through membrane processes. The most important advantages of using membrane processes are the ease of operation and little or no requirement of chemicals. Based on pore size, the membrane processes could be classified into Microfiltration (MF), Ultrafiltration (UF) and Nanofiltration (NF). The membranes are also classified as organic, inorganic and composite membranes. The primary disadvantage of using membranes is fouling. Irreversible and reversible foulings occur while treating produced water. The usage of appropriate pre-treatment process reduces the membrane fouling to a greater extent. Commercial treatment methods based on reverse osmosis and ion exchange processes are also discussed.

2. Characteristics of produced water

The physical and chemical properties of produced water depend on the geographic location of the field, the geological formation with which the produced water has been in contact for thousands of years, and the type of hydrocarbon product being produced.

The main constituents of produced water are as follows:

- dissolved and dispersed oil compounds
- dissolved formation minerals
- production chemical compounds
- production solids (formation, corrosion, scale, bacteria, waxes, and asphaltenes)
- dissolved gases

Produced waters discharged from gas/condensate platforms are about 10 times more toxic than the produced waters discharged from oil wells, but, the volumes from gas production are much lower; hence the total impact may be less.

2.1. Constituents in produced water

1. **Dispersed oil:** Oil is an important contaminant in produced water since it can create potentially toxic effects near the discharge point. It can significantly contribute to Biological Oxygen Demand (BOD) and hence affects the aquatic or marine ecosystem. Usually the size of dispersed oil droplets would be 4-6 microns, but it may vary from 2-30 microns. The current treatment systems could recover oil droplets of size up-to 10 microns.
2. **Dissolved Organic Compounds:** They include organic acids, polycyclic aromatic hydrocarbons (PAHs), phenols and volatiles. Volatile hydrocarbons can occur naturally in produced water. Concentrations of these compounds are usually higher in produced water from gas-condensate-producing platforms than in produced water from oil-producing platforms.
3. **Treatment Chemicals:** They include biocides, reverse emulsion breakers, and corrosion inhibitors. Corrosion inhibitors can form stable emulsions. Some chemicals are highly toxic even at low concentrations such as 0.1 ppm.
4. **Produced Solids:** They consist of precipitated solids (scales), sand and silt, carbonates, clays, corrosion products and other suspended solids produced from the formation and from well bore operations.
5. **Bacteria:** Anaerobic bacteria present in produced water may lead to corrosion.
6. **Metals:** Zinc, Lead, Manganese, Iron and Barium are the metals usually present in produced water. They are in general less toxic when compared to organic constituents. But they may precipitate to form undesired solids which hinder the treatment processes.

An example of key parameters of produced water is listed below in Table 1.

3. Produced water management[1]

a. **Injection into oil wells:** The produced water is injected into the same oil well from where it is obtained or transported to the discharge well at another location. The cost varies from $0.70 to $4.00.

b. **Direct discharge:** The produced water is discharged directly as per the regulation norms. The cost varies from $0.03 to $0.05.

c. **Reuse in oil and gas operation:** The produced water could be treated and used in the oil and gas processing industries. The cost varies from $0.04 to $0.07.

d. **Consumed in beneficial use:** Treating produced water to convert it into an asset. The cost varies from $0.25 to $2.00.

Parameter	Natural Gas Produced Water	Oil field Produced Water
Oil/grease(ppm)	40	560
pH	4.4-7.0	4.3–10
TSS(ppm)	5500	1000
TDS(ppm)	360,000	6554
TOC(ppm)	67-38,000	1500
COD(ppm)	120,000	1220
Density(kg/m³)	1020	1140
Arsenic(ppm)	0.005-151	0.005–0.3
Lead(ppm)	0.2-10.2	0.008-8.8
Chromium(ppm)	0.03	0.02–1.1
Mercury(ppm)	--	0.001–0.002
Oil droplet size(μm)	2 to 30	

Table 1. Key parameters of importance in produced water treatments [1]

4. Treatment methods

4.1. Physical treatment

4.1.1. Physical adsorption

Activated carbon, organoclay, copolymers, zeolite, resins are widely used to treat produced water. The combination of activated carbon and organoclays proved to be more efficient in removing total petroleum hydrocarbons (TPH).[2] Copolymers reduce the oil content up to 85%.[3] Zeolites are efficient in removing BTEX compounds.[4] A multi-stage adsorption and separation system was developed, for example, by EARTH Canada Corporation to recover dispersed oil droplets in water, whose size is greater than 2 microns.[5]

4.1.2. Sand filters

They are generally used to remove metals from produced water. Process requires series of pre-treatment steps such as pH adjustment, an aeration unit and a solid separation unit. The removal efficiency is as high as 90%.[6]

4.1.3. Cyclones

A compact floatation unit (CFU) could remove dispersed oil from 50% to 70% using a centrifugal force. [7]The major drawback of using a cyclone is its low efficiency and inability to remove dissolved components.[8]

4.1.4. Evaporation

Evaporation does not require chemical treatment which eliminates the risk of secondary sludge handling. It also does not require highly skilled labor. On the other hand, the requirement of energy is very high which increases the operating cost. The energy consumption could be brought down by reusing hot vapor to heat the fresh feed.[9]

4.1.5. Dissolved air precipitation (DAP)

In this process, water at 500 kPa(for example) is saturated with air in a packed column separator. The pressure is released into the water column which causes the formation of air bubbles. It induces the flotation of aliphatic and aromatic hydrocarbons, and removes the aliphatic compounds more efficiently than aromatic compounds.[10]

4.1.6. C-TOUR

It is a patented technology that uses liquid condensate to extract dissolved components from produced water. In field trials, the removal efficiency of dispersed oil was found to be 70%.[7]

4.1.7. Freeze-thaw/evaporation

This technology uses the principle of solubility dependency on temperature. When the solution is cooled below the freezing point of the solvent but not below the depressed freezing point of the solution, relatively pure crystals of solvent and unfrozen concentrated solutions are obtained. If we couple this process with conventional evaporation, large volumes of clean solvent could be obtained. The process is capable of removing 90% of Total Recoverable Petroleum Hydrocarbons (TRPH). But it has several limitations like the requirements of sub-zero ambient temperatures and large land surface.[11]

4.2. Chemical treatment

4.2.1. Chemical precipitation

The suspended solids and colloidal particles could be removed by coagulation and flocculation. Several coagulants like modified hot lime, FMA (a mixed metal polymer), Spillsorb, calcite and ferric ions were used as coagulant to treat produced water. The disadvantages of this process are its ineffectiveness for dissolved components and the increased concentration of metals in the sludge formed.[12, 13]

4.2.2. Chemical oxidation

It uses a combination of strong oxidants (e.g: O3 and H2O2), irradiation (e.g: UV) and a catalyst (e.g: photocatalyst), and oxidizes the organic components to their highest stable oxidation states.[14]

4.2.3. Electrochemical process

Almost 90% of BOD and COD could be removed from produced water in a short time (of the order of 6 minutes) by using an active metal and graphite as an anode and iron as a cathode. During the process, Mn^{2+} is formed, which oxidizes and coagulates the organic contaminants.[15]

4.2.4. Photocatalytic treatment

The pH of the solution is increased to a value of 11 by adding soda. The photochemical reaction was then carried out on the supernatant obtained from the flocculation unit. Titanium dioxide is usually used as photocatalyst. The COD removal efficiency and toxicity reduction were found to be higher in photoelectrocatalysis than that in photocatalysis.[16]

4.2.5. Fenton process

Nearly 95% of COD and dispersed oil content can be reduced by combining flocculation with the Fenton oxidation adsorption process. The flocculent used is poly-ferric sulfate. [17]

4.2.6. Treatment with ozone

Sonochemical oxidation could destroy BTEX compounds but the addition of hydrogen peroxide does not improve the efficiency. The process requires high initial and operating cost. [18]

4.2.7. Room temperature ionic liquids

The hydrophobic room temperature ionic liquids remove certain soluble organic components efficiently, but not much of the other contaminants. Hence, the screening of ionic liquids depends on the constituents of produced water.[19]

4.2.8. Demulsifiers

Some surfactants used as production chemicals are responsible for the stabilization of oil-water emulsions. They reduce the oil-water interfacial tension. Demulsifiers are surface-active agents that would disrupt the effects of surfactants. But a number of solids like silts, iron sulphide and paraffin, etc., present in the crude oil complicate the process.[20]

4.3. Biological treatment

The produced water could be treated with aerobic as well as anaerobic microorganisms. The microorganisms disintegrate the organic and ammonia compounds, but could not treat dissolved solids.[21] The COD removal efficiency increased up to 90% while treating produced water with Bacillus sp.[22]

5. Membrane treatment processes

Conventional treatment methods are capable of removing suspended particles with particle size of 5.0 μm or above.The disposal and reinjection regulations are becoming more stringent and the conventional methods are not able to treat produced water which can meet these regulations.[23] The general specification for acceptable quality of oil-fields produced water for discharging into surface water (or re-injection) are less than 42 mg/L of oil/water, and less than 10 mg/L of Total Suspended Solids (TSS).[24] Conventional treatment processes are not able to meet these water effluent standards. New technologies should be utilized to separate both fine particles and dissolved components.[23]

Membrane processes are a rather new separation process for treatment of produced water. Membrane separation processes, including microfiltration (MF), ultrafiltration (UF), nanofiltration (NF) and reverse osmosis (RO), are able to treat produced water and generate water with high standards to meet regulations. The driving force of the above mentioned membranes processes is pressure gradient.[23]

5.1. Advantages of membrane technology

Membrane technologies have some advantages that make them popular for produced water treatment processes:[25, 26]

- sludge reduction
- high quality of permeate
- smaller space needed
- ease of operation
- minimal impact on permeate quality with variation in feed water quality
- little or no chemicals required
- possibility for recycling of waste streams
- possibility for having an automated plant
- moderate capital costs
- ability to be combined easily with other separation processes
- low energy consumption
- continuous separation

But there may be some drawbacks for using membrane processes including concentration, polarization/membrane fouling, low selectivity or low flux and low membrane lifetime.

According to the above mentioned advantages, that membrane separation processes can, in some circumstances, be viable for treatment of produced water.[25]

5.2. Membrane properties

There are different types of membrane processes, membrane materials, and feed water compositions, but the main goal of preparing the membranes is the same. An ideal membrane should:

- be mechanically resistant
- have a high permeate flowrate
- have a high selectivity for a specific component

Having high permeate flowrate means having large pore sizes. A high level of selectivity for a certain component is achievable with small pore sizes and the range of pore sizes should be narrow. The last two parameters present a dilemma, as one is in conflict with the other. The membrane mechanical resistance depends on the membrane thickness. Therefore the membrane should have a thin layer of material (the selective layer), narrow pore sizes, and high porosity. [27]

According the type of materials and mechanism of separations, membranes may be categorized as porous or dense. Separation of dense membranes is based on physicochemical interaction of permeate and the membrane material. Separation mechanism of porous membranes is based on the mechanical separation by size of permeates and pore sizes of membrane (sieving).[27]

5.3. Types of membranes

Membranes can be generally classified based on their structure or morphology. The detailed classification of membranes is reported in Table2. Symmetric and asymmetric membranes are two classes. Symmetric membranes have different types including isotropic microporous, nonporous dense membrane, and electrically charged membranes. Asymmetric membranes are divided into Loeb-Sourirajan anisotropic, thin-film composite anisotropic, and supported liquid membranes.[28]

Membranes can also be classified based on the type of materials like ceramic, inorganic, and composite membranes.

5.3.1. Polymeric membranes

Polymeric membranes have some advantages including high efficiency for the removal of particles, emulsified and dispersed oil; small size; low energy requirements, and being cheaper than ceramic membranes. They also have some disadvantages including the inability to separate volatile and low molecular weight compounds, fouling problems due to oil, sulfide or bacteria which may be required to be cleaned daily, an inability to be used at temperatures above 50 °C, and they also create the possibility of having radioactive by-product in the effluent and need for some pre-treatment processes. [26]

Process	Mechanism of separation	Material/Type	Typical Objective
Microfiltration (MF)	Separation by sieving through macropores (>50 nm)	Polymeric and inorganic / Porous	Removal of suspended solids, large organic molecules, and large colloidal particles including microorganisms (used for reducing colloidal suspensions and turbidity)
Ultrafiltration (UF)	Separation by sieving through mesopores (2-50 nm)	Polymeric and inorganic / Porous	Removal of large dissolved solute molecules and suspended colloidal particles, including bacteria and macromolecules such as proteins
Nanofiltration (NF)	Separation through combination of charge rejection, solubility-diffusion and sieving through micropores (<2 nm)	Polymeric / Dense	Removal of multivalent ions and specific charged or polar molecules
Reverse osmosis (RO)	Separation is based on the difference in solubility and diffusion rates of water and solutes	Polymeric / Dense	Removal of low molecular weight components such as inorganic ions

Table 2. Membrane processes [26, 27]

Polymeric membranes are prepared from materials like polyacrylonitrile (PAN) and Polyvinylidenediflouride (PVDF). These membranes are relatively cheap. Polymeric membrane should be tested via integrity testing to be sure that they are not damaged. Their life cycle is approximately 7 years. Polymeric membranes can be used to treat feed streams containing high TDS contents. Their efficiency for dead-end and cross-flow operations are 85% and 100%, respectively.[29]

5.3.2. Inorganic membranes

Inorganic membranes have better chemical and thermal stability than polymeric membranes. There are four different types of inorganic membranes, such as ceramic membranes, glass membranes, metallic membranes (including carbon), and zeolitic membranesm are utilized in MF and UF processes. Metallic membranes are prepared from metal powders like stainless steel, tungsten or molybdenum. Ceramic membranes are synthesized from a combination of metals like aluminium, titanium, silicium, or zirconium with non-metals like oxides, nitrides, or carbides. Aluminium oxide (γ-Al$_2$O$_3$) and zirconium oxide (ZrO$_2$) are the most important materials for ceramic membranes. Glass membranes (silica or SiO$_2$) can also be considered as ceramic membranes. Zeolite membranes are a new class of membranes which have a narrow pore size.[30]

a. Ceramic membranes

Ceramic membranes are prepared from nitrides, oxides, or carbides of metals like zirconium, titanium, or aluminum. Tubular modules have been the most widely used, and in which the feed flows inside the membrane channel. Tubular membranes consist of a porous support layer (typically α-alumina), one or more decreasing pore diameter layers and an active layer responsible for separation (α-alumina, zirconia, etc.).[29, 26]

Ceramic membranes have some advantages including higher flux due to their higher porosity, more hydrophilic surface than organic membranes, better recovery performance of the membrane due to better resistance against mechanical, thermal, and chemical stress than organic membranes.[31, 32]The main advantage of ceramic membranes is its capability to meet the current water treatment effluent standards with no chemical pre-treatment.[33]Ceramic membranes have some disadvantages including sealing problems because of thermal expansion of ceramic membrane, and module housing. Ceramic membranes should be handled carefully as they are brittle.[31, 32]Another advantage of ceramic membranes is a greater removal of the particles at higher flux than polymeric membranes due to well-defined pore distribution of the ceramic membranes.

Membrane fouling is one of the problems associated with usage of membranes in produced water treatment processes, but the high chemical and thermal stability of the ceramic membranes will make the chemical and thermal cleaning methods possible; this is not the case for the polymeric membranes.[26]

Thermal and chemical stability of mullite ceramic membranes are very high compared to other ceramic membranes. They are very cheap as they can be produced by extruding kaolin clay.[32]

Organic matter, oil and grease, and metal oxides can be removed using ceramic membranes but dissolved ions and dissolved organics cannot be separated and some pre-treatment processes like pre-coagulation, straining or cartridge filtration should be utilized. Feed streams containing high amount of total dissolved solids (TDS) can be treated using ceramic membranes, but high ion-concentration may cause irreversible fouling. Based on several studies, it was concluded that ceramic membranes have better performance, lower energy requirement, longer life cycle (more than 10 years), but it requires higher capital cost, than polymeric membranes.[29]

Abadi et al. [31]used a tubular ceramic (α-Al$_2$O$_3$) microfiltration (MF) membrane with a pore size of 0.2 μm and a stainless steel housing for the treatment of oily wastewater from API effluent of Tehran refinery. Effects of the transmembrane pressure (TMP), cross flow velocity (CFV) and temperature on permeate flux; total organic compound (TOC) and fouling resistance (FR) were investigated. The recommended working conditions were 1.25 bar for the transmembrane pressure (TMP), 2.25 for the cross flow velocity (CFV) and 32.5 °C for temperature. Backwashing was stated to be useful to prevent the declination in permeate flux. Oil and grease content was reduced to 4 mg/L therefore meeting the National Discharge Standard, and total organic compound removal efficiency was more than 95%. These systems were recommended to replace the conventional wastewater treatment method.

In a different study, ceramic microfiltration was used for treating produced water from two onshore and two offshore pilot plants. Dispersed oil and suspended solids concentration were less than 5 mg/L and 1 mg/L, respectively, in permeate stream. Produced water was pretreated to avoid membrane fouling. At suitable membrane pore size and cross velocity, flux was increased up to 1750 gal/ft^{2}D. [34]

b.   Zeolite membranes

Zeolite membranes have attracted much research in separation of ions from aqueous solutions by reverse osmosis. The separation mechanism includes electrostatic repulsion (Donnan exlusion) at intercrystalline pore entrances and size exclusion of hydrated ions. These special kinds of separations have made the zeolite membrane a unique separation process for the removal of organics and electrolytes from water by RO processes. [35, 36]

Zeolite membranes are mostly used in composite structure due to their high fragility. Composite zeolite membranes consist of a thin film of zeolite supported on a porous material like ceramics, metal glasses, and porous alumina. Among different support materials,alumina is the most widely used. Some studies for zeolite composite membranes will be presented in the following section.[37]

c.   Composite membranes

Cui et al. [38] reported in a study the application of zeolite/ceramic membranes in MF of oil-water emulsion. NaA/α-Al$_2$O$_3$ microfiltration membranes were produced by hydrothermal synthesis on porous ceramic tubes with inter-particle pore sizes of 0.2 to 1.2 μm. They were prepared and utilized for the separation of oil-water emulsion. Feed water containing 100-500 mg/L oil was treated to produce clean water (<3 mg/L oil). Fouling resistance of hydrophilic NaA membrane was better than tubular ceramic membrane. In general, zeolite membranes are cheaper than ceramic membranes. Silica and alumina reagents like TEOA and sodium aluminate may be used for preparing inexpensive zeolite membranes. The energy-intensive sintering process, which is widely used for preparing ceramic MF membranes, is not needed for preparing zeolite membranes. Zeolite membranes demonstrated good stability against oil fouling and caustic cleaning solutions. Oil rejection efficiency and flux were more than 99%, and 85 Lm^{-2}h^{-1}flux was obtained for water containing 100 mg/L oil.Backwashing with hot water and alkali solution did not affect the membrane performance.

Li et al. [39] used inorganic nano-sized alumina particles for the modification of tubular UF module equipped with Polyvinylidenedifluoride (PVDF) membranes. The PVDF-Al$_2$O$_3$ membrane was used to treat oily wastewater from an oil field.Chemical oxygen demand and total organic carbon retention efficiencies were 90% and 98%, respectively. The type of process was cross-flow. It was seen that permeation performance was improved and flux was increased to twice that of unmodified one. The antifouling performance of modified membrane was favorable and flux recovery ratio increased to 100% by washing with 1 wt% (commercial) OP-10 surfactant solution. The permeation water quality met the requirement for oil field injection or drainage. Oil content and suspended solids content were both 1 mg/L after membrane treatment, and was therefore demonstrated that PVDF-Al$_2$O$_3$ composite membrane can be used in oily wastewater treatment.

Four commercial thin-film composite polyamide membranes including RO membrane, two ultra-low pressure RO membranes, and one NF membrane were used for treatment of produced water to reach the standards of potable and irrigation waters. Produced water from sandstone aquifers was treated using a two-stage membrane unit. TOC concentration was less than 200 μg/L. The system can recover 60-80% iodide which causes the final concentration of iodide in retentate to be more than 100 mg/L and can be used for commercial iodide recovery purposes.[40]

Molecular sieve zeolite membranes were prepared on the inner surface of tubular α-alumina substrate, and produced water was treated in a RO separation process. Good ion rejection and chemical and mechanical stabilities led to the conclusion that molecular sieve zeolite membrane can be used in produced water purification, while polymeric membranes have many problems with fouling and structure instability. The overall ion rejection was 98.4% for synthetically produced water. [36]

Li et al. [35] looked into the separation of NaCl solutions in presence of counter-ions or at increased pressure and concentration from produced water employing a pure silicate zeolite membrane synthesized on a commercial tubular α-alumina substrate (PALL, pore diameter = 0.2 μm). Ion and water transport are highly dependent on the operating pressure, feed ion concentration and solution chemical composition. Exponential increase of ion flux was due to increasing feed ion concentration. Reductions in ion rejection rate and water flux were seen in the presence of high valence cations. It was suggested that the MFI-type zeolite membrane can be used for the desalination of produced water.

d.   Membrane bioreactor (MBR)

A membrane bioreactor (MBR) has two steps including the activated sludge process in which produced water is treated biologically and the membrane filtration process which biomass (activated sludge) is separated from treated water using the membrane. MF and UF are used for the separation process. Better effluent quality will be obtained using an MBR compared to conventional activated sludge process. [41]

Use of membrane bioreactors has some advantages over conventional methods for treating produced water which includes lower energy costs, compactness, no need for chemical additives, low sludge production, high loading rate capacity and high quality of the treated water. [42]There are few studies available for treatment of produced water using an MBR.

e.   Membrane distillation

Membrane distillation (MD) is a separation process based on a thermally driven membrane. Vapour pressure gradient is the driving force for mass transfer through the membrane. MD is the only membrane separation process where the performance does not change with changing of feed TDS content. Operating cost of MD is lower than conventional distillation processes. The following materials are typically used for MD: Polyvinylidenedifluoride (PVDF), Polypropylene (PP), and Polytetrafluoroethylene (PTEE). Flat-sheet and hollow-fiber are two typical modules for MD. Direct contact MD (DCMD), vacuum MD (VMD), seeping gas MD (SGMD), and air gap MD (AGMD) are four different methods of MD operation. [29]

Direct contact membrane distillation (DCMD) separation process utilizes a hydrophobic microporous membrane which in one side has a hot brine feed and on the other side a cold distillate stream. Water vapor passes through the membrane pores from the hot brine section. For treating hot brines, if reverse osmosis (RO) process was to be used, the feed should first be cooled which requires some energy, but DCMD process can treat hot feed streams without cooling which is an advantage of DCMD over RO process. Furthermore, the application of DCMD process above 100°C eliminates porous substrates which are required for low temperatures. [43]

Produced water feeds with TDS content of more than 35000 mg/L can be processed using MD and all non-volatile solutes (like Na, B and heavy metals) are rejected with a theoretical efficiency of 100%, but the diffusivity of compounds having higher volatility than water, diffuse faster through the membrane. As a pre-treatment, pre-filtration of feed is required to remove all compounds which may wet the hydrophobic surface of the membrane. MD systems have larger footprint than nanofiltration or reverse osmosis systems. [29]

5.4. Fouling

In the treatment of produced water using membranes, the permeate flux may be decline. This declination may be due to concentration polarization or fouling[26]. Two general types of membrane fouling exist for oily waste water treatment including reversible fouling and irreversible fouling. Reversible fouling is a result of deposited colloidal particles or solutes on the membrane surface and in the membrane pores. Pure water backwashing may be utilized to reverse the flux declination in reversible fouling. Irreversible fouling is a result of strong chemical or physical sorption of particles and solutes on the membrane surface and in the pores. The only method for recovering the flux declination by irreversible fouling may be washing with acid or alkali solutions. However, aggressive cleaning methods may not be able to recover the initial permeability of irreversibly fouled membranes [39].

Membrane fouling may be due to biofouling, scaling, organic fouling, and colloidal fouling: [26, 44]

- microbial contamination of water may produce a biofilm on the surface of the membrane which will decrease water permeation through the membrane (biofouling).
- salts may precipitate on the membrane surface which will cause scaling.
- coating the surface or plugging the pores in the porous support layer by hydrocarbons is called organic fouling.
- accumulation of clay and silica on the surface of the membrane will cause colloidal fouling.

The mechanism of fouling has not been clearly identified yet because measurements taken from laboratory-scale processes cannot truly describe or explain the fouling phenomenon in pilot-scale or full-scale membrane processes. [45] Many feed streams which are processed using membranes have particles with sizes from nanometers to micrometers which cause membrane fouling and because of this problem membrane fouling may not be stopped. [32]

Different strategies can be utilized to mitigate the effect of fouling: [46]

- oil, solids, and gels can be reduced using some pre-treatment processes
- keeping high cross-flow velocity on the feed side
- a cleaning cycle may be utilized
- backwashing or reverse flow washing may be used
- operating at higher temperatures

Coagulation, granular activated carbon and low pressure membrane processes (e.g., MF and UF membranes) are some of the pre-treatment methods which can be utilized before high pressure membranes to prevent fouling. [45] Addition of disinfectants and anti-scaling agents may control fouling. [26]The most possible foulant for NF and RO membranes after a UF treatment process could be calcium bound to inorganic materials and silica bound to organic materials. [45]

Membrane fouling is a function of feed properties (ionic strength, pH, particle concentration, and particle size), membrane properties (pore size, charge, and hydrophibicity) and hydrodynamics (transmembrane pressure and cross-flow velocity). One of the most important parameters in controlling the flux is the cross-flow velocity. As pressure increases, flux increases linearly. A cross-flow velocity of 3 m/s was reported as normal velocity, operation above it may reduce the rate of fouling, however utilization of higher pressures needs more energy and very high pressures (consequently high velocities) may result in severe fouling or membrane compaction. [26]

Hydrophilic membranes experience reduced fouling problems because their adsorptivity of hydrocarbons are less than hydrophobic membranes. Many methods including surface segregation, surface coating, and surface graft polymerization may be used for increasing the hydrophilicity of membranes. Some polymers including polyvinylidene fluoride (PVDF), polymethyl methacrylate (PMMA), polymethyacrylate (PMA), polyvinylacetate (PVAc), and cellulose acetate (CA) may be used for blending as hydrophilic polymers to increase the hydrophilicity of membranes. Blending of inorganic materials was shown to increase the membrane permeability and the membrane surface properties was better controlled than blending of organic materials[44, 39].

Maguire-Boyle et al. [44]used alumina nanoparticles (alumoxanes) stabilized with hydrophilic cysteic acid to prepare the surface functionalized alumina fabric composite membranes. It was observed that surface chemistry of the membrane may be changed by coating. They consequently have better permeability of water compared to hydrocarbons.

Hydrophilicity of a membrane was increased in a study [46] but it was seen that adhesion of some components of oil and chemicals in produced water changed the wettability of the surface and degradation in membrane performance was observed over time.

Nanofiltration (NF) membranes can be used at lower pressures than reverse osmosis (RO) membranes but fouling is a problem for NF membranes which will increase the transmembrane pressure and reduce the flux. A trial-and–error method can be used to control fouling and clean the membranes. Membrane fouling may be controlled by changing the surface chemistry of the membrane. Mondal and Wickramasinghe used polyamide based thin film composite (TFC) membrane surfaces and modified them by photo-induced grafting of N-isopropyl acrylamide. It was seen that the hydrophobicity of the membrane surface was increased substantially which would control fouling [47].

Pedenaud et al. [48] reported that if produced water contained sufficient level of barium and sulfate, it may cause irreversible fouling on ceramic membranes and have very low long term filtration, and in order to prevent precipitation, a suitable anti-scalant should be used.

Another surface modification process study was conducted by Wandera et al. [49] to reduce membrane fouling. A low molecular weight cut-off cellulose ultrafiltration membrane was modified utilizing surface-initiated atom transfer radical polymerization. Utilizing this method limited the fouling and, when it occurred, foulants were removed without using chemicals. The rate of decrease in water flux was slowed and finally flux recovery got better than in the unmodified membrane. It was stated that this modification process can be used for other membrane materials to reduce the fouling problem.

5.5. Cleaning methods

There are different methods to clean membranes in order to reduce fouling problems which includes cross-flushing, back-flushing, use of chemicals, and use of ultrasound:[26]

- air bubbles in forward washing (AirFlush) may be utilized for membrane cleaning (cross-flushing)
- reversion of flow direction through permeate channel to feed channel may be utilized in back-washing method but in cases where adsorbents were strongly adsorbed to the membrane surface this method may not be very useful
- some chemicals may be utilized including acids (calcium salts and metal oxides may be dissolved), alkalis (silica, organic/biological foulants, and inorganic colloids may be removed), surfactants (displace foulants, dissolve hydrophobic foulants, and emulsify oils), oxidants (bacteria and organic compounds may be oxidized), sequestrates (metal cations may be removed from a solution), and enzymes (foulants may be degraded)

- utilization of ultrasound method may be useful in increasing membrane flux by breaking cake layer or by decreasing solute concentration at membrane surface

Among the cleaning methods, backwashing has been used widely but is not recommended as the membrane usually experiences a reduction in permeate flux after each backwash. Another problem with backwashing is that the operation must be stopped to backwash the membrane. [26] Chemicals used for membrane cleaning were lye solutions (1 % (w/w) NaOH solution, Ultrasil P3-14, Ultrasil P3-10 for 30 to 60 min). Washing intervals depends on the oil and grease concentration in the feed, permeate flow rate, success of washing, and the period of membrane degradation. [46]

5.6. Modelling

Membrane fouling crucially depends on the permeate flux. Usually there is a rapid decrease in permeate flux and then flux decreases gradually with time. As a result, modeling of the fouling process is important. Three different types of models exist for the description of the permeate flux decline. Empirical models are accurate but cannot explain the fouling mechanisms. Fouling mechanisms can be described by theoretical models but these models cannot give accurate results for flux reduction with time. Semi-empirical models which have their parameters estimated from experimental data can accurately predict the flux decline and interpret the fouling mechanism. [32]

Blocking models may be used for prediction of filtration flux from analyses of blocking chart and resistance coefficients. Pore blocking is caused by bulk phase particles that are small enough to enter and deposit inside the pore. There are three types of blocking models. The standard pore blocking happens when the particle/droplet size is smaller than the mean pore size of the membrane. It is called adsorptive fouling or pore narrowing. Complete pore blocking occurs when particle/droplet size is approximately equal to the mean pore size of the membrane. Intermediate pore blocking occurs when more particles/droplets settle over already deposited particles/droplets. Standard pore blocking, Complete pore blocking and Intermediate pore blocking models can be used to obtain the permeate flux. Blocking chart is a graph drawn between the permeate flux (q) and the particle accumulation (cv), from which the blocking index(i) can be obtained. The resistance coefficient (K) can be found using the following equation.

$$\frac{d^2t}{dv^2} = K\left(\frac{dt}{dv}\right)^i \qquad (1)$$

Where v is the filtrate volume, t is time, and i is the blocking index.

Abbasi et al. [32] studied the fouling mechanism in cross-flow microfiltration (MF) of a synthesized oily wastewater. They stated that the cake formation model had the best results when compared to experimental data. The intermediate pore blocking model had the best results for the prediction of the flux after the cake formation model, and the worst result pertained to the pore blocking model.

6. Combined systems

Membrane treatment processes can be very efficient for the treatment of oily waste water but the utilization of membranes as the only unit for treatment of produced water may cause severe fouling of the membranes due to the presence of high level of oil, suspended solids, and bacteria. Therefore, some pre-treatment processes are required before membrane processes.[50]Removal of salts and many of the associated organics and inorganics can be achieved using the RO process. However, a pre-treatment step such as acidification is needed for the removal of low molecular weight organics and inorganics like boron, and the utilization of RO process without a pre-treatment step cannot be effective and some post treatment processes may also be needed after the RO process to meet the required standard. [51]

Treatment processes depend on the characteristics of produced water and these differ from well to well. Some experimental studies for determination of characteristics of produced water are required for designing a treatment process, relying solely on the literature is not recommended. [52]

Lee and Frankiewicz [46] recommended using some pre-treatment processes to reduce the oil and solids contents to less than 50 ppm and 15 ppm respectively, before utilizing a hydrophilic UF membrane with a pore size of 0.01 μm to reduce the fouling problem. It was stated that if the oil concentration can be reduced to less than 50 ppm, the membrane permeability can be maintained in an acceptable value for 4 days or longer with washing at several intervals. A daily washing was necessary for treating a feed stream containing 200 ppm as oil concentration but the washing intervals extended to 6 to 8 days for an oil concentration of 25-50 ppm by utilizing some pre-treatment processes. De-sanding and de-oiling hydrocyclones and pre-filtering are examples of pre-treatment processes used. The removal efficiency of hydrocyclone for solids and oils were 73% and 54% respectively in the pre-treatment processes. Total oil and grease content was less than 2 mg/L in the permeate stream and the efficiency of water recovery was 98%.

Some experiments were conducted to treat produced water for irrigation quality standards and to model the oil and salt separation processes. The proposed process included microfiltration, utilizing sorption pellets made of a modified clay material (organoclay PS12385) and reverse osmosis (RO) units for separation of salts. The average loading capacity of clay pellets was better than the activated carbon (more than 60%) and it was seen that packed bed can separate more than 90% of the oil. The RO membrane can separate more than 95% of TDS. This process may be used to recover up to 90% of water from produced water. Utilizing two membranes in series with lower surface area for the second membrane than the first membrane may give similar results to membranes in parallel at a lower capital cost.[53] Note that, for produced water, Benko puts the cost of material for ceramic membranes at $180/ft^2 and that of polymeric membranes at $ 40/ft^2. If productivity is the criteria, then the cost becomes $ 60 and $ 20 for ceramic and polymeric membranes, respectively. [60]

In a process, boron and solubilized hydrocarbons were separated from aqueous liquids such as produced water. In this study divalent ions were removed by adding a water softener to the water and raising the pH to about 9.5. The liquid was then passed through a composite polyamide reverse osmosis (RO) membrane. The efficiency of method was able to reduce boron concentration to less than 2 mg/L. [54]

A pilot plant including aeration tank, air floatation, sand filter and UF membrane was used for treating produced water for discharge or injection into an oil-well purposes. It was shown that oil and suspended solid contents were reduced to less than 0.5 mg/L and 1.0 mg/L. The concentration of Fe and some bacteria also met the required standard for injection and discharging purposes. [50]

Çakmakci et al. [52] utilized dissolved air floatation (DAF), acid cracking (AC), coagulation (CA) with lime and precipitation, cartridge filters (CDF), microfiltration (MF) and ultrafiltration (UF) as pre-treatment processes and nanofiltration (NF) and reverse osmosis (RO) were used to reduce the salt content. To obtain the best effluent quality and high permeate flux different combinations of treatment methods were tested. A combination was suggested to reduce COD to less than 250 mg/L.

In a pilot study, produced water was treated for industrial, irrigation and potable water use. The proposed unit included warm softening, coconut shell filtration, cooling (fin-fan), trickling filter, ions exchange and reverse osmosis. Silica level was reduced to 3 mg/L by adding 400 mg/L $MgCl_2$. Hardness, TDS, boron and ammonia were removed up to 95%, 95%, 90% and 80%, respectively. [55]

A configuration of the process was suggested including wash tanks, dissolved gas floatation, walnut shell filtration, warm lime softening, membrane bioreactor and reverse osmosis to reach potable and irrigation water standards. [56]

Doran et al. [57] used a combination of warm precipitate softening at pH 9.7, cooling, fixed-film biological organics oxidation, pressure filtration, ion-exchange softening and reverse osmosis to treat produce potable water. [57]

A pilot-scale hybrid reverse osmosis process was used to treat produced water for irrigation or discharge to surface waters. It was suggested that decreasing the conductivity may reduce the salt concentration. After treatment, the conductivity was reduced by 98% and TDS by 96% which was found to be an acceptable level for irrigation or discharge to surface waters. [51]

7. Commercial treatment processes

7.1. RO based processes

7.1.1. CDM Technology

CDM Smith produces a technology that is a combination of three major processes such as ion exchange process, reverse osmosis and evaporation. It is mostly used to treat high TDS coal bed methane (CBM) produced water. UV disinfection is also included to reduce the bacterial activity. The total cost of treatment per barrel of produced water was found to be $ 0.30. The inclusion of pre-treatment processes reduced the membrane fouling to a great extent, and water recovery ranged from 50% to 90%. [62, 29]

7.1.2. Veolia: OPUS™ – Optimized pre-treatment and separation technology

It is designed to treat sparingly soluble solutes (e.g., SiO_2, $CaSO_4$, and $Mg(OH)_2$), organics, and boron. The raw produced water is acidified and degasified. It is followed by Multiflo™ chemical softening, which is a series of coagulation, flocculation and sedimentation. Decant from sedimentation is fed into packed-bed media filtration column. The microorganisms present would be removed by IX resin. Water is then pressurized and treated by BWRO (Brackish Water Reverse Osmosis) membrane at high pH. The entire process system could fit on a cargo trailer. Produced water recovery is estimated to be greater than 90%. [62, 29]

7.1.3. Eco-sphere: Ozonix™

Ozonix™ is primarily used for the treatment of frac flow-back water, but it could also be used for produced water treatment. The feed water is mixed with supersaturated ozonized water in a reaction vessel. The hydroxyl radicals, formed from ozone, readily oxidize metals, and decompose soluble and insoluble organic compounds and microorganisms. The reaction vessel had two electrodes to induce precipitation of hard salts. Water is then treated with activated carbon cartridge filter and a RO membrane. Water recovery approaches 75%.[29]

7.1.4. GeoPure water technologies

The GeoPure desalination process is a combination of pre-treatment, ultrafiltration and reverse osmosis. This technology was developed for the treatment of oil and natural gas produced waters. Water recovery was reported to be 50%.[29]

7.2. Ion-Exchange (IX) based processes

7.2.1. EMIT: Higgins Loop

EMIT Higgins Loop technology is widely used for CBM produced water treatment. The Higgins Loop is a continuous counter current ion exchange contactor for liquid phase separations of ionic components. The IX resin adsorbs sodium ions in exchange of hydrogen ions. Hence the pH of water is reduced which eventually reduces bicarbonate levels. The resin saturated with sodium ions are regenerated by 4.11 M HCl. Product water recovery typically exceeds 99%.[29]

7.2.2. Drake: Continuous selective IX process

The Drake system is a three-phase, continuous fluidized bed system to remove monovalent cations. A strong acid cation exchange resin is used. Energy requirements are slightly less than that required for the EMIT Higgins Loop system. The maximum product water recovery is reported to be 97%.[29]

7.2.3. Eco-Tech: Recoflo® compressed-bed IX process

The Eco-Tech compressed bed systems are an extension of conventional packed bed IX processes. One system has two separate compressed-bed columns for anion and cation removal. Another system has three separate compressed-bed columns that contain a primary cation bed and anion bed followed by a polishing cation bed. Recoflo® systems are primarily used for recovering metals from effluent electrolytes. These are more mobile than conventional and Higgins Loop processes. A system has been installed in Powder River Basin to treat 1.5 Mgd of CBM produced water. [62, 29]

7.2.4. Catalyx Fluid Solutions/RGBL IX process

It was designed to minimize resin wastage during regeneration. The sodium and bicarbonate ions are removed by ion exchange chemical reaction.

$$Na^+ + HCO_3^- + R...H^+ \rightarrow R...Na^+ + H_2O + CO_2 \tag{2}$$

Waste minimization is done by the use of three tanks that are responsible for shuffling regenerating agent and rinse waters of various qualities during IX resin regeneration cycles.[29]

8. Conclusion

Produced water may be treated using different methods of operation. The criteria used to compare the technologies are in general, robustness, reliability, mobility, flexibility, modularity, cost, chemical and energy demand, and brine or residual disposal requirements. Many process and water quality specific factors should be taken into account when selecting a produced water treatment process. Temperature of the feed water may help determine which type of desalination treatment process should be employed since many technologies work more efficiently at high temperatures, while others use feed stream at low temperature. On the other hand if ion removal is necessary, it is important to consider the type of ions that need to be removed. Membrane processes most often remove divalent ions to a greater extent than monovalent ions, which may make the sodium adsorption ratio higher and render the water less suitable for beneficial use as irrigation water or surface discharge.[62] Membrane processes can treat produced water to meet many water quality requirements. Fouling is one of the major drawbacks of membranes which depends on permeate flux and its stability on time, but can be minimized by using hydrophobic membranes.

Modeling of permeate flux decline in crossflow filtration of oily wastewater is important from both the economical and technological points of view. Empirical models are the most accurate, but need experimentation and are not capable of making accurate predictions. Theoretical models are useful for the prediction of permeate flux under different operating conditions without the need for time-consuming experiments but there is no theoretical model that can accurately describe the crossflow filtration process. Therefore, it is required to work on the improvement of the theoretical models to make more accurate predictions and to better understand the fouling mechanisms.

Some pre-treatment methods should be utilized before membrane processes to increase the membrane life cycle. Different membrane units can be used as a polishing step. Each method has some advantages as well as some disadvantages and a combination of methods may be more useful for treatment of produced water to meet the different water quality requirements. Utilization of a certain type of process or combination of processes is highly dependent on the characteristics of the produced water.

Characteristics of produced water differ from well to well and the determination of the produced water characteristics is required. The design of a treatment process wholly based on the literature is not possible.Finally, the unpredictable and rapid onset of upsets in the produced water treatment process often results in unplanned maintenance and production losses.

Ceramic membranes may be a viable treatment technology for many produced water applications. Ceramic membranes will most likely be needed as a pre-treatment technology if desalination is required. While the capital cost of ceramic membrane is presently higher than polymeric, a ceramic membrane offers advantages over a polymeric membrane such as increased chemical, mechanical and thermal stability, and therefore higher lifespan and higher productivity.

International standards demand more efficient separation systems than those now in common use.More research and development is required in membrane development as, although many new products show promises, membrane filtration is still considered at the development stages. Effluent streams that once were treated as "waste" will then be considered a valuable "resource" but, unknown toxic effects and public acceptance are also important barriers for potable reuse of all wastewaters.

Author details

Rangarajan T. Duraisamy, Ali Heydari Beni and Amr Henni
*Industrial/Process Systems Engineering , Faculty of Engineering and Applied Sciences,
University of Regina, Regina, Saskatchewan, Canada*

Acknowledgement

The authors would like to acknowledge the financial help in a form of two grants (equipments and operations) from Western Economic Diversification Canada, Entreprise Saskatchewan (Saskatchewan Government) and the Petroleum Technology Research Centre (PTRC-Regina).

9. References

[1] Ahmadun F.-R., Pendashteha A., Abdullaha L. C., Biaka D. R. A., Madaenic S. S., Abidina Z. Z. (2009) Review of technologies for oil and gas produced water treatmen. *Journal of Hazardous Materials.* 170 (2-3): 530-551.

[2] Doyle D. and Brown A. (2000) Produced Water treatment and hydrocarbon removal with organoclay. in *SPE Annual Technical Conference and Exhibition*, Dallas, Texas.

[3] Carvalho M., Clarisse M., Lucas E. and Barbosa C. (2002) Evaluation of the polymeric materials (DVB copolymers) for produced water treatment. in *SPE International Petroleum Exhibition and Conference*, Abu Dhabi, 2002.

[4] Janks J. and Cadena F. (1992) Investigations into the use of modified zeolites for removing benzenes, toluene and xylene from saline produced water. in *Produced Water: Technological/Environmental Issues and Solutions,* New York, Plenum Publishing Corp.: 473–488.

[5] Plebon M., Saad M. and Fraser S. (2005) Further Advances in Produced Water De-oiling Utilizing a Technology that Removes and Recovers Dispersed Oil in Produced Water 2 micron and Larger. Available: http://www.ipec.utulsa.edu/.

[6] Adewumi M., Erb J. and Watson R.W. (1992) Design considerations for a cost effective treatment of stripper oi lwell produced water. In *Produced Water: Technological/Environmental Issues and Solutions,* New York, Plenum Publishing Corp.: 511–523.

[7] Knudsen B., Hjelsvold M., Frost T., Grini P., Willumsen C. and Torvik H. (2004) Meeting the zero discharge challenge for produced water. *Proceeding of the Seventh SPE International Conference on Health, Safety, and Environment in Oil and Gas Exploration and Production,* Calgary, 2004.

[8] Broek W. V. d. and Zande M. V. d. (1998) Comparison of plate separator, centrifuge and hydrocyclone. *SPE International Oil and Gas Conference and Exhibition*, Beijing, 1998.

[9] Becker R. B. (2000) Produced and process water recycling using two highly efficient systems to make distilled water. *SPE Annual Technical Conference and Exhibition*, Dallas , 2000.

[10] Thoma G., Bowen M. and Hollensworth D. (1999) Dissolved air precipitation solvent sublation for oilfield producedwater treatment. *Separation and Purification Technology.* 16: 101–107.

[11] Boysen J. and Boysen D. (2008) The Freeze-Thaw/Evaporation (FTE) process for produced water treatment, disposal and beneficial uses. Available: http://ipec.utulsa.edu/Conf2008/Manuscripts%20&%20presentations%20received/Boyse n_37_FreezeThaw.pdf.

[12] Garbutt C. (1997) Innovative treating processes allow steam flooding with poor quality oilfield water. *SPE Annual Technical Conference and Exhibition*, San Antonio, 1997.

[13] Zhou F., Zhao M., Ni W., Dang Y., Pu C. and Lu F. (2000) Inorganic polymeric flocculent FMA for purifying oilfield produced water: preparation and uses. *Oilfield Chemistry.* 17, 256–259.

[14] Renoua S., Givaudana J., Poulaina S., Dirassouyanb F. and Moulin P. (2008) Landfill leachate treatment: Review and opportunity. *Journal of Hazardous Materials.* 150(3): 468–493.

[15] Ma H. and Wang B. (2006) Electrochemical pilot-scale plant for oil field produced wastewater by M/C/Fe electrodes for injection.*Journal of Hazardous Materials*. B132: 237–243.

[16] Li G., An T., Nie X., Sheng G., Zeng X., Fu J., Lin Z. and Zeng E. (2007) Mutagenicity assessment of produced water during photoelectrocatalytic degradation.*Environmental Toxicology and Chemistry*. 26: 416–423.

[17] Yang Z. and Zhang N. (2005) Treatment of produced wastewater by flocculation settlement-Fenton oxidation–adsorption method.*Xi'an Shiyou University(Natural Science)*. 20: 50-53.

[18] Morrow L., Martir W., Aghazeynali H. and Wright D. (1999) Process of treating produced water with ozone". USA Patent 5,868,945.

[19] McFarlane J., Ridenour W., Luo H., Hunt R., Paoli D. D. and Ren R. (2005) Room temperature ionic liquids for separating organics from produced water.*Separation Science and Technology*. 40: 1245–1265.

[20] Deng S., Yu G., Jiang Z., Zhang R. and Ting Y. (2005) Destabilization of oil droplets in produced water from ASP flooding.*Colloids and Surfaces*. 252: 113-119.

[21] Palmer L., Beyer A. and Stock J. (1981) Biological oxidation of dissolved compounds in oilfield produced water by a field pilot biodisk.*J. Petrol. Technol*. 8308-PA: 1136–1140.

[22] Li Q., Kang C. and Zhang C. (2005) Waste water produced from an oilfield and continuous treatment with an oil-degrading bacterium.*Process Biochem*. 40: 873–877.

[23] Li L. and Lee R. (2009) Purification of Produced Water by Ceramic Membranes: Material Screening, Process Membranes. *Separation Science and Technology*. 44: 3455-3484.

[24] Bader M. (2007) Seawater versus producedwater in oil-fields water injection operations.*Desalination*. 208: 159–168.

[25] Mondal S. and Wickramasinghe S. R. (2008) Produced water treatment by nanofiltration and reverse osmosis membranes.*Journal of Membrane Science*. 322: 162-170.

[26] Ashaghi K. S., Ebrahimi M. and Czermak P. (2007) Ceramic Ultra- and Nanofiltration Membranes for Oilfield Produced Water Treatment: A Mini Review.*The Open Environmental Journal*. 1: 1-8.

[27] Judd S. and Jefferson B. (2003) Membranes for Industrial Wastewater Recovery and Re-use. Oxford: Elsevier Advanced Technology.

[28] Baker R. W. (2004) Membrane Technology and Applications, Second Edition. England: John Wiley & Sons, Ltd.

[29] Technical Assessmentof Produced Water Treatment Technologies. Colorado School of Mines. Available:
http://aqwatec.mines.edu/produced_water/treat/docs/Tech_Assessment_PW_Treatment_Tech.pdf. Accessed November 2009.

[30] Mulder M. (1996) Basic Principles of Membrane Technology, second edition. Dordrecht, The Netherlands: Kluwer Academic Publishers.

[31] Abadi S. R. H., Sebzari M. R., Hemati M., Rekabdar F. and Mohammadi T. (2011) Ceramic membrane performance in microfiltration of oily wastewater.*Desalination*. 265: 222-228.

[32] Abbasi M., Sebzari M. R., Salahi A. and Mirza B. (2012) Modeling of Membrane Fouling and Flux Decline in Microfiltration of Oily Wastewater Using Ceramic Membranes.*Chemical Engineering Communications*. 199: 78-93.

[33] Ebrahimi M., Willershausen D., Ashaghi K. S., Engel L., Placido L., Mund P., Boldaun P. and Czermak P. (2010) Investigations on the use of different ceramic membranes for efficient oil-field produced water treatment.*Desalination. 250:* 991-996.

[34] Chen A., Flynn J., Cook R. and Casaday A. (1991) Removal of oil, grease, and suspended solids from producedwater with ceramic crossflow microfiltration.*SPE Prod. Eng. 6:* 131–136.

[35] Li L., Liu N., McPherson B. and Lee R. (2008) Influence of counter ions on the reverse osmosis through MFI zeolite membranes: implications for producedwater desalination.*Desalination. 228:* 217-225.

[36] Liu N., Lu J., Li L. and Lee R. (2007) Factors determining the reverse osmosis performance of zeolite membranes on producedwater purification. in *Factors determining the reverse osmosis performance of zeolite membranes on producedwater purification,* Houston, Texas, USA, 2007.

[37] Valtchev V. (2005)Verified Syntheses of Zeolitic Materials, 2nd Revised Edition, Preparation of zeolite membranes," Laboratoire de Matériaux, Minéreaux. Available: http://www.iza-online.org/synthesis/VS_2ndEd/Membranes.htm. Accessed 2005 Nov 16.

[38] Cui J., Zhang X., Liu H., Liu S. and Yeung K. L. (2008) Preparation and application of zeolite/ceramic microfiltration membranes for treatment of oil contaminated water.*Journal of Membrane Science. 325:* 420-426.

[39] Li Y. S., Yan L., Xiang C. B. and Hong L. J. (2006) Treatment of oily wastewater by organic–inorganic composite tubular ultrafiltration (UF) membranes.*Desalination. 196:* 76-83.

[40] Xu P., Drewes J. E. and Heil D. (2008) Beneficial use of co-producedwater through membrane treatment: technical-economic assessment.*Desalination. 225:* 39–155.

[41] Li N. N., Fane A. G., Ho W. S. W. and Matsuura T. (2008) Advanced Membrane Technology and Applications, New Jersey: A John Wiley & Sons, Inc., Publication.

[42] Kose B., Ozgun H., Ersahin M. E., Dizge N., Koseoglu-Imer D., Atay B., Kaya R., Altınbas M., Sayılı S., Hoshan P., Atay D., Eren E., Kinaci C. and Koyuncu I. (2012) Performance evaluation of a submerged membrane bioreactor for the treatment of brackish oil and natural gas field produced water.*Desalination. 285:* 295-300.

[43] Singh D. and Sirkar K. K. (2012) Desalination of brine and producedwater by direct contact membrane distillation at high temperatures and pressures.*Journal of Membrane Science. 389:* 380-388.

[44] Maguire-Boyle S. J. and Barron A. R. (2011) A new functionalization strategy for oil/water separation membranes.*Journal of Membrane Science. 382:* 107-115.

[45] Chon K., Kim S. J., Moon J. and Cho J. (2012) Combined coagulation-disk filtration process as a pretreatment of ultrafiltration and reverse osmosis membrane for wastewater reclamation: An autopsy study of a pilot plant. *Water Research. 46:* 1803-1816.

[46] Lee J. and Frankiewicz T. (2005) Treatment of Produced Water With an Ultrafiltration (UF) Membrane - A Field Trial. in *SPE Annual Technical Conference and Exhibition,* Dallas, Texas, 2005.

[47] Mondal S. and Wickramasinghe S. (2012) Photo-induced graft polymerization of N – isopropyl acrylamide on thin film composite membrane: Produced water treatment and

antifouling   properties.   *Separation   and   Purification   Technology*.doi: 10.1016/j.seppur.2012.02.024.

[48] Pedenaud P., Heng S., Evans W. and Bigeonneau D. (2011) Ceramic Membrane and Core Pilot Results for Produced Water Management. in *Offshore Technology Conference*, Rio de Janeiro, Brazil, 2011.

[49] Wandera D., Wickramasinghe S. R. and Husson S. M. (2011) Modification and characterization of ultrafiltration membranes for treatment of produced water.*Journal of Membrane Science. 373:* 178-188.

[50] Qiao X., Zhang Z., Yu J. and Ye X. (2008) Performance characteristics of a hybrid membrane pilot-scale plant for oilfield produced wastewater.*Desalination. 225:* 113-122.

[51] Murray-Gulde C., Heatley J. E., Karanfil T., Rodgers Jr. J. H. and Myers J. E. (2003) Performance of a hybrid reverse osmosis-constructed wetland treatment system for brackish oil field producedwater.*Water Res. 37:* 705-713.

[52] Çakmakci M., Kayaalp N. and Koyuncu I. (2008) Desalination of producedwater from oil production fields by membrane processes.*Desalination. 222:* 176-186.

[53] Barrufet M., Burnett D. and Mareth B. (2005) Modeling and operation of oil removal and desalting oilfield brines with modular units. in *SPE Annual Technical Conference and Exhibition*, Dallas, Texas, USA, 2005.

[54] Tao F. T., Pilger P. F. and Dyke C. A. (1993) Reducing aqueous boron concentrations with reverse osmosis membranes operating at a high pH. US Patent 5,250,185, 5 October 1993.

[55] Funston R., Ganesh R. and Leong L. Y. (2002) Evaluation of Technical and Economic Feasibility of Treating Oilfield ProducedWater to Create a "New" Water Resource. Available: http://www.kennedyjenks.com/News_Pubs/GWPC2002_Roger_Funston.pdf.

[56] Tsang P. B. and Martin C. J. (2004) Economic evaluation of treating oilfield producedwater for potable use. in *Thermal Operations and Heavy Oil Symposium*, Bakersfield, USA, 2004.

[57] Doran G. F., Carini F. H., Fruth D. A., Drago J. A. and Leong L. Y. (1997) Evaluation of technologies to treat oil field producedwater to drinking water or reuse quality. in *SPE Annual Technical Conference and Exhibition*, San Antonio, Texas, USA, 1997.

[58] Liangxiong L., Whitworth T. and Lee R. (2003) Separation of inorganic solutes from oil-field producedwater using a compacted bentonite membrane.Journal of Membrane Science. 217: 215-225.

[59] Kojima S., Okada S., Sasaki T., Oba T., Fujise A. and Kusuyama K. (2011)Reliability Study for Mebrane-Processed water for Injection (WFI).*PDA Journal of GMP and Validation in Japan.*13: 47-55.

[60] Dallbauman L. and Sirivedhin T. (2005) Reclamation of producedwater for beneficial use.Separation Science and Technology.40: 185–200.

[61] Benko, L. K. (2009) Ceramic membranes for produced water treatment. World oil J. 230 (4). Access: http://www.worldoil.com/April-2009-Ceramic-membranes-for-produced-water-treatment.html

[62] Guerra K., Dahm, K. Dundorf, S. (2011) Oil and Gas Produced Water Management and Beneficial Use in the Western United States. U.S. Department of the Interior, Bureau of Reclamation.

Treatment of Organic-Contaminated Water

Chapter 11

Decontamination of Wastewaters Containing Organics by Electrochemical Methods

Florica Manea and Aniela Pop

Additional information is available at the end of the chapter

1. Introduction

The ultimate goal of the sustainable wastewater management is the protection of the environment in a manner commensurate with public health and socio-economic concerns [1]. Industrial effluents containing toxic and refractory organic pollutants cause severe environmental problems and their removal or degradation is required. Industries like petrochemical, chemical, pulp and paper, dyeing and pharmaceutical generate a variety of toxic organic compounds, and among them, phenol derivates have received an increased attention in the last years [2-4]. From phenol derivates, the most toxic compounds are the chlorinated and nitro-substituted phenols, which are used as pesticides and bactericides [3]. In most countries, the maximum allowable concentration of phenol derivates in the effluent streams is required to be less than 1 ppm [5], because these derivates cause several serious health problems: *e.g.,* liver and kidney damage and blood pressure drop, cardiac toxicity including weak pulse, cardiac depression and reduced blood pressure [6]. Hence, phenol derivates must be treated to satisfy the stringent water quality regulations and the demand for recycling of water in the process [2]. The objective of Integrated Pollution Prevention and Control (IPPC) Directive [7] regarding the industrial effluent reuse as raw water into production process represents a challenge for the wastewater treatment technologies. Usually, for the treatment of the industrial effluents containing refractory organics, the conventional wastewater treatment for organics removal or destruction are ineffective to reach the water quality appropriate for its reusing, and new alternative methods are required.

During the last years, the electrochemical processes have been developed as alternative options for the remediation of wastewaters containing organic pollutants, due to the main advantages of the electrochemical methods, *e.g.,* environmental compatibility, versatility, high energy efficiency, amenability of automation and safety, and cost effectiveness. The electrochemical technology currently offers promising approaches for the prevention of pollution problems from industrial effluents, acting in principal as „end-of-pipe" technology [8-11].

Electrochemical processes should be regarded as viable alternatives for enhanced conventional wastewater treatment technology. Based on the wastewater type, composition and their further usage, the various electrochemical processes could be applied individually or/and integrated into a variety of technological flows involving physical, chemical and biological methods. Thus, the electrocoagulation process could replace the conventional coagulation method integrated into a technological flow containing or not the electrooxidation process as advanced water treatment method. Also, the electrooxidation process could be applied as advanced oxidation process suitable especial for degradation/mineralization of the recalcitrant organic pollutants. However, there are many important aspects for the optimization of an electrochemical process, starting with the electrochemical process type, continuing with the electrode materials and not the least, the operating variables, like the current density and pH.

2. Electrochemical process for organics separation

Electrocogulation represents an important technique for investigation and application of the electrochemical treatment of wastewaters containing organics as alternative for the coagulation process [8,12,13]. In general, during the electroflotocoagulation process the coagulant is generated in-situ by "sarcrificial anodes", the most usual being aluminium or iron electrodes. The electrolytic dissolution of aluminium anode lead to the generation of Al^{3+} and $Al(OH)^{2+}$ at low pH, which are converted into $Al(OH)_3$ at appropriate pH values and finally polymerized to $Al_n(OH)_{3n}$, which acts the same as in the coagulation process. Also, $Fe(OH)_n$, n=2,3 is produced by electrolytic dissolution of iron anodes are dissolved by electrolysis, which can remove the pollutants from wastewaters. The metal hydroxides act as coagulant species, which destabilise and aggregate the suspended particles and precipitates and adsorb dissolved and suspended pollutants (e.g., suspended solids, dissolved organic matter). The metal anode dissolution is accompanied by hydrogen gas evolution at cathodes, the bubbles capturing and floating the suspended solids formed and thus removing pollutants [14,15]. The presence of organic and suspended matter activates or inhibits these main electrode processes partially by adsorption on electrode and involvement in surface film formation. Also, the presence of Cl^-, SO_4^{2-} anions influences the aluminium dissolution, with further involvement in the electrocoagulation process [12]. Good process efficiencies have been reported for electrocoagulation applying on wastewaters [12,14,16,17], and also, greater efficiency for the removal of chemical oxygen demand (COD) parameter and suspended solids in comparison with conventional coagulation process [15]. The main advantages of electrocoagulation versus coagulation are simple equipment and easy operation, no chemical added and a smaller amount of precipitate and sludge [8].

3. Electrochemical oxidation of organics

Electrochemical oxidation represents a very promising way for the treatment of wastewaters containing organic pollutants. Taking into consideration the conventional wastewater treatment flow, the electrooxidation methods exhibit great potential either for replacement of conventional process with electrooxidation one or its integration within the technological flow [18]. Two approaches have been proposed by Comninellis [19] of the electrooxidation process in wastewater treatment:

- so-called electrochemical conversion method, in which the bio-recalcitrant organics are transformed into biodegradable compounds;
- the electrochemical combustion or incineration method, considered as advanced electrooxidation process that allows a complete degradation or mineralization of organic pollutants.

The electrochemical conversion process should be integrated prior to the biological stage of the conventional wastewater treatment to enhance the biodegradability degree and as consequence, the biological treatment performance.

The advanced electrooxidation process is based on the electrochemical generation of hydroxil radicals (•OH), which is a very high powerful oxidant able to mineralize the organic pollutants from water [20].

Two different oxidation mechanisms for the electrochemical oxidation of pollutants have been proposed: direct anodic oxidation, where the pollutants are destroyed at the anode surface, and indirect oxidation where the oxidation process occurred via a mediator that is electrochemically generated. In general, during the electrooxidation of wastewaters, both oxidation mechanisms may coexist with different proportions in direct relation with the electrode material and operation conditions [21].

The key of the electrochemical process performance is given by the electrode material, which should to exhibit several properties, $e.g.$, high physical and chemical stability, resistance to corrosion and formation of passivation layers, high electrical conductivity, catalytic activity and selectivity, low cost/life ratio. Also, a very important aspect for the electrode material assessment is the potential value at which the side reaction of oxygen evolution occurs. High O_2 overvoltage electrodes with large potential window are suitable for an effective electrooxidation of organics from wastewater. Carbon-based electrodes represent an attractive electrode material for electrooxidation with respect to the above-discussed features. A large variety of carbon electrodes suitable for the degradation of organics from aqueous solution have been reported, $e.g.$, glassy carbon [22-24], carbon pastes [25], and boron-doped diamond electrode [26-29].

The performance of new carbon based electrode materials for organics electrooxidation are presented and discussed in this chapter, pointing out: their preparation, the electrochemical characterization of new carbon based composite electrode materials comparative with the new commercial boron doped diamond electrode (BDD) and the conventional glassy carbon (GC) electrode, the assessment of electrochemical activities of each material for the target pollutants, *i.e.,* 4-chlorophenol, 4-nitrophenol, 4-aminophenol, 2,4-dichlorophenol and pentachlorophenol by the chronoamperometry and multiple-pulsed amperometry technique application. Also, the results of 4-aminophenol degradation by photocatalytically-assisted electrochemical process using BDD electrodes are presented in this chapter. In addition, the dual role of the electrode material and electrochemical method for both organics degradation and detection is briefly presented.

4. Experimental procedures

4.1. Preparation of carbon composite materials

Carbon-based composite electrodes presented in this chapter, *i.e.,* expanded graphite-epoxy (EG-Epoxy), carbon-nanofiber expanded graphite-epoxy (CNF-EG-Epoxy), expanded graphite–silver-zeolite-epoxy (AgZEG) and expanded graphite-polystyrene (EG-PS) composite electrodes were obtained by film casting and respective, hot press methods.

4.1.1. Epoxy matrix

The composite electrodes, *i.e.,* expanded graphite-epoxy (EG-Epoxy), carbon-nanofiber expanded graphite-epoxy (CNF-EG-Epoxy), expanded graphite–silver-zeolite-epoxy (AgZEG), made using epoxy matrix were obtained from two-component epoxy resin (LY5052, Araldite / 5020 Aradur) mixed with conductive expanded graphite (EG) fillers powder and silver modified zeolite (clinoptilolite) [30-32]. The ratio was chosen to reach 20 weight percent (w/w) for each component, *i.e.,* expanded graphite (EG), carbon nanofibers (CNF) (PS-447 BOX), silver- zeolite ion-exchanged function of the electrode composition. The mixing was performed in a two roll-mill at room temperature. For the EG-Epoxy and CNF-EG-Epoxy electrodes, the epoxy was cured in a hot press at 80 º C for 40 minutes. Simultaneously, the material was shaped in a plate of 1 mm thickness. The plate was slowly cooled down (for about 12 h) to the room temperature without removing the applied pressure. The electrodes were made cutting plates of 9 mm^2 and put on a glass supports and electrical contacts were made using a silver paint. For AgZEG electrode, the resulting pastes from the two roll-mill procedure were embedded in polyethylene and electrical contacts were made using copper wire, and discs with a surface area of 19.63 mm^2 were obtained. Prior to use, the working electrode was gradually cleaned, first polished with abrasive paper and then on a felt-polishing pad by using 0.3 μm alumina powder (Metrohm, Switzerland) in distilled water for 5 minutes and rinsing with distilled water.

4.1.2. Polystyrene matrix

The expanded graphite-polystyrene (EG-PS) composite electrode was prepared from low viscosity polystyrene pellets (PS N2000, Crystal from Shell) mixed with conductive expanded graphite (EG) filler powder (Conductograph, SGL Carbon). EG was added to a toluene (Fluka) solution of PS N2000 with constant stirring. Ultrasound was used for about 2h to mix well the components. The obtained viscous mixture was cast into a film using a

Teflon mold and the toluene was left to evaporate at room temperature for 48 h, and then in a vacuum oven at 50 º C for 24 h. The final thickness of the obtained film was about 1 mm [33]. Plates with a surface area of 9 mm² were cut from the composition and put on a glass supports and electrical contacts were made using a silver paint. The electrodes were isolated on the sides by epoxy resin.

4.2. Structural and electrical characterization

The carbon-based composite electrodes were characterized morphologically by scanning electron microscopy and a homogeneous distribution of carbon filler within the insulating matrix was found [30-32]. The electrical resistance of each electrode was measured by a four-point resistance method [34] and the results are gathered in Table 1.

Electrode	Electrical resistance (Ωcm^{-1})
EG-Epoxy	5.3
CNF-EG-Epoxy	11.6
EG-PS	3.2

Table 1. Electrical resistance of the carbon composite electrodes

4.3. Electrochemical characterization of the carbon electrodes

The carbon composite electrodes have been studied in comparison with commercial glassy carbon (GC) and boron-doped diamond (BDD) electrodes provided by Metrohm, Switzerland, and respective, Windsor Scientific Ltd., UK. The electrochemical characterization of the electrodes was carried out by cyclic voltammetry (CV). All measurements were carried out using an Autolab potentiostat/galvanostat PGSTAT 302 (Eco Chemie, The Netherlands) controlled with GPES 4.9 software and a three-electrode cell, with a saturated calomel electrode as reference electrode, a platinum counter electrode and carbon based working electrode.

4.4. Bulk electrolysis and photocatalytically-assisted electrolysis experiments

The electrochemical degradation experiments were carried out by batch process using an undivided cell of 1 dm³ volume, for different organics concentrations in 0.1 M Na₂SO₄ supporting electrolyte. The BDD/Nb electrodes (100 mm x 50 mm x 1 mm) with 280 cm³ geometric area provided by CONDIAS, Germany were used as anodes, and stainless steel plates (100 mm x 50 mm x 1 mm) were employed as cathodes under vertical arrangement. A regulated DC power supply (HY3003, MASTECH) was used under galvanostatic regime at 5 and 10 mA cm⁻² current densities.

The photocatalytically-assisted electrolysis experiments were performed under the similar working conditions as the electrolysis experiments in the presence of 1g dm⁻³ TiO₂-supported zeolite catalyst [35,36]. The suspension was illuminated with a 6 W UV lamp emitting 254-365 nm wavelengths. The intensity of UV lamp was about 2100 μWcm^{-2}, the illumination surface area of 32.15 cm^2, and the distance between UV lamp and cell surface was 5 cm.

After each 30 minutes, samples were drawn from the cell and organics degradation was monitored by UV-vis spectroscopy. Also, total organic carbon (TOC) parameter was used to check organics mineralization. UV-vis spectrometric measurements were performed on a Varian 100 Carry using 1 cm quartz cell, and TOC content of the samples was analyzed at a Shimadzu TOC analyzer. Measurements of pH were done by an Inolab WTW pH meter.

4.4.1. Analytical procedures

The degradation process efficiency (η, %) and electrochemical degradation efficiency (E, mg/C· cm^2) for organics degradation was determined based on equations (1) and (2) [37,38]:

$$\eta(\%) = \frac{\left(C_{org,i} - C_{org,f}\right)}{C_{org,i}} x100 \tag{1}$$

$$E = \frac{\left(C_{org,i} - C_{org,f}\right)}{C*S} \times V (mg\ /\ C\,cm^2) \tag{2}$$

where $C_{org,i}$-$C_{org,f}$ is the change in the organic concentration determined by spectrophotometry during experiments for a charge consumption of C corresponding to various electrolysis time, V is the sample volume (700 cm^3) and S is the area of the electrode surface (cm^2).

The electrochemical efficiency for organics mineralization was determined based on equation (2) modified as (3) taking into consideration the change in TOC measurements during experiments, determining (TOC_0 – TOC)

$$E_{TOC} = \frac{\left(TOC_0 - TOC\right)}{C*S} \times V (mg\ /\ C\ cm^2) \tag{3}$$

The mineralization current efficiency (MCE) for each electrolyzed solution was calculated based on equation (4) [39]:

$$MCE = \frac{nFV_s\Delta(TOC)_{exp}}{4.32 \times 10^7\ mIt} \times 100\,(\%) \tag{4}$$

where n is the number of electrons consumed in the mineralization process of 4-AP, F is the Faraday constant (=96 487 C mol^{-1}), V_s is the solution volume (dm^3), $\Delta(TOC)_{exp}$ is the experimental TOC decay (mg dm^{-3}), 4.32 x 10^7 is a conversion factor for units homogenization (=3 600 s h^{-1} x 12 000 mg of carbon mol^{-1}), m is the number of carbon atoms in organic, I is the applied current (A), and t is time (h). The number of electrons consumed is determined based on the overall mineralization reaction of organics to CO_2

The specific energy consumption, W_{sp}, was calculated with the relation (5):

$$W_{sp} = C_{sp} \times U \quad \left(kWh\ dm^{-3} \right) \tag{5}$$

where C_{sp} represents the specific charge consumption of C corresponding to 1 dm³ and U is the cell voltage (V).

5. Results of organics degradation by electrochemical and photocatalytically-assisted electrochemical oxidation

5.1. Electrochemical behaviours of the carbon based electrodes in the presence of organics

5.1.1. Cyclic voltammetry studies

The electrochemical behaviours of the carbon based electrodes in 0.1 M Na2SO4 supporting electrolyte were studied by cyclic voltammetry (CV) to explore the organics oxidation process on the electrodes in order to clarify the relationship between experimental variables and the electrode response, and also, to determine the potential window and the oxygen evolution reaction potential value. As can be seen in Figure 1, the AgZEG electrode exhibited the oxidative/reductive peak corresponding to redox peaks of the Ag/Ag(I) couple. For this electrode, the background current is similar with CNF-EG-Epoxy, and higher than for the EG-Epoxy electrode, also a common aspect for the electrocatalytic electrode type. In the presence of organic pollutant, 4-chlorophenol, chosen as an example of monosubstituted phenol derivates, an anodic oxidation peak appeared at the potential values presented in Table 2, which are prior to the oxygen evolution potential. No evidenced oxidation peak occurred at GC electrode even if a slight current increase was found in the presence of 4-chlorophenol, which informed that there is no limiting current characteristics to the transport control. Except BDD electrode that can be classified as high O2 overvoltage electrode (Eo2 about 1.5 V vs. Ag/AgCl), the others shlould be regarded as low O2 overvoltage electrodes (Eo2 about 1 V vs. Ag/AgCl) characterized by a high electrochemical activity towards oxygen evolution and low chemical reactivity towards oxidation of organics. Effective oxidation of pollutants at these electrodes may occur at low current densities due to at high current densities, significant decrease of the current efficiency is expected due to the oxygen evolution. In contrast, at high O2 overvoltage anodes, higher current densities may be applied with minimal contribution from the oxygen evolution side reaction.

Except 2, 4-dinitrophenol, which exhibited the oxidation potential of about 1.2 V vs. Ag/AgCl, similar results regarding the oxidation peak appearance at the potential value before oxygen evolution were found for other phenolic derivates, i.e., 4-aminophenol, 4-nitrophenol, 2,4-dinitrophenol and pentachlorophenol (the results are not shown here). However, for mono- and di-nitro substituted phenol derivates, the cathodic current increased at the potential value of about -0.4 V Ag/AgCl, which was ascribed to the reduction of –NO2 [40,41]. In addition, in the cathodic potential range a polymer of aminophenol formed cathodically could occur. The polymer resulted from 2,4-dinitrophenol is less stable than the polymer resulting from mono-substituted nitrophenol, and can be further oxidized and mineralized [42].

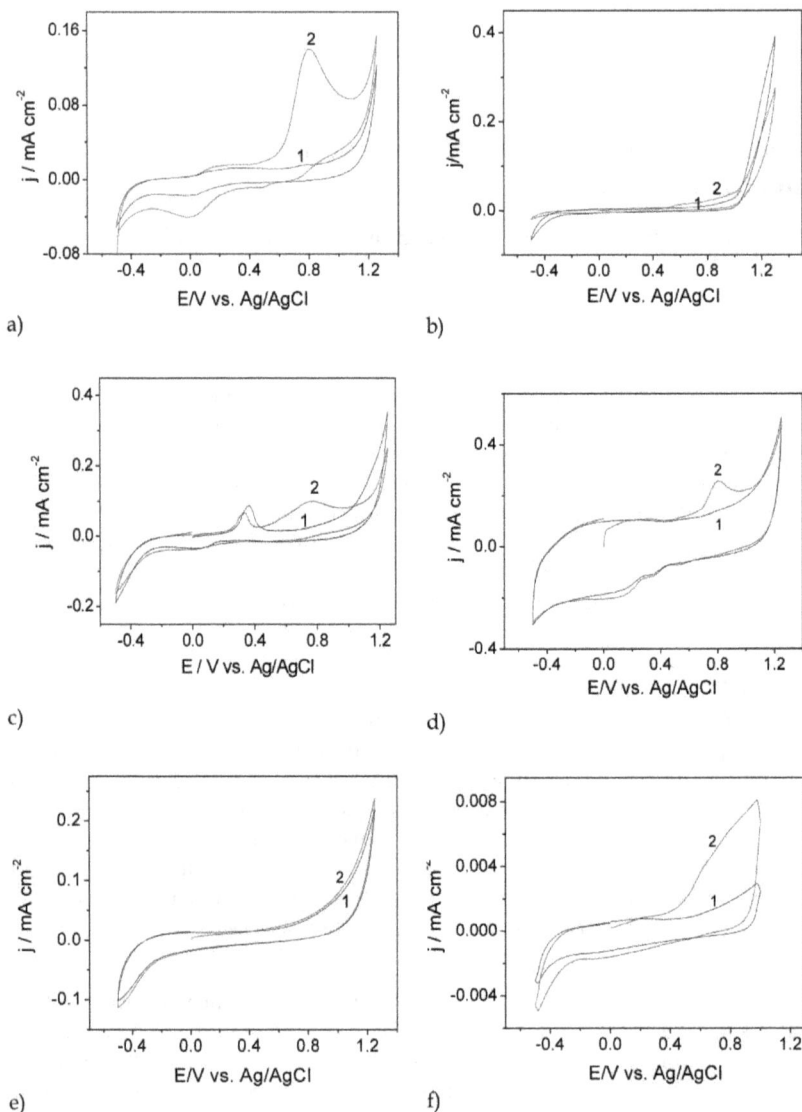

Figure 1. Cyclic voltammograms recorded with a potential scan rate 0.05 Vs^{-1} between 1.25 and -0.5V vs. SCE in a 0.1M Na$_2$SO$_4$ supporting electrolyte (curve 1) and in the presence of 25 mg·L^{-1} 4-CP (curve 2) at a) EG-Epoxy, b) EG-CNF-Epoxy, c) AgZEG. d) EG-PS; e) GC; f) BDD electrode.

Electrode material	Scan number	Δj (mA·cm^{-2})	Anodic peak current densities reduction degree (%)	E (V)
EG-Epoxy	1	0.12323	-	0.8
	2	0.07492	39.2	0.8
	3	0.05743	53.4	0.8
CNF-EG-Epoxy	1	0.01184	-	0.8
	2	0	100	0.8
	3	0	100	0.8
AgZEG	1	0.07147	-	0.8
	2	0.04467	37.5	0.8
		0.03495	51.1	0.8
EG-PS	1	0.24299	-	0.8
	2	0.15065	38.00	0.8
	3	0.12344	49.20	0.8
BDD	1	$2.720 \cdot 10^{-3}$	-	0.6
	2	$0.955 \cdot 10^{-3}$	64.88	0.6
	3	$0.893 \cdot 10^{-3}$	67.17	0.6

Table 2. Voltammetric parameters for 25 mgL^{-1} 4-CP oxidation on carbon based electrodes.

It can be noticed that starting with second CV the peak current decreased by quite 100 % for CNF-EG-Epoxy composite electrode, which is due to the electrode fouling. For other electrodes the peak current decreases by the similar order starting with the second CV. The effect of the scan rate on the anodic peak current of the first CV has been investigated for each carbon-based electrode in 0.1 M Na$_2$SO$_4$ supporting electrolyte and in the presence of 25 mg/L 4-CP (the results are not shown here). The linear proportionality of the anodic peak current with the square root of the scan rate obtained for all tested electrodes indicated process control by mass transport. No intercept of 0 suggested that the adsorption steps and specific surface reaction cannot be neglected. The most strong adsorption effects were noticed for EG-CNF-Epoxy electrode, these results being in concordance with those obtaining by three consecutive scanning using CV. Even the oxidation potential value is almost constant the 4-CP oxidation is not reversible because no corresponding reduction cathodic peak appeared by backward scanning. In general, the electrode fouling owes the complex mechanism of phenols oxidation on carbon based electrode, which involves both the adsorption of reactant/intermediate or final oxidation products and the formation of passive, nonconductive layer of oligomer products of oxidation process on its surface [43-46]. Silver ions introduction within the electrode composition in order to act as catalyst is not justified based on CV results regarding the electrode activity towards 4-CP at various concentration. AgZEG electrode exhibited a slight lower electrochemical activity in comparison with EG-Epoxy electrode (see Figure 2).

For all tested phenolic derivates the similar behaviours of the carbon-based electrodes regarding the presence of the limiting current density (i$_{lim}$) corresponding to the phenolic derivates oxidation peak were found by cyclic voltammetry. CV results must be considered as reference for electrooxidation applying under galvanostatic regime (constant current). Thus, depending on the applied current density (i_{appl}), two different operating regimes have been identified [43,47] *i.e.*, $i_{appl} < i_{lim}$ the electrolysis is under current control with the current efficiency of 100% and $i_{appl} > i_{lim}$ the electrolysis is under mass-transport control and the secondary reactions of oxygen evolution are involved with a decreasing of current efficiency. However, the choice of i_{appl} must consider the process or electrochemical efficiency, which is limited by the electrode fouling and also, the electrical charge consumption.

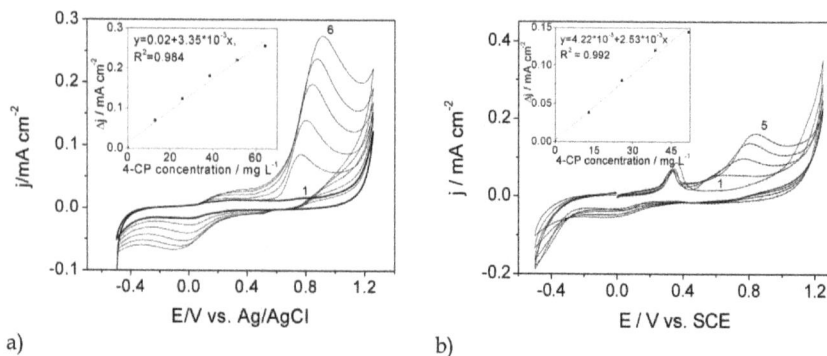

Figure 2. Cyclic voltammograms of a) EG-Epoxy and b) AgZEG in the presence of different 4-CP concentrations: 1-0 mM; 2-12.5 mgL^{-1}; 3-25 mgL^{-1}; 4-37.5 mgL^{-1}; 5-50 mgL^{-1}; 6-62.5 mgL^{-1} in a) 0.1 M Na$_2$SO$_4$ supporting electrolyte; potential scan rate 0.05 Vs^{-1}. Inset: the calibration plots of the current densities vs. 4-CP concentration.

5.1.2. Effect of UV irradiation on voltammetric behaviour of BDD electrode [35]

Figure 3 shows the cyclic voltammograms of BDD electrode in the absence and in the presence of 4-aminophenol without and under UV irradiation. In the absence of 4-AP no difference between the cyclic voltammograms was noticed, thus UV irradiation does not influence the BDD electrode behaviour. In the presence of 4-AP without UV irradiation, the potential where the direct electrochemical oxidation of 4-AP is established. Under UV irradiation, the photoresponse of anodic current is higher and more two oxidation peaks at about +0.5 V and +0.9 V vs. Ag/AgCl appear. This aspect could be attributed to changing the mechanism of the electrochemical oxidation of 4-AP, involving the further oxidation occurring, which was similar with the above-mentioned results found at the high potential scan rate. Also, the higher current under UV irradiation could be attributed to increase in the degradation rate of 4-AP.

Based on the above-presented results and taking into account the open circuit potential behavior evolution of BDD electrode under UV irradiation and the presence of TiO$_2$-supported catalyst [35], it may be postulated that the 4-AP oxidation exhibited a complex mechanism. 4-AP could be oxidised directly on the anode and indirectly by hydrogen peroxide and hydroxyl radicals. When Z-TiO₂ particles are illuminated at a wavelength shorter than 400 nm, the generation of electron-hole pairs takes place. The photoelectrons act as reducing agents and peroxide radical ($O_2^{\bullet-}$) is generated, which may become a source of other oxidising species, such as hydrogen peroxide and hydroxyl radicals. The photogenerated holes have a high positive oxidation potential to oxidise organic compounds and water to form hydroxyl radicals (HO•), which are the main oxidising species in the photocatalysis. In an electric field, the major electrochemical reactions occur as follows: anode-direct oxidation, under water stability potential region, equation (6), under water decomposition potential region equation (7), cathode-under water decomposition potential region, equation (8).

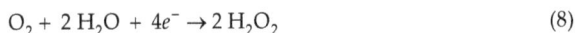

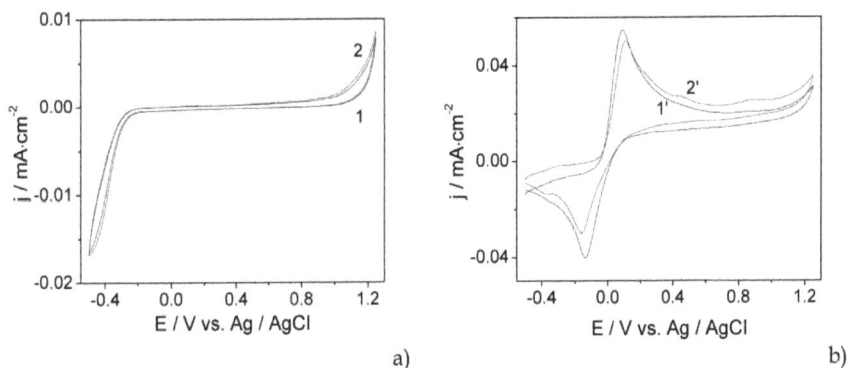

$$4 - AP - n\ e^- \rightarrow \text{Products} \tag{6}$$

$$2\,H_2O - 2\,e^- \rightarrow 2\,OH^{\bullet} + 2\,H^+ \tag{7}$$

$$O_2 + 2\,H_2O + 4e^- \rightarrow 2\,H_2O_2 \tag{8}$$

Figure 3. Cyclic voltammograms recorded at BDD electrode in: (a) 1M Na₂SO₄ supporting electrolyte under dark (1) and UV irradiation (2); and (b) in the presence of 50 mgL⁻¹ 4-AP under dark (1') and UV irradiation (2') at the potential scan rate of 0.1 Vs⁻¹.

5.1.3. Chronoamperometric and multiple-pulsed amperometry results

To establish the best operating conditions, chronoamperometry measurements were performed to determine the electrode performance for electrochemical and photocatalytically-assisted electrochemical degradation of organics. The chronoamperograms recorded for each electrode in the presence of 25 mgL⁻¹ 4-CP during 2 h of electrolysis are presented in Figure 4.

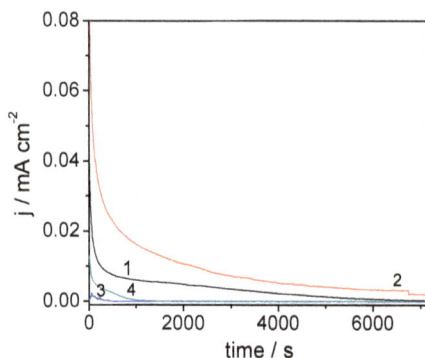

Figure 4. Chronoamperograms recorded at +1.1 V vs. Ag/AgCl for: 1-EG-Epoxy, 2-CNF-EG-Epoxy, 3-EG-PS, 4-BDD in the presence of 25 mgL^{-1} 4-CP, with the background substracted.

Chronoamperometric results have been used to determine the serie of electroactivities for 4-CP oxidation at the tested electrodes. For a more electroactive material having a diffusion coefficient D, the current corresponding to electrochemical reaction is described by the Cottrell equation [48]:

$$I(t) = nFAC^* (D/_{\pi t})^{1/2} \tag{9}$$

where D and C* represent the diffusion coefficient (cm^2·s^{-1}), and the concentration of the solution (mol·cm^{-3}), respectively. The dependency between I and t$^{-1/2}$ is linear, and the diffusion coefficient can be determined from the slope. Accoding to Cottrell equation, the slope is given by the following term: $nFAC^*(D/\pi)^{1/2}$, and the apparent diffusion constant can be calculated (see Table 3):

Electrode	Apparent diffusion constant, D' (s^{-1})
EG-Epoxy	21.20·10^{-6}
CNF-EG-Epoxy	58.97·10^{-6}
PS-EG	4.269·10^{-9}
BDD	1.378·10^{-6}
GC	194.61 10^{-6}

Table 3. Apparent diffusion constant of 4-CP

From the above presented results, the electroactivity serie for 4-CP oxidation at the tested electrodes was established: S$_{EG-PS}$>S$_{BDD}$>S$_{EG-Epoxy}$>S$_{CNF-EG-Epoxy}$>S$_{GC}$. Also, this aspect was proved by process efficiency and electrochemical efficiency determination based on the 4-CP concentration monitoring using UV-VIS spectrophotometry measurements (see Table 4).

Electrode	Process efficiency (η_{CF}) (%)	Electrochemical efficiency (E_{CF}) ($g \cdot C^{-1} \cdot cm^{-2}$)
EG-Epoxy	17.26	$4.74 \cdot 10^{-4}$
CNF-EG-Epoxy	20.14	$2.73 \cdot 10^{-4}$
EG-PS	23.09	$3.20 \cdot 10^{-2}$
GC	4.47	$0.12 \cdot 10^{-4}$
BDD	9.14	6.73

Table 4. Process efficiency (η_{CF}) and electrochemical efficiency (E_{CF}) for 4-CP degradation by chronoamperometry

In general, the dependence of the electrochemical reaction rates on the concentration of organic pollutants has been described by a pseudo first-order equation based on the relation $\ln (C_0/C_t)=k_{app}(t)$. Also, the apparent rate constant was calculated based on the relation of $\ln(C_0/C_t)=k'_{app}(C)$, in which C represents the electrical charge passed during electrochemical oxidation of 4-CP. The kinetic results were obtained by monitoring changes in 4-CP concentration as a function of the reaction time. Based on the results gathered in Table 4 can be concluded that 4-CP oxidation process occurred with a fast kinetics on carbon composite electrode. However, in terms of charge consumption, the BDD electrode exhibited the best apparent rate constant, an expected result takes into account that no oxygen evolution occurred at the potential value applied on this electrode and lower electrical charge was involved.

Electrode	Rate constant, k	
	K'_{app}, C^{-1}	K_{app}, s^{-1}
EG-Epoxy	2.041	$2.638 \cdot 10^{-5}$
CNF-EG-Epoxy	1.2056	$3.125 \cdot 10^{-5}$
PS-EG	1139.47	$3.646 \cdot 10^{-5}$
GC	0.0538	$6.39 \cdot 10^{-6}$
BDD	1415.3	$1.33 \cdot 10^{-5}$

Table 5. Electrochemical reaction rate constant of 4-CP at the carbon-based electrodes

Even if the carbon composite electrode exhibited better performances expressed in terms of the process efficiency, BDD electrode exhibited superiority and a real potential for electrochemical degradation based on its electrochemical efficiency in relation with lower energy consumption. An example of optimization of operating conditions using chronoamperometry (CA) for the electrochemical oxidation of 4-aminophenol (4-AP) using boron-doped diamond (BDD) electrode is presented in Figure 5. Various oxidizing potential levels, *i.e.*, 0.5, 0.9, 1.25 and 1.75 V vs. Ag/AgCl were applied for optimization. It can be noticed that higher oxidation potential value better degradation process efficiency but with higher energy consumption that decrease the electrochemical efficiency and limits for the practical applications.

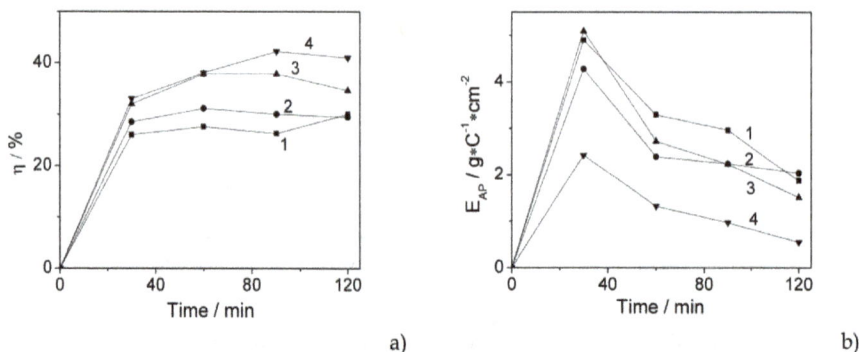

Figure 5. 4-AP removal efficiency (a) and electrochemical efficiency (b) by electrooxidation process using BDD electrode in 1 M Na₂SO₄ and in the 64 mg·L⁻¹ 4-AP; the applied techniques are: curve 1 – E=0.5 V vs Ag/AgCl, curve 2 – E=0.9 V vs Ag/AgCl, curve 3 – E=1.25 V vs Ag/AgCl, curve 4 – E=1.75 V vs Ag/AgCl.

Multiple-pulsed amperometry (MPA) is a technique appropriate for in-situ electrochemical cleaning of an electrode during the oxidation process. The effect of MPA applying is presented here in comparison with optimized CA technique for the electrochemical oxidation of 4-AP using boron-doped diamond (BDD) electrode. The operating conditions for MPA, a cleaning potential of +1.75 V and an oxidation potential value of +1.25 V vs. Ag/AgCl were selected, taking into account that the oxidation potential is similar to that of CA and that the cleaning potential should be higher but very close the oxygen evolution potential.

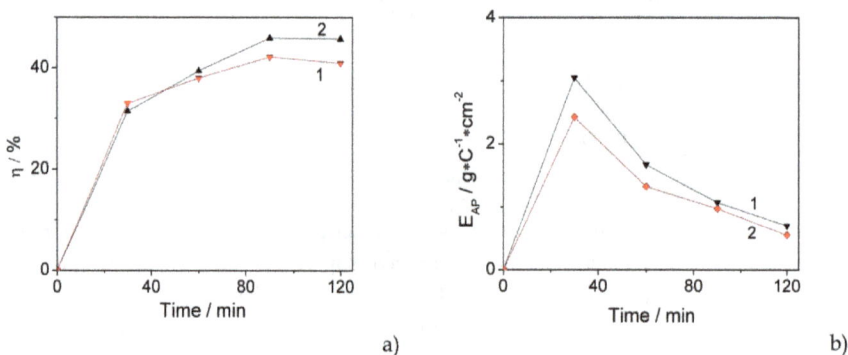

Figure 6. 4-AP removal efficiency (a) and electrochemical efficiency (b) by electrooxidation process using BDD electrode in 1 M Na₂SO₄ and in the 64 mg·L⁻¹ 4-AP; the applied techniques are: curve 1 – CA, curve 2 – MPA.

Even if MPA application as an electrochemical process provided better results in terms of the 4-AP degradation process efficiency (η, %) (Figure 6a), better electrochemical activity in terms of the electrochemical degradation efficiency was reached in CA application (Figure 6b). This is a reasonable result because during pulse application of +1.75 V a proportion of the current was consumed for oxygen evolution as a side reaction. Applying MPA operating conditions by selecting potential values of +1.25 and +1.75V, respective vs. Ag/AgCl enables the electrochemical oxidation of 4-AP to take place, through both direct oxidation on the electrode surface and indirect oxidation by electrogeneration of the oxidants under the water decomposition potential.

5.2. Bulk electrolysis for the electrochemical degradation of 4-aminophenol using boron-doped diamond electrode

As example of bulk electrochemical degradation is given for 4-AP performed at two current densities, 5 and 10 mA cm^{-2}, which were selected in according with the literature data [42] to generate hydroxyl radicals with low energy consumption, and the results are presented in Figures 7 and 8.

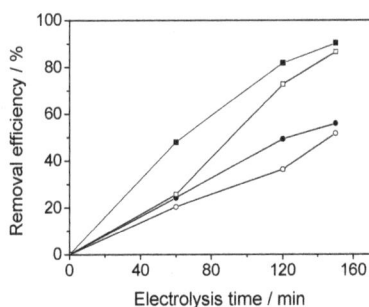

Figure 7. Evolution of removal efficiency of 4-AP determined by spectrophotomertry (solid) and TOC (open) by electrooxidation at the current density of 10 mA cm^{-2} (■) and 5 mA cm^{-2} (•).

It can be observed that both process efficiencies determined by spectrophotometry and total organic carbon (TOC) parameter increase with electrolysis time and respective, with specific electrical charge passed in the electrochemical oxidation (Figure 7). TOC parameter was checked to assess the mineralization degree. Based on the literature data concerning the phenol based organics electrooxidation under water decomposition potential range, 4-AP electrooxidation occurred in two stages, e.g., the first stage of 4-AP transformation into carboxylic acids and the second stage of the further oxidation of carboxylic acid to carbon dioxide [42]. The difference between 4-AP concentration and TOC evolution with the electrolysis time for both applied current densities proved the existence of the two oxidation stages. Better results were obtained by application of higher current density (10 mA cm^{-2}) for the same electrolysis time but it has to take in consideration that for higher current density higher specific electrical charge passed in the electrochemical oxidation, the aspect that limited the application from the economic consideration. The results of the electrochemical performance presented in Figure 8 show that the electrolysis cell operating at the current density of 5 mA cm^{-2} lead to the better results for 4-AP degradation and mineralization due to more part of electrical charge is consumed for the abundant oxygen evolution at higher current density applying. Also, an electrolysis time of 60 minutes should be considered as optimum.

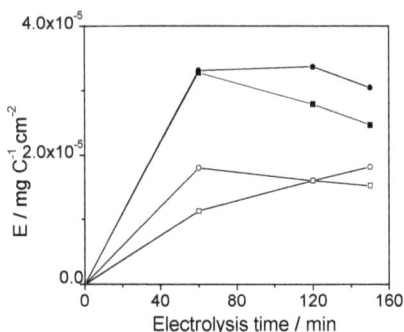

Figure 8. Evolution of removal electrochemical efficiency of 4-AP determined by spectrophotometry (solid) and TOC (open) by electrooxidation at the current density of 10 mAcm^{-2} (■) and 5 mA cm^{-2} (•).

During electrolysis, the cell voltage remains constant that informs that no electrode fouling occurred. Also, the pH of the residual solution first decreased slightly and later increased to reach a value close to 8. These changes can be attributed to both anodic and cathodic processes which occur in the cell. On the cathode the main reaction is the electrochemical generation of hydroxyl species by hydrogen peroxide reduction at the cathode surface [49], which led to the alkaline pH. At the same time, on the anode several reactions coexist, i.e., organic oxidation and oxygen evolution. The oxygen evolution led a pH decrease. The organic oxidation with the carboxylic acids formation in the first stage and subsequently its oxidation to carbon dioxide are supported by pH evolution, thus, the presence of the carboxylic acids lead to a pH decrease and by its conversion the pH is increased [36].

In addition, the relationship between mineralization current efficiency, specific energy consumption and electrolysis time for each current density is shown in Figure 9. For the same electrolysis time, at higher current density the charge consumed and implicit, specific energy consumption increased.

The initial pH is one of the most important operating parameter in the electrooxidation process. The performance of 4-AP degradation and mineralization process by electrooxidation was slightly improved under more alkaline versus acidic conditions. The influence of the pH mainly acts on the level of the oxidation mechanism under water decomposition region via the action of hydroxyl radicals physisorbed on BDD surface [50]. Alkaline media may favor the generation of hydroxyl radical and the electrooxidation performance is improved. Also, the electroactive species in alkaline medium is more easily oxidized than that of acidic medium (protonated form) [49].

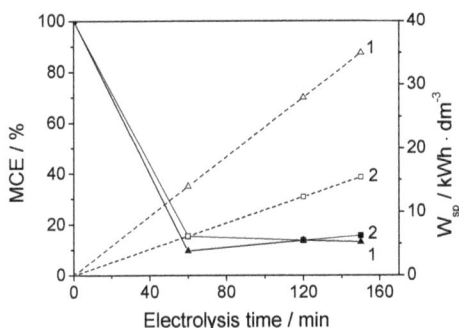

Figure 9. Evolution of the mineralization current efficiency (solid) and the specific energy consumption (open) with the electrolysis time in the electrochemical oxidation of 4-AP in 0.1 M Na₂SO₄ supporting electrolyte at the current densities of 10 mA cm⁻² (curve 1) and 5 mA cm⁻² (curve 2).

5.2.1. Photocatalytically-assisted electrochemical oxidation of 4-AP using BDD electrode and TiO2-supported zeolite catalyst

Based on the above-presented results, a high current density applying is limited by economic criterion. In according with the literature data [51-56] the presence of a catalyst in the electrical field can enhance the treatment efficiency with lower energy consumption. In order to improve the treatment performance of the 4-AP degradation and mineralization at lower current density (5 mAcm⁻²), the application of photocatalysis using TiO₂-supported zeolite catalyst on the bulk electrolysis using BDD electrode was performed.

UV spectra of 4-AP evolution presented in Figure 10 were recorded comparatively with initial 4-AP after the application of each oxidation process, i.e., electrooxidation, photocatalytycally -assisted electrooxidation at the same current density in comparison with photocatalysis using 1g dm⁻³ TiO₂-supported zeolite catalyst. The photocatalytically-assisted electrooxidation seems to be the most efficient in comparison with the photocatalysis and electrooxidation processes.

The progression of 4-AP degradation by above-presented processes indicated different oxidation efficiencies, and the profiles are presented in Figure 11a. The degradation assessment at 60 minutes showed that the lowest 4-AP degradation efficiency was reached by photocatalysis. Thus, after the reaction time of 120 minutes, about 23, 56 and 88 % of 4-AP was degraded in the photocatalysis, electrochemical oxidation and respective, photocatalytically-assisted electrooxidation. Thus, it can be seen that the net efficiency of combined process was greater than the sum of both electrooxidation and photocatalysis processes. This can be regarded as a synergetic effect of the photocatalytically-assisted electrooxidation process.

To assess the 4-AP mineralization by all above-presented processes, the progression of 4-AP degradation expressed in terms of TOC is shown in Figure 11b. After the reaction time of

120 minutes, about 16, 36 and 72 % of TOC was removed by the application of photocatalysis, electrochemical oxidation and respective, photocatalysis-assisted electrooxidation. The lower values of TOC removal degrees compared with 4-AP removal ones supported the fact that for all applied oxidation processes the complete mineralization of 4-AP to CO_2 was occurred via intermediaries as carboxylic acids.

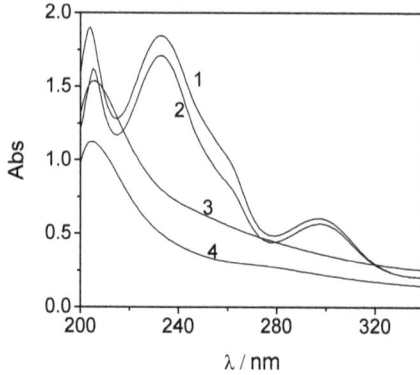

Figure 10. UV spectra profiles of 4-AP after degradation by photocatalysis (2), electrooxidation (3), photocatalytically-assisted electrooxidation (4), in comparison with initial 4-AP (1).

In general, the dependence of electrochemical reaction rates on the concentration of organic pollutants has been described by a pseudo first-order equation. Also, the photocatalysis reaction rate follows the Langmuir-Hinshelwood kinetic model, which can be simplified to a pseudo first-order equation [39]. For all applied oxidation processes the pseudo first-order equation was chosen to determine the apparent rate constant (k_{app}), which was calculated based on the relation of $\ln(C_0/C_t) = k_{app}(t)$. The kinetic results were obtained by monitoring of changes in 4-AP concentration determined by spectrophotometry and TOC parameter as a function of the reaction time. Table 6 shows the apparent rate constant for each applied oxidation process.

The apparent rate constant determined by spectrophotometry analysis of photocatalytically-assisted electrooxidation was higher by 8.1 and 2.5 times than photocatalysis and electrooxidation, respectively. Also, the apparent rate constants determined by TOC analysis are higher by 9.6 and 2.6 times versus photocatalysis and electrooxidation, respectively. The kinetics of TOC reduction in all applied processes showed that TOC reduction was slower in the photocatalytic process, suggesting that photocatalysis process was inefficient in the mineralization and quite degradation of 4-AP.

According to the definition of the synergetic effect (SF)= $k_{app, \text{ photocatalytically-assisted electrooxidation}}$ - ($k_{app,\text{photocatalysis}} + k_{app,\text{electrooxidation}}$) [57], the SF was calculated from the kinetic constants determined by spectrophotometry and TOC analysis. Thus, (SF)$_{\text{spectrophotometry}}$ of 0.0068 min^{-1} and (SF)$_{\text{TOC}}$ of 0.00627 min^{-1} indicated the apparent kinetic synergetic effects in the photocatalytically-assisted electrooxidation process for both degradation and mineralization of 4-AP.

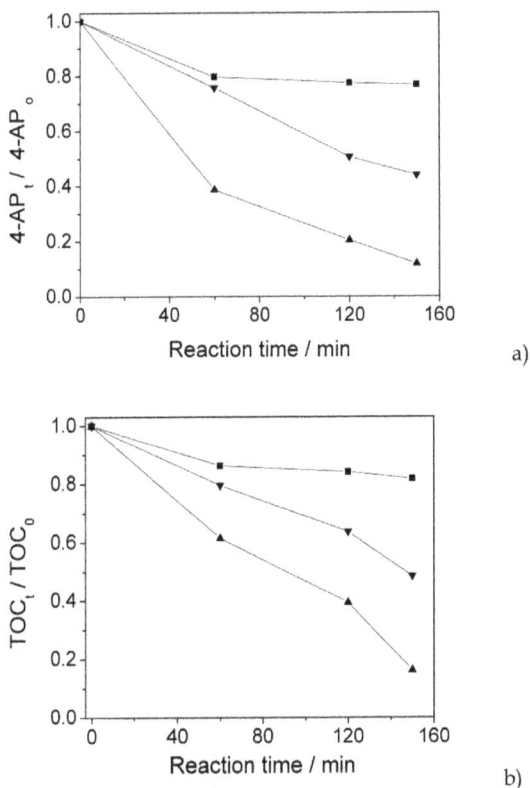

Figure 11. Evolution of 4-AP degradation with the reaction time, expressed as: a) 4-AP$_t$/a-AP$_0$ and b) TOC$_t$/TOC$_0$; (■)photocatalysis; (▼)electrooxidation; (▲)photocatalytically-assisted electrooxidation.

Type of 4-AP determination	Applied oxidation process	Apparent rate constant, k$_{app}$ (x 10^{-3} min^{-1})	Correlation factor, R^2
spectrophotometric	photocatalysis	1.68	0.9118
	electrooxidation	5.6	0.9968
	photocatalytically-assisted electrooxidation	13.6	0.9961
TOC	photocatalysis	1.26	0.9353
	electrooxidation	4.57	0.9821
	photocatalytically-assisted electrooxidation	12.1	0.9886

Table 6. The pseudo-first order kinetic apparent constant for the degradation of 4-AP determined by spectrophotometry and TOC analysis

For the electrochemical and combined processes, the apparent rate constant determined by spectrophotometry and TOC analysis was calculated based on the relation of $\ln(C_0/C_t)=k'_{app}(C)$, in which C represents the electrical charge passed during electrochemical and combined oxidation of 4-AP. The results are gathered in Table 7, and it can be noticed that the apparent rate constants determined for 4-AP concentration and TOC measurements for photocatalysis-assisted electrooxidation process are the best. Also, the effective mineralization rate was assessed as the ratio between the apparent rate constant calculated for TOC analysis and the apparent rate constant calculated for spectrophotometry analysis ($k'_{app,TOC}/k'_{app,4-AP}$). The value of this ratio closer to 1 indicated that the effective mineralization rate was achieved. The value of 0.88 compared with 0.82 for electrooxidation confirmed the enhancement effect of photocatalysis application on the electrooxidation process of 4-AP using BDD electrode to reach the mineralization.

Type of 4-AP determination	Applied oxidation process	Apparent rate constant, k'_{app} (C^{-1})	Correlation factor, R^2
spectrophotometric	electrooxidation	0.240	0.9968
	photocatalytically-assisted electrooxidation	0.586	0.9961
TOC	electrooxidation	0.196	0.9633
	photocatalytically-assisted electrooxidation	0.518	0.9886

Table 7. The apparent rate constant for the degradation of 4-AP determined by spectrophotometry and TOC analysis function of electrical charge

The enhancement effect of the combination of photocatalysis and electrooxidation processes operating at high voltage or current density may be attributed to two major factors, i.e., reducing the recombination of hole-electron pair resulted from photocatalysis, and indirect electrochemical oxidation of 4-AP. In according with literature data [57] it is supposed that the degradation mechanism is very complex, involving many oxidation pathways, e.g., anodic oxidation involving oxygen evolution, oxidation by electrogenerated H_2O_2 and OH•, oxidation by photogenerated hole and OH• and the photoelectrocatalytic synergetic effect.

It is well-known that the operating cost of an electrochemical oxidation depends strongly on the cell voltage as the specific energy consumption (W_{sp}). By the application of photocatalysis-assisted electrooxidation process for 4-AP degradation a higher mineralization current efficiency at low specific energy consumption was reached (Figure 12). It was found that for the same energy consumption of 12 kWhm^{-3}, photocatalysis-assisted electrooxidation process allowed removing 72% TOC in comparison with the electrochemical process for which 32% TOC removal was achieved. Thus, both technical and economic aspects regarding degradation process of 4-AP from water suggested the superiority of photocatalytically-assisted electrooxidation process in comparison with the electrochemical process [36].

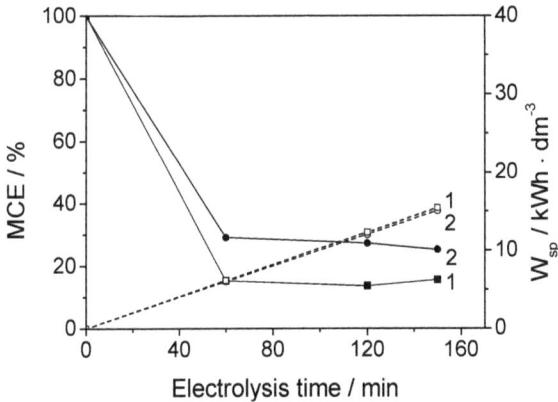

Figure 12. Evolution of the mineralization current efficiency (solid) and the specific energy consumption (open) with the electrolysis time in the electrochemical oxidation of 4-AP in 0.1 M Na₂SO₄ electrolyte for the application of: 1-electrooxidation; 2- photocatalytically-assisted electrooxidation.

It is clear that the (photo)electrochemical processes are energy consumption, being more expensive in comparison with the conventional biological process, but it must be taken into consideration that the (photo)electrochemical processes are considered advanced treatment one and should be integrated within the conventional treatment technological flow. When the electrochemical oxidation is the stage prior to biological one, biodegradability of the organic pollutants is enhanced, improving the performance of the biological process. In contrast, application of electrochemical oxidation as a finishing step envisaging completely mineralizing the refractory organic matter after biological treatment. A reduction of 25% of the energy consumption was reported by Panizza after coupling the electrooxidation and the biological processes [58].

5.3. Dual role of the electrochemical methods for organics degradation and determination

It is noteworthy that electrode materials for water quality monitoring by voltammetric/amperometric detection are conceptually very similar to those used in the degradation of pollutants from water, so the development of suitable electrode materials and electrochemical techniques involves actually relatively easy adaptation from a field to another, taking into account the application peculiarities, e.g., electrode geometry and design for degradation application. As example, the assessment of 4-CP concentration after chronoamperometry applying using EG-Epoxy composite electrode at the potential value of +1.1 V vs. Ag/AgCl after two hours of electrolysis was performed by using the cyclic voltammetric detection with the same electrode in comparison with spectrophotometric and classical chemical oxygen demand (COD) method. Figure 13 shows CVs recorded for initial 4-CP concentration and after chronoamperometry applying. The reduction degrees of 4-CP concentration by applying the chronoamperometry, determined by cyclic voltammetric detection, spectrophotometric and classical COD method are shown in Table 8.

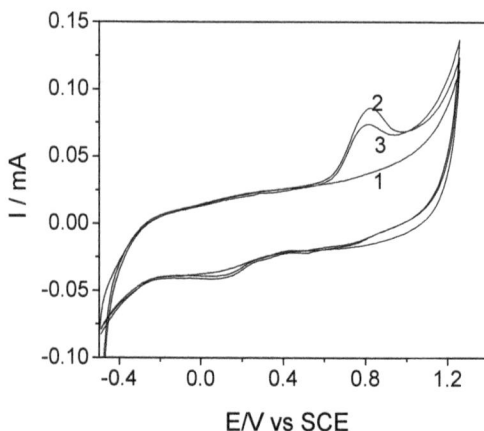

Figure 13. Cyclic voltammograms of EG-Epoxy in 0.5 M Na$_2$SO$_4$ supporting electrolyte (1) and in the presence of 25 mgL^{-1} 4-CP concentrations: before (2) and after (3) chronoamperometry applying at +1.1 V vs. Ag/AgCl for 2 h.

4-CP reduction degree (%)	COD reduction degree (%)	Amperometric signal reduction degree (%)
17.26	15	18

Table 8. The reduction degree of 4-CP concentration by applying the chronoamperometry at the potential value of +1.1 V vs. Ag/AgCl for two hours

Compared with the initial concentration of 4-CP, cyclic voltammogram shape recorded after chronoamperometry applying is changed, the current at the higher potential value but below oxygen evolution increased. This aspect could be explained by the intermediate formation of 4-CP oxidation which is oxidizable at higher potential value, *e.g.*, carboxilic acid that is oxidized on composite electrode at the potential value of 1.1 V vs. SCE [59]. This behaviour is in concordance with spectrophotometric and COD results, denoting a good accuracy of the voltammteric method and suitability for degradation process control.

6. Conclusions

Electrochemical processes should be regarded as viable alternatives for enhanced conventional wastewater treatment technology. Based on the wastewater characteristics and further usage, the various electrochemical processes could be applied individually or/and integrated into a variety of technological flows involving physical, chemical and biological methods. The electrocoagulation process could replace the conventional coagulation method integrated into a technological flow containing or not the electrooxidation process as advanced water treatment method. Also, the electrooxidation process could be applied as advanced oxidation process suitable especial for degradation/mineralization of the recalcitrant organic pollutants from wastewaters. The efficiency of electrochemical oxidation system is given mainly by the electrode material, which represents the key of the process performance. Carbon-based electrodes represent very promising electrode materials for degradation and mineralization of phenol derivates from wastewaters. Pulsed voltage application allows in-situ electrochemical cleaning the electrode materials leading to enhanced electrode performance, but its common usage is limited by higher energy consumption. Very promising results in relation with technical-economic criteria were achieved using photocatallytically-assissted electrooxidation process for degradation and mineralization of organics from wastewaters in comparison with simple electrooxidation process. Moreover, it must be kept in mind that the economic aspects will be improved by integration of the (photo)electrochemical process within the conventional treatment technological flow before or after the biological stage in accordance with the practical requirements, to improve the biodegradability of the recalcitrant organics or to mineralize them. Another important aspect that is not exploited is represented by dual role of the electrode material and electrochemical method in wastewater treatment and quality control, which could allow to set-up the electrochemical process control-integrated wastewater electrochemical treatment flow. A major area for future research is to establish the methodologies for the integration of the electrochemical treatment process involving more cost-effective electrodes in wastewaters decontamination and quality control.

Author details

Florica Manea[*] and Aniela Pop
Faculty of Industrial Chemistry and Environmental Engineering
"Politehnica" University of Timișoara, Romania

Acknowledgement

This work was partially supported by the strategic grants POSDRU/89/1.5/S/57649, Project ID 57649 (PERFORM-ERA), POSDRU/21/1.5/G/13798 co-financed by the European Social Fund – Investing in People, within the Sectoral Operational Programme Human Resources Development 2007-2013 and partially by the PN-II-ID-PCE 165/2011 and PNII-RU-PD129/2010 Grants.

7. References

[1] Metcalf & Eddy Inc. (1991) Wastewater Engineering: Treatment, Disposal and Reuse. New York: McGraw-Hill. 1334 p.

[*] Corresponding Author

[2]   Kim KH, Ihm SK (2011) Heterogeneous Catalytic Wet Air Oxidation of Refractory Organic Pollutants in Industrial Wastewaters: A Review. J. hazard. mat. 186(1): 16-34.

[3]   Liotta LF, Gruttadauria M, Di Carlo G, Perrini G, Librando V (2009) Heterogeneous Catalytic Degradation of Phenolic Substrates: Catalysts Activity. J. hazard. mat. 162(2–3): 588–606.

[4]   Busca G, Berardinelli S, Resini C, Arrighi L (2008) Technologies for The Removal of Phenol from Fuid Streams: A Short Review of Recent Developments. J. hazard. mat. 160: 265–288.

[5]   Bapat PS, Gogate PR, Pandit AB (2008) Theoretical Analysis of Sonochemical Degradation of Phenol and Its Chloro-Derivatives. Ultrason. sonochem. 15(4): 564–570.

[6]   Barrios-Martinez A, Barbot E, Marrot B, Moulin P, Roche N (2006) Degradation of Synthetic Phenol-Containing Wastewaters by MBR. J. membr. sci. 281(1–2): 288–296.

[7]   ***Directive 2008/1/EC of the European Parliament and of the Council of 15 January 2008 concerning integrated pollution prevention and control.

[8]   Rajeshwar K, Ibanez JG (1997) Environmental Electrochemistry. San Diego: Academic Press. 776 p.

[9]   Rajeshwar K, Ibanez JG, Swain GM, (1994) Electrochemistry and Environment. J. appl. electrochem. 24: 1077-1091.

[10] Walsh FC (2001) Electrochemical Technology for Environmental Treatment and Energy Conversion. Pure appl. chem. 73: 1819-1837.

[11] Chen G (2004) Electrochemical Technologies in Wastewater Treatment. Sep. puri. technol. 38: 11-41.

[12] Bebeselea A, Pop A, Orha C, Danielescu C, Manea F, Burtica G (2006) Aspects Regarding The Wastewater Treatment by Electroflotocoagulation. Environ. eng. manag. j. 5(5): 1071-1077.

[13] Mouli PC, Mohan SV, Reddy SJ (2004) Electrochemical Processes for The Remediation of Wastewater and Contaminated Soil: Emerging Technology, J. sci. ind. res. 63: 11-19.

[14] Ciorba GA, Radovan C, Vlaicu I, Masu S (2002) Removal of Nonylphenol Ethoxylates by Electrochemically-Generated Coagulants. J. appl. electrochem. 32: 561-566.

[15] Jiang JQ, Graham N, Andre C, Kelsall GH, Brandon N (2002) Laboratory Study of Electro-Coagulation–Flotation for Water Treatment. Wat. res. 36: 4064-4078.

[16] Chen X, Chen G, Yue PL (2000) Separation of Pollutants From Restaurant Wastewater by Electrocoagulation, Sep. purif. technol. 19:65-76.

[17] Pouet MF, Grasmick A (1995) Urban Wastewater Treatment Electrocoagulation and Flotation. Water sci. technol. 31: 275-283.

[18] Anglada A, Urtiaga A, Ortiz I (2009) Contributions of Electrochemical Oxidation to Waste-Water Treatment: Fundamentals and Review of Applications. J. chem. technol. biotechnol. 84: 1747–1755.

[19] Comninellis Ch (1994) Electrocatalysis in The Electrochemical Conversion/Combustion of Organic Pollutants for Waste Water Treatment. Electrochim. acta. 39(11-12): 1857-1862.

[20] Brillas E, Arias C, Cabot PL, Centellas F, Garrido JA, Rodriguez RM (2006) Degradation of Organic Contaminants by Advanced Electrochemical Oxidation Methods. Port. Electrochim. Acta. 24: 159-189.

[21] Yoshida K, Yoshida S, Seki Y, Takahashi T, Ihara I, Toyoda K (2007) Basic Study Of Electrochemical Treatment of Ammonium Nitrogen Containing Wastewater Using Boron-Doped Diamond Anode. Environ. res. 65: 71–73.

[22] Motoc S, Manea F, Pop A, Pode R, Teodosiu C (2012) Electrochemical Degradation of Pharmaceutical Effluent on Carbon-Based Electrodes. Environ. eng. manag. j. 11(3): 627-634.

[23] Hegde RN, Shetti NP, Nandibewoor ST (2009) Electro-Oxidation and Determination of Trazodone at Multi-Walled Carbon Nanotube-Modified Glassy Carbon Electrode. Talanta. 79: 361–368.

[24] Patil RH, Hegde RN, Nandibewoor ST (2011) Electro-Oxidation and Determination of Antihistamine Drug, Cetirizinen Dihydrochloride at Glassy Carbon Electrode Modified with Multi-Walled Carbon Nanotubes. Coll. surf. B: biointerf. 83: 133–138.

[25] Shahrokhian S, Souri A, Khajehsharifi H (2004) Electrocatalytic Oxidation of Penicillamine at a Carbon Paste Electrode Modified with Cobalt Salophen. J. electroanal. chem. 565: 95–101.

[26] Brillas E, Garcia-Segura S, Skoumal M, Arias C (2010) Electrochemical Incineration of Diclofenac in Neutral Aqueous Medium by Anodic Oxidation using Pt and Boron-Doped Diamond Anodes. Chemosphere. 79: 605-612.

[27] Dominguez JR, Gonzalez T, Palo P, Sanchez-Martin J (2010) Anodic Oxidation of Ketoprofen on Boron-Doped Diamond (BDD) Electrodes. Role of Operative Parameters. Chem. eng. j. 162: 1012–1018.

[28] Garrido JA, Brillas E, Cabot PL, Centellas F, Arias C, Rodriguez RM (2007) Mineralization of Drugs in Aqueous Medium by Advanced Oxidation, Port. electrochim. Acta. 25: 19-41.

[29] Klavarioti M, Mantzavinos D, Kassinos D (2009) Removal of Residual Pharmaceuticals from Aqueous Systems by Advanced Oxidation Processes, Environ. int. 35: 402–417.

[30] Manea F, Radovan C, Pop A, Corb I, Burtica G, Malchev P, Picken S, Schoonman J (2009) Carbon Composite Electrodes Applied for Electrochemical Sensors. In: Baraton MI editor. Sensors for Environment, Health and Security. NATO Science for Peace and Security Series C: Environmental Security. Springer. pp. 179-189.

[31] Pop A, Manea F, Radovan C, Malchev P, Bebeselea A, Proca C, Burtica G, Picken S, Schoonman J (2008) Amperometric Detection of 4-Chlorophenol on Two Types of Expanded Graphite based Composite Electrodes. Electroanal. 20(22): 2460-2466.

[32] Orha C, Manea F, Pop A, Burtica G, Fazakas-Todea I (2008) Obtaining and Characterization of Zeolitic Materials with Antibacterial Properties, Rev. chim. 59(2): 173-177.

[33] Corb I, Manea F, Radovan C, Pop A, Burtica G, Malchev P, Picken S, Schoonman J (2007) Carbon-based Composite Electrodes: Preparation, Characterization and Application in Electroanalysis. Sensors. 7: 2626-2635.

[34] Mironov VS, Kim JK, Park M, Lim S, Cho WK (2007) Comparison of Electrical Conductivity Data Obtained by Four-Electrode and Four-Point Probe Methods for Graphite-based Polymer Composites. Polym. test. 26(4): 547-555.

[35] Ratiu C, Manea F, Lazau C, Orha C, Burtica G, Grozescu I, Schoonman J (2011) Zeolite-Supported TiO_2 based Photocatalysis-Assisted Electrochemical Degradation of p-Aminophenol from Water. Chem. pap. 65(3): 289-298.

[36] Ratiu C, Manea F, Lazau C, Grozescu I, Radovan C, Schoonman J (2010) Electrochemical Oxidation of p-Aminophenol from Water with Boron-Doped Diamond Anodes and Assisted Photocatalytically by TiO_2-Supported Zeolite. Desalination. 260: 51–56.

[37] Wang YH, Chan KY, Li XY, So SK (2006) Electrochemical Degradation of 4-Chlorophenol at Nickel-Antimony Doped Tin Oxide Electrode. Chemosphere. 65: 1087-1093.

[38] Vlaicu I, Pop A, Manea F, Radovan C (2011) Degradation of Humic Acid from Water by Advanced Electrochemical Oxidation Method. Wat. sci. technol.: Wat. supp. 11.1: 85-95.

[39] Guinea E, Centellas F, Garrido JA, Rodriguez RM, Arias C, Cabot PL, Brillas E (2009) Solar Photoassisted Anodic Oxidation of Carboxylic Acids in Presence of Fe^{3+} using a Boron-Doped Diamond Electrode. Appl. catal. B: environ. 89: 459-468.

[40] Wang XG, Wu QS, Liu WZ, Ding YP (2006) Simultaneous Determination of Dinitrophenol Isomers with Electrochemical Method Enhanced by Surfactant and Their Mechanisms Research. Electrochim. acta. 52: 589-594.

[41] Canizares P, Saez C, Lobato J, Rodrigo MA (2004) Electrochemical Treatment of 4-Nitrophenol-Containing Aqueous Wastes using Boron-Doped Diamond Anodes. Ind. eng. chem. res. 43, 1944–1951.

[42] Canizares P, Lobato J, Paz R, Rodrigo MA, Saez C (2005) Electrochemical Oxidation of Phenolic Wastes with Boron-Doped Diamond Anodes. Water res. 39: 2687–2703.

[43] Gherardini L, Michaud PA, Panizza M, Comninellis Ch, Vatistas N (2001) Electrochemical Oxidation of 4-Chlorophenol for Waste Water Treatment. Definition of Normalized Current Efficiency. J. electrochem. soc. 148: D78–D84.

[44] Ureta-Zanartu MS, Bustos P, Berrios C, Diez MC, Mora ML, Gutierrez C (2002) Electrooxidation of 2,4-Dichlorophenol and other Polychlorinated Phenols at a Glassy Carbon Electrode. Electrochim. acta. 47: 2399-2406.

[45] Skowronski JM, Krawczyk P (2007) Improved Electrooxidation of Phenol at Exfoliated Graphite Electrodes. J. solid state electrochem. 11: 223-230.

[46] Manea F, Burtică G, Bebeşelea A, Pop A, Corb I, Schoonman J (2007) Degaradation of 4-Chlorophenol from Wastewater by Electrooxidation using Graphite based Composite Electrodes. Environmental Science and Technology. Starrett SK, Hong J, Wilcock RJ, Li Q, Carson JH, Arnold S, editors. Houston: American Science Press. vol. 1. p.281.

[47] Rodrigo MA, Michaud PA, Duo I, Panizza M, Cerisola G, Comninellis Ch (2001) Oxidation of 4-Chlorophenol at Boron-Doped Diamond electrodes for Wastewater Treatment. J. electrochem. soc. 148: D60-D64.

[48] Bard AJ, Faulkner LR (1980) Electrochemical Methods, Fundamentals and Applications, New Zork: Wiley.

[49] Martinez-Huitle CA, Brillas E (2009) Decontamination of Wastewater Containing Synthetic Organic Dyes by Electrochemical Methods: A General Review. Appl. catal. B: environ. 87: 105-145.

[50] Lissens G, Pieters J, Verhaege M, Pinoy L, Verstraete W (2003) Electrochemical Degradation of Surfactants by Intermediates of Water Discharge at Carbon-Based Electrodes. Electrochim. acta. 48: 1655-1663.

[51] An T, Li G, Xiong Y, Zhu X, Xing H, Liu G (2001) Photoelectrochemical Degradation of Methylene Blue with Nano TiO_2 under High Potential Bias. Mater. phys. mech. 4: 101-106.

[52] Chen J, Liu M, Zhang J, Ying X, Jin L (2004) Photocatalytic Degradation of Organic Wastes by Electrochemically Assisted TiO_2 Photocatalytic System, J. environ. manag. 70: 43-47.

[53] Lei Y, Shen Z, Chen X, Jia J, Wang W (2006) Preparation and Application of Nano-TiO_2 Catalyst in Dye Electrochemical Treatment. Water. SA. 32(2): 205-210.

[54] Shen ZM, Wu D, Yang J, Yuan T, Wang WH, Jia JP (2006) Methods to Improve Electrochemical Treatment Effect of Dye Wastewater. J. hazard. mater. 131: 90-97.

[55] Wang B, Gu L, Ma H (2007) Electrochemical Oxidation of Pulp and Paper Making Wastewater Assisted by Transition Metal Modified Kaolin. J. hazard. mater. 143: 198-205.

[56] Gomathi Devi L, Kottam N, Girish Kumar S, Anantha Raju KS (2009) Mechanismof Charge Transfer in The Transition Metal Ion Doped TiO_2 with Bicrystalline Framework of Anatase and Rutile: photocatalytic and photoelectrocatalytic activity. Catal. lett. 131(3-4): 612-617.

[57] An T, Zhang W, Xiao X, Sheng G, Fu J, Zhu X (2004) Photoelectrocatalytic Degradation of Quinoline with a Novel Three-Dimensional Electrode-Packed Bed Photocatalytic Reactor. J. photochem. photobiol. A: chem. 161: 233-242.

[58] Panizza M, Zolezzi M, Nicolella C (2006) Biological and electrochemical oxidation of naphtalenesulfonates. J chem technol biotechnol 81:225–232.

[59] Manea F, Radovan C, Corb I, Pop A, Burtica G, Malchev P, Picken S, Schoonman J (2008) Simultaneous Determination of 4-Chlorophenol and Oxalic Acid using an Expanded Graphite-Epoxy Composite Electrode. Electroanal. 20(5): 1719-1722.

Chapter 12

Treatment Technologies for Organic Wastewater

Chunli Zheng, Ling Zhao, Xiaobai Zhou, Zhimin Fu and An Li

Additional information is available at the end of the chapter

1. Introduction

1.1. Resources of organic wastewater

There are several contaminants in wastewater, with organic pollutants playing the major role. Many kinds of organic compounds, such as PCBs, pesticides, herbicides, phenols, polycyclic aromatic hydrocarbons (PAHs), aliphatic and hetercyclic compounds are included in the wastewater, and industrial and agricultural production as well as the people living could be the source of organic wastewater endangering the safety of the water resource [1]. The wastewater of the farmland may contain high concentration of pesticides or herbicides; the wastewater of the coke plant may contain various PAHs; the wastewater of the chemical industry may contain various heterogeneity compounds, such as PCB, PBDE; the wastewater discharged by the food industry contains complex organic pollutants with high concentration of SS and BOD; and the municipal sewage contains different type of organic pollutants, such as oil, food, some dissolved organics and some surfactants. These organic pollutants in water can harm the environment and also pose health risks for humans.

1.2. Common poisonous substances in organic wastewater

The organic pollutants in the wastewater could be divided into two groups according to their biological degradation abilities. The organic pollutants with simple structures and good hydrophilicity are easy to be degraded in the environment. These organic pollutants, such as polysaccharide, methanol could be degraded by the bacteria, fungus and algae. However, some of them, such as acetone and methanol, could cause acute toxicity when existed in wastewater at a high concentration. On the other hand, the persistent organic pollutants, such as PAHs, PCBs, and DDT, are very slowly metabolized or otherwise degraded. Some of them, for example, the pesticides were widely used for several years. Although their concentration as well as the cute toxicity in the wastewater is lower than the soluble organic pollutant, they can be sequestered in sediment and exist for decades, and transport into the wastewater and then the food chain. The POPs are lipid soluble, and many of them mentioned above are carcinogenic, teratogenic, and neurotoxic. Since they are persistent, long way transported and toxic, these organic pollutants draw more attentions.

The classic poisonous substances in organic wastewater are as follow:

1.   Water organic matter

Water organic matter is the genetic name of the organic compounds in the sediment and wastewater. Generated from the residues of the animal, plants and microorganisms, the water organic matters could be divided into two categories: one is non - humic, which is composed of the various organic compounds of organisms, such as protein, carbohydrate, organic acids, etc., the other is a special organic compounds named humus. Water organic matter could affect the physical and chemical properties of the water, and could also influence the self-purification, degradation, migration and transformation process in the water.

2.   Formaldehyde

Formaldehyde is an organic compound with the formula CH_2O. The main sources of formaldehyde are organic synthesis, chemical industry, synthetic fiber, dyestuff, wood processing and the paint industry emissions of wastewater. With a strong reducibility, formaldehyde could easily combine with a variety of material, and is easy to be together. Formaldehy is a stimulus to skin and mucous membrane. It could enter the central nervous of human body and cause retinal damage

3.   Phenols

Phenols are a class of chemical compounds consisting of a hydroxyl group (-OH) bonded directly to an aromatic hydrocarbon group. The phenol in the wastewater mainly comes from the coking plant, refining, insulation material manufacturing, paper making and phenolic chemical plant. Phenol is known human carcinogen and is of considerable health concern, even at low concentration. Phenol also has potential to decrease the growth and the reproductive capacity of the aquatic organisms.

4.   Nitrobenzene

Nitrobenzene is an organic compound with the chemical formula $C_6H_5NO_2$. It is produced on a large scale as a precursor to aniline. In the laboratory, it is occasionally used as a solvent, especially for electrophilic reagents. Prolonged exposure may cause serious damage to the central nervous system, impair vision, cause liver or kidney damage, anemia and lung irritation[2]. Recent research also found nitrobenzene as a potential carcinogenic substance.

5.   PCBs

PCBs are biphenyl combined with 2 to 10 chlorine atoms. PCBs are widely used as dielectric and coolant fluids, for example in transformers, capacitors, and electric motors, and various kinds of PCBs could be found the wastewater of this factories[3]. PCBs are carcinogenic, and could accumulate in adipose tissue, causing brain, skin and the internal organs disease, and influence nerve, reproductive and immune system. PCBs also have shown toxic and mutagenic effects by interfering with hormones in the body. PCBs, depending on the specific congener, have been shown to both inhibit and imitate estradiol.

6.   PAHs

PAHs are recalcitrant organic pollutants consisting of two or more fused benzene rings in linear, angular, or cluster arrangements. PAHs occur in oil, coal, and tar deposits, and PAHs in the aquatic system could come from accidently leaking, atmosphere deposition and contaminated sediment release. The concentration of PAHs, especially the PAHs with high molecular weight, in the water is usually low in the water owing to their hydrophobia property, but they are still among the most problematic substances as they could accumulate in the environment and threaten the development of living organisms because of their acute toxicity, mutagenicity or carcinogenity [4].

7.   Organophosphorus pesticide

The wastewater of organophosphorus pesticide manufacturers often contains a high concentration of organophosphorus pesticide, intermedia productions and degradation productions, and the wastewater from the farmland could contain some of this pesticide since this substance could exist in the environment for a period of time. The discharge of water contained organophosphorus pesticide could cause serious environmental pollution. Some organophosphorus pesticides have acute poison on the people and livestock. In spite of the severe toxicity of the organophosphorus pesticide, it is easy to be degraded in the environment [5].

8.   Petroleum hydrocarbons

The petroleum hydrocarbons in the water system mainly come from the industrial wastewater and municipal sewage. The industry, such as oil exploration, oil manufacture, transportation and refining could produce the wastewater with a mixture of various petroleum hydrocarbons. The petroleum hydrocarbons are toxic towards aquatic living things, and they could also aggravate the water quality by forming a layer of oil film, which could decrease the oxygen exchange of the air and water body.

9.   Atrazine

Atrazine is the most widely used herbicide in conservation tillage systems, which are designed to prevent soil erosion. This chemical herbicide could stop pre- and post-emergence broadleaf and grassy weeds in dry farmland, and increase the production of the major crops [6]. The wastewater contained atrazine mainly comes from the chemical industry manufacturing this product and the farmlands which are over loaded. This substance could remain in the environment for a period of time, and it has been detected in the surface water and groundwater of many countries and regions. Atrazine could volatilize at high temperature and release poisonous gas such as carbon monoxide, nitrogen oxides, which could irritate people's skin, eyes and respiratory tract. Besides, atrazine also has a potential cause of birth defects, low birth weights and menstrual problems when consumed at concentrations below federal standards.

1.3. Environmental hazards of organic wastewater

High mount of hydrophilic organic pollutants, such as organic matters, oil could consume a large amount of soluble oxygen. The acute toxicity and high quantity of oxygen demand could worsen the water quality and lead to great damage to the aquatic ecological system. However, their bad influence towards the environment will not last long, since they could easily be degraded by microorganisms.

The situation is different for the POPs, which have low water solubility, high accumulation capacity and potential carcinogenic, teratogenic, and neurotoxic properties. For example, many of the organochlorine pesticides cited above are carcinogenic, teratogenic, and neurotoxic. The dioxins and benzofurans are highly toxic and are extremely persistent in the human body as well as the environment. Several of the POPs, including DDT and its metabolites, PCBs, dioxins, and some chlorobenzene, can be detected in human body fat and serum years after any known exposures. Lindane (hexachlorocyclohexane), which was used for the treatment of body lice and as a broad-spectrum insecticide, could cause very high tissue levels, and could cause acute deaths when improperly used.

Many factors, such as the characters of the pollutants, the environmental factors (PH value, temperature etc.), aging process could affect the toxicity of the organic wastewater, and their long-term influence to the ecosystem deserve further investigation.

1.4. Monitoring analysis method of poisonous substances

1. Gross analysis

The amount of organic compounds in wastewater is generally evaluated by chemical oxygen demand (COD) test, biological oxygen demand (BOD) test, and (TOC) test.

The basis for the COD test is that nearly all organic compounds can be fully oxidized to carbon dioxide with a strong oxidizing agent under acidic conditions. The COD value is always measured by the acidic potassium permanganate method and potessium dichromate method, and could reflect the pollution degree of reducing matter in water, including ammonia and reducing sulfide, so in wastewater with high quantity of reducing matter, the COD value will overestimate the organic pollutants in the water.

BOD value is the amount of dissolved oxygen needed by aerobic biological organisms in a body of water to break down organic material present in a given water sample at certain temperature over a specific time period. The BOD value is most commonly expressed in milligrams of oxygen consumed per liter of sample during 5 days of incubation at 20 °C and is often used as a robust surrogate of the degree of organic pollution of water. This is not a precise quantitative test, although it is widely used as an indication of the organic quality of water.

TOC value is the mount of total carbon (water soluble and suspended in water) in the water. Using combustion during the assessment, this method could oxidize all the organic pollutants, and value reflects the amount of organic matter more directly than BOD_5 or COD.

The COD, BOD and TOC test could quickly reflect the organic pollution in the wastewater, however, they can't reflect the kinds of organic matter and composition of the water, and therefore cannot reflect the total amount of the same total organic carbon pollution caused by different consequences.

2. Chromatography-mass spectrometry method

Chromatography-mass spectrometry method is an advanced method to separate and define the organic pollutants in the waste water. Spectrometry is the collective term for a set of laboratory techniques for the separation of mixtures. The separation is based on differential partitioning between the mobile and stationary phases. The structure diversity of different components in the wastewater results in a different retention on the stationary phase and thus changing the separation. The mobile phase of the chromatography can be gas or liquid, so the chromatography can be divided into gas chromatography (GC) and liquid chromatography (LC).

The mass spectrometer could ionize the organism and shoot it through an electric field. Since the electric field could bend the path (trajectory) of lighter molecules more than that of heavy molecules, the organic matter of different mass would strike at different position (the position is fixed for each organic matter) in the detector. This method could identify and quantify organic pollutants. The combination of chromatography and mass spectrometry could offer complete information on the type of organic pollutants in a sample and the concentration of each pollutant in the sample.

2. Biological treatment technology of organic wastewater

2.1. Principle of the biodegradation

Biodegradation is a process using microorganisms, fungi, green plants and their enzymes to remove the pollutants from natural environment or transform them harmless. Biodegradation could happen in nature world and is used in wastewater treatment in recent years since humanity strives to find sustainable ways to clean up contaminated water economically and safely.

2.2. Biodegradation of organic compounds

Chemical, physical and biological methods have been used to remove the organic compounds from the wastewater, and biological method has been paid much attention owing to its economic and ecologic superiority. The biodegradation rate and biodegradation degree of the organic substance partly depended on the characters of the substance. Some of the organic pollutants like organic matters, organophosphorus pesticide, which have relativity high water solubility and low acute toxicity, are bioavailable and easy to be degraded [7]. However, for some POPs and xenobiotic organic pollutants, such as polychlorinated biphenyls (PCBs), polyaromatic hydrocarbons (PAHs), heterocyclic compounds, pharmaceutical substances, which possess a higher bioaccumulation, biomagnification and biotoxicity properties, are reluctant to biodegradation in the nature condition. Organic material can be degraded aerobically with oxygen, or anaerobically, without oxygen [8].

2.3. Aerobic biodegradation

The principle of the aerobic biodegradation is as follow: oxygen is needed by degradable organisms in their degradation at two metabolic sites, at the initial attack of the substrate and at the end of respiratory chain [9]. Bacteria and fungi could produce oxygenases and peroxidases they could help with the pollutant oxidization and get benefits from observing the energy, carbon and nutrient elements released during this process. A huge number of bacterial and fungal general possess the capability to release non-special oxidase and degrade organic pollutants. There are generally two types of relationships between the microorganism and organic pollutants: one is that the microorganisms use organic pollutant as sole source of carbon and energy; the other is that the microorganisms use a growth substrate as carbon and energy source, while another organic compound in the organic substrate which could not provide carbon and energy resource is also degraded, namely cometabolism.

The classic aerobic biodegradation reactors include activated sludge reactor and membrane bioreactor.

2.3.1. Activated sludge reactor

Activated sludge is a process for treating sewage and industrial wastewaters using air and a biological floc composed of bacteria and protozoans. This technique was invented by Ardern and Lockett at the beginning of last century and was considered as a wastewater treatment technique for larger cities as it required a more sophisticated mode of operation (Fig. 1) [10].

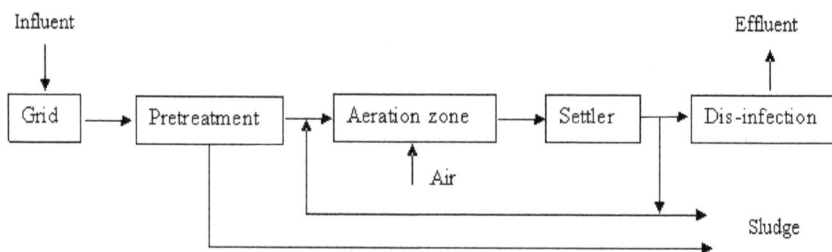

Figure 1. The scheme of the activated sludge reactor [1]

This process introduced air or oxygen into a mixture of primary treated or screened wastewater combined with organisms to develop a biological floc which reduces the organic content of the sewage, which is largely composed of microorganisms such as saprotrophic bacteria, nitrobacteria and denitrifying bacteria. With this biological floc, we could degrade the organic pollutant and bio-transform the ammonia in wastewater. Generally speaking, the process contained two steps: adsorption followed by biological oxidation.

The technique could effectively remove the organic matters, nitrogeneous matters, phosphate in the wastewater, when there is enough oxygen and hydraulic retention time. However, the contaminated water is always short of oxygen, which could cause sludge bulking, a great problem decrease the water quality of the effluent. The oxygen concentration could be increased by including aeration devices in the system, but research need to be done to find out the optimal value since aeration would cause an increase of the costs of the wastewater treatment plants. Researches are also required to deal with the excess activated sludge, the by-product of this process, with a relatively low cost.

2.3.2. Membrane bioreactor

Membrane bioreactor (MBR) is the combination of a membrane process like microfiltration or ultrafiltration with a suspended growth bioreactor, and is now widely used for municipal and industrial wastewater treatment. The scheme of the reactor is showed in Fig. 2 [11].

The Principle of this technique is nearly the same as activated sludge process, except that instead of separation the water and sludge through settlement, the MBR method uses the membrane which is more efficient and less dependent on oxygen concentration of the water.

The MBR has a higher organic pollutant and ammonia removal efficiency in comparison with the activated sludge process. Besides, the MBR processes is capable to treat waste water with higher mixed liquor suspended solids (MLSS) concentrations compared to activated sludge process, thus reducing the reactor volume to achieve the same loading rate.

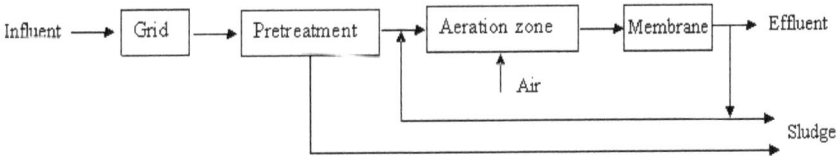

Figure 2. The scheme of the MBR reactor

However, membrane fouling greatly affects the performance of this technique, since fouling leads to significantly increase trans-membrane pressure, which increased the hydraulic resistance time as well as the energy requirement of this reactor. Alternatively frequent membrane cleaning and replacement is therefore necessary, but it significantly increases the operating cost.

2.4. Anaerobic biodegradation

Anaerobic degradation is a series of processes in which microorganisms break down biodegradable material in the absence of oxygen. The principle of the anaerobic degradation is as follow: first, the insoluble organic pollutant brakes down the into soluble substance, making them available for other bacteria; second, the acidogenic bacteria convert the sugars and amino acid into carbon dioxide, hydrogen, ammonia and organic acid; third, the organic acids convert into acetic acid, ammonia, hydrogen and carbon dioxide; finally, the methanogens convert the acetic acid into hydrogen, carbon dioxide and methane, a kind of gaseous fuel [9].

Anaerobic degradation processes have always been considered to be slow and inefficient, in comparison to aerobic degradation. However, the anaerobic degradation not only decreases the COD and BOD in the waste water, but also produces renewable energy. Moreover, the anaerobic bacteria could break down some persistent organic pollutants, such as lignin and high molecular weight PAH, which show little or no reaction to aerobic degradation. Besides, anaerobic processes could treat the wastewater with high loads of easy-to-degrade organic materials (wastewaters of the sugar industry, slaughter houses, food industry, paper industry, etc.) efficiently and costly. These advantages make investigation and application of anaerobic microbial mineralization in organic polluted water important.

Generally speaking, anaerobic reactor could be divided into anaerobic activated sludge process and anaerobic biological membrane process. The anaerobic activated sludge process includes conventional stirred anaerobic reactor, upflow anaerobic sludge blanket reactor, and anaerobic contact tank. The anaerobic biological membrane process includes fluidized bed reactor, anaerobic rotating biological contactor, anaerobic filter reactor. Upflow anaerobic sludge blanket reactor and anaerobic filter reactor are selected as the representative of the two kinds of reactors mentioned above.

2.4.1. Upflow anaerobic sludge blanket reactor (UASB)

The UASB system was developed in 1970s. No carrier is used to in the UASB system, and liquid waste moves upward through a thick blanket of anaerobic granular sludge suspended in the system. As shown in Fig. 3, mixing of sludge and wastewater is achieved by the generation of methane within the blanket as well as by hydraulic flow. And the triphase separator (gas, liquid, sludge biomass) could prevent the biomass loss of the sludge through the gas emission and water discharge. The advantage of this system are that 1) it contains a high concentration of naturally immobilized bacteria with excellent settling properties, and could remove the organic pollutants from wastewater efficiently; 2) a high concentrations of biomass can be achieved without support materials which reduces the cost of construction. These advantages would increase the efficient and stable performance of this system [12].

2.4.2. Anaerobic biofilter

Anaerobic biofilter, so called anaerobic fixed film reactors, is a kind of high efficient anaerobic treatment equipment developed in 1960 s. These reactors use inert support materials to provide a surface for the growth of anaerobic bacteria and to reduce turbulence to allow unattached populations to be retained in the system (Fig 4). The organic matter of wastewater is degraded in the system, and produce methane gas, which will be released from the pool from the top [13].

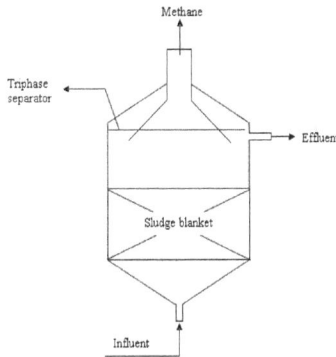

Figure 3. The scheme of upflow anaerobic sludge blanket reactor [1]

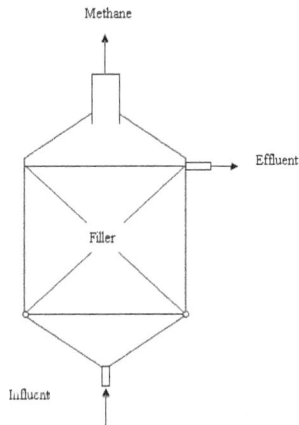

Figure 4. The scheme of the anaerobic biofilter [1]

The advantages of this system are as follow: 1) the filler provides a large surface area for the growth of the microorganisms, and the filler also increases hydraulic retention time of the wastewater; 2) the system provides a large surface area for the interaction between the wastewater and film; 3) the fact that microorganisms grow on the filler reduces the run of the degraders. These advantages could increase the efficiency of this treatment, and guarantee the water quality of the effluent. The backward of this system is that the system could be blocked when dealing with high concentration organic water, especially in the water inlet parts. And no simple and effective way for filter washing has been developed yet.

2.5. Combination of the aerobic and anaerobic biodegradation

Compared with the single anaerobic and aerobic reactors, the combination of the anaerobic and aerobic reactor is more efficient in organic pollutants degradation. The advantages of the combined system are as follow: 1) the anaerobic process could get rid of the organic matters and suspended solid from the wastewater, reduce the organic load of the aerobic degradation as well as the production of aerobic sludge, and finally reduce the volume of the reactors; 2) wastewater pretreated by anaerobic technology is more stable, indicating that anaerobic process could reduce the load fluctuation of the wastewater, and therefore decrease the oxygen requirement of the aerobic degradation; 3) the anaerobic process could modify the biochemical property of the wastewater, making the following aerobic process more efficient. Investigation showed that the wastewater from aerobic-anaerobic combined reactor are more stable and ready for degradation, indicating that this technical have a huge potential for application. The classic aerobic-anaerobic reactors include A/O reactor, A2/O reactor, oxidation ditch, constructed wetland.

Two classic aerobic biodegradation reactors, oxidation ditch and constructed wetland are introduced.

2.5.1. Oxidation ditch

The oxidation ditch is a circular basin through which the wastewater flows. Activated sludge is added to the oxidation ditch so that the microorganisms will digest the organic pollutants in the water. This mixture of raw wastewater and returned sludge is known as mixed liquor. The rotating biological contactors could add oxygen into the flowing mixed liquor, and they could also increase surface area and create waves and movement within the ditches. Once the organic pollutant has been removed from the wastewater, the mixed liquor flows out of the oxidation ditch. Sludge is removed in the secondary settling tank, and part of the sludge is pumped to a sludge pumping room where the sludge is thickened with the help of aerator pumps [14]. Some of the sludge is returned to the oxidation ditch while the rest of the sludge is sent to waste.

The oxidation ditch is characterized by simple process, low maintain consumption, steady operation, and strong shock resistance. The effluent of the system has high water quality effluent with low concentration of organic pollutants, nitrogen and phosphorus. However, the problems of this reactor, such as sludge expansion, rising sludge and foam, are important factors which confines the development of this technique.

2.5.2. Constructed wetland

A constructed wetland is an artificial wetland which could wetlands act as a biofilter, removing sediments and pollutants such as heavy metals and organic pollutants from the water. Constructed wetland is a combination of water, media, plants, microorganisms and other animals. Constructed wetlands are of two basic types: subsurface-flow and surface-flow wetlands [15].

Physical, chemical, and biological processes combine in wetlands to remove contaminants from wastewater. Besides absorbing heavy metals and organic pollutants (especially POPs) on the filler of the constructed wetland, plants can supply carbon and other nutrients such as nitrogen through their roots to for the growth and reproduction of the microorganisms. Plants could also pump oxygen to form an aerobic and anaerobic area in the deep level of constructed wetland to assist the breaking down of organic materials. The major reactor in constructed wetland was supposed to be microorganisms, and microorganisms and natural chemical processes are responsible for approximately 90 percent of pollutant removal, while, the plants remove about 7-10 percent of pollutants. In addition to organic pollutants, this device could remove the nitrogen and phosphorous in the wastewater and prevent eutrophication.

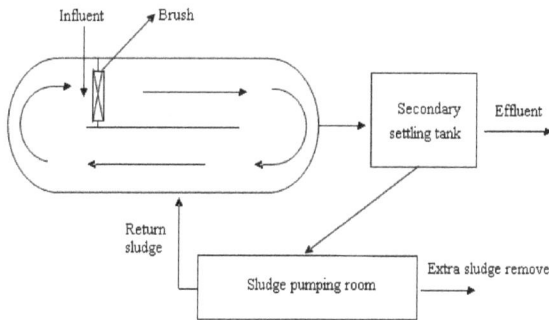

Figure 5. The scheme of the oxidation ditch [1]

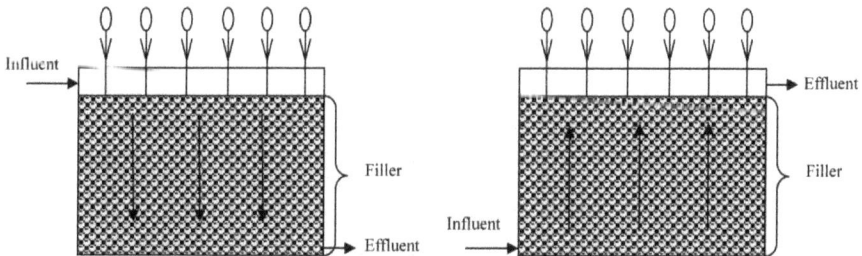

Figure 6. The scheme of the subsurface flow constructed wetland of different feeding pattern

As an economical, easy management and ecological friendly reactor, constructed wetland is supposed to be a promising technique to treat the wastewater in developing country. However, this technique was not widely used up to now for 1) the plants couldn't adapt to heavy contaminated wastewater, which strikes its application scope; 2) the device of this technique demands large area of land; 3) the efficiency of this device relativity lower than other biological device such as activated sludge process and membrane bioreactor. Thus, efforts should be made in plants selection, device structure modification and multiple devices combination to enhance the adaption and efficiency of this technique.

3. Chemical oxidation technologies

Nowadays, due to the increasing presence of molecules, refractory to the microorganisms in the wastewater streams, the conventional biological methods cannot be used for complete treatment of the effluent and hence, introduction of newer technologies to convert it into less harmful or lower chain compounds which can be then treated biologically, has become imperative. Chemical oxidation technology is one of these newer technologies which use chemical oxidant (H_2O_2, O_3, ClO_2, K_2MnO_4, K_2FeO_4 and so on) oxide pollutant to slightly toxic, harmless substances or transform it into manageable form. However, Chemical oxidation technologies constitute the use of oxidizing agents such as ozone and hydrogen peroxide, exhibit lower rates of degradation. Therefore, advanced oxidation processes (AOPs) with the capability of exploiting the high reactivity of hydroxyl radicals in driving oxidation have emerged a promising technology for the treatment of wastewaters containing refractory organic compounds. Several technologies like Fenton, photo-Fenton, wet oxidation, ozonation, photocatalysis, etc. are included in the AOPs and their main difference is the source of radicals.

3.1. Chemical oxidation technologies under normal temperature and pressure

This part aims at highlighting three different oxidation processes operating at ambient conditions viz. Fenton's chemistry (belonging to the class of AOPs) and ozonation, use of hydrogen peroxide (belonging to the class of chemical oxidation technologies).

3.1.1. Classification and principle

3.1.1.1. Hydrogen peroxide

Hydrogen peroxide (H_2O_2) is an environment friendly oxide which could oxidate organic pollutants efficiently and economically. The standard reduction potentials (1.77V, 0.88v) of hydrogen peroxide imply that it is a strong oxidant in both acidic and basic solutions [16]. It can oxidize many kinds of organic contaminants in wastewater directly. The very slow decomposition rate of hydrogen peroxide in the drinking water treatment, with mild operation conditions, can ensure a longer disinfection. Also, it can be utilized as a dechlorination agent (reductant) without organic halogen compounds production. Therefore, the hydrogen peroxide is the ideal drinking water pre-oxidant and disinfectant.

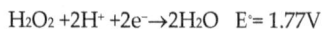

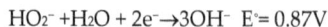

$H_2O_2 + 2H^+ + 2e^- \rightarrow 2H_2O$ $E^\circ = 1.77V$

$HO_2^- + H_2O + 2e^- \rightarrow 3OH^-$ $E^\circ = 0.87V$

However, considering for the removal of organic compounds in wastewater, the reactivity of hydrogen peroxide is generally low and largely incomplete due to kinetics, in particular in acidic media. It can be enhanced by homogeneous and/or heterogeneous catalysts, the progress named wet hydrogen peroxide catalytic oxidation (WHPCO). WHPCO operates at temperatures in the 20-80⊜ range and atmospheric pressure.

3.1.1.2. Fenton

The Fenton's process has its origin in the discovery reported in 1894 that ferrous ion strongly promotes the hydrogen peroxide oxidation of tartaric acid. The mechanism of the Fenton's process is quite complex, and some papers can be found in the literature where tens of equations are used for its description. Nevertheless, it can be summarized by the following steps: first, a mixture of H_2O_2 and ferrous iron in acidic solution generates the hydroxyl radicals which will subsequently attack the organic compounds present in the solution [17].

$Fe^{2+}+H_2O_2 \rightarrow Fe^{3+}+HO^-+HO\bullet$

As iron (II) acts as a catalyst, it has to be regenerated, which seems to occur through the following scheme:

$Fe^{3+}+H_2O_2 \leftrightarrow Fe\text{-}OOH_2^{+}+H^{+}$

$Fe\text{-}OOH_2^{+} \rightarrow Fe^{2+}+HO_2 \bullet$

The important mechanistic feature of the Fenton reaction is that in the outer-sphere single electron transfer from Fe^{2+} to H_2O_2 and generates hydroxyl radicals and hydroxide anions. Hydroxyl radicals are after fluorine atoms the most oxidizing chemical species. They are extremely powerful species to abstract one electron from an electron rich organic substrate or any other species present in the medium to form hydroxide anion. The oxidation potential of hydroxyl radicals has been estimated as +2.8 and +2.0V at pH 0 and 14, respectively. The high reactivity of $HO\bullet$ ensures that it will attack a wide range of organic compounds. Fenton reaction gives rise to CO_2 and the heteroatoms also form the corresponding oxygenated species such as NOx, SOx and POx, meaning that the carbons and heteroatoms of the organic substrate are converted to inorganic species. Equations illustrate the cyclic processes occurring in Fenton chemistry under aerobic conditions leading to the formation of CO_2.

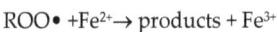

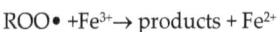

$RH+HO\bullet \rightarrow R\bullet + H_2O$

$R\bullet +Fe^{3+} \rightarrow R^{+}+Fe^{2+}$

$R^{+}+H_2O \rightarrow ROH + H^{+}$

$R\bullet +Fe^{2+} \rightarrow$ products $+ Fe^{3+}$

$R\bullet +O_2 \rightarrow ROO\bullet$

$R\bullet + \bullet OOH \rightarrow RO\bullet + \bullet OH$

$ROO\bullet +RH \rightarrow ROOH + R\bullet$

$ROO\bullet +Fe^{2+} \rightarrow$ products $+ Fe^{3+}$

$ROO\bullet +Fe^{3+} \rightarrow$ products $+ Fe^{2+}$

The performance of Fenton oxidation application to wastewater treatment was based on the following parameters: operating pH, amount of ferrous ions, concentration of hydrogen peroxide, initial concentration of the pollutant, type of buffer used for pH adjustment, operating temperature and chemical coagulation. The optimum pH has been observed to be 3 in the majority of the cases. The pollutant removal efficiency increases with an increase in the dosage of ferrous ions and hydrogen peroxide. However, care should be taken while selecting the dosage, for high dosage leasing environmental question and high treatment cost. The optimum dosage is available in the open literature or required to establish in laboratory scale studies under similar conditions.

The conventional Fenton reaction that hydrogen peroxide in conjunction with an iron(II) salt to produce high fluxes of hydroxyl radicals is homogeneous catalytic reaction. Therefore, the application of conventional Fenton reaction is complicated by the problems typical of homogeneous catalysis, such as catalyst separation, regeneration, etc. It is necessary to control pH carefully to prevent precipitation of iron hydroxide. Thus, heterogeneous catalysts Fenton reaction, i.e., solids containing transition metal cations (mostly iron ions) have been developed and tested [18].

3.1.1.3. Ozonation

Ozone is one of the most powerful oxidants with an oxidation potential of 2.07 V. In acidic conditions, ozone undergoes selective electrophilic attack which occurs at particular parts of complexing agent with high electronic density. Under alkaline environment, ozone is catalyzed by OH^- in basic conditions to intermediate compounds such as superoxide, HO radicals and HO_2 radicals which are highly reactive. Apart from pH, the degradation of target compounds in the liquid phase corresponds to the amount and form (species) of oxidants present in a reactor [19].

$O_3 + OH^- \rightarrow HO_2^- + O_2$

$O_3 + OH^- \rightarrow HO\bullet + O_3^-$

$O_3 + HO_2^- \rightarrow HO_2\bullet + O_3\bullet$

The applications of ozonation for water treatment offer various advantages. Due to its short half-life of less than 10 min, the oxidant degrades most of pollutants rapidly. However, at pH 10, the half-life of ozone in solutions is less than 1 min. As a result, ozonation extensively consumes energy, thus reducing its treatment efficiency. Due to the improvement in ozone production from pure oxygen and the increase of its concentration in the feeding gas, an ozone generation with less cost may be economically attractive.

The performance of ozonation application to wastewater treatment was based on the following parameters: operating pH, ozone partial pressure, contact time and interfacial area, presence of radical scavengers, operating temperature, presence of catalyst, combination with other oxidation processes.

Very low reaction rates have been observed for the degradation of complex compounds or mixture of contaminants by ozonation alone. Catalyst such as BST catalyst TiO_2 fixed on alumina beads, Fe (II), Mn (II) can be used to increase the degradation efficiency. Heterogeneous catalytic ozonation has received increasing attention in recent years due to its potentially higher effectiveness in the degradation and the mineralization of refractory organic pollutants and a lower negative effect on water quality. The major advantage of a heterogeneous over a homogeneous catalytic system is the ease of catalytic retrieval from the reaction media. Results suggest that catalytic ozonation with MnOx/MZ, CoOx/MZ and CuOx/Al₂O₃ is a promising technique for the mineralization of refractory organic compounds in water [20].

3.1.2. Reactors

3.1.2.1 Typical reactor used for Hydrogen peroxide

Introduction of hydrogen peroxide into the waste stream is critical due to lower stability of hydrogen peroxide. An addition point should give large residence time of H_2O_2 in the pollutant stream, but due to the practical constraints and poor mixing conditions, it is not always possible to inject H_2O_2 in line and an additional holding tank is required. The simplest, faster and cheapest method for injection of hydrogen peroxide is gravity feed system. Pump feed systems can also be used, but it requires regular attention.

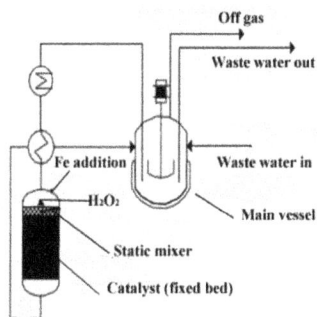

Figure 7. Typical reactor used for WHPCO technology.

Figure 7 reports a simplified flow diagram of the WHPCO technology for the treatment of olive oil milling waste water using Fe-ZSM-5 solid catalysts. H_2O_2 is added progressively at the top of a fixed bed catalytic reactor (before a static mixer), in order to maximize its local concentration. An iron solution is added on the top of the reactor to maintain catalyst activity constant. The feed solution is recirculated to and from a tank in order to have good turbulence in the catalyst bed, but also to guarantee the necessary total residence time to obtain the required level of removal of phytotoxic chemicals.

3.1.2.2. Typical reactor used for fenton oxidation

A batch Fenton reactor essentially consists of a nonpressurized stirred reactor with metering pumps for the addition of acid, base, a ferrous sulfate catalyst solution and industrial strength (35-50%) hydrogen peroxide. The reactor vessel should be coated with an acid-resistant material, because the Fenton reagent is very aggressive and corrosion can be a serious problem. pH of the solution must be adjusted at 6, usually iron hydroxide is formed. For many organic pollutants, the ideal pH for the Fenton reaction is between 3 and 4, and the optimum catalyst to peroxide ratio is usually 1:5 wt/wt. Addition of reactants are done in the following sequence: dilute sulfuric acid catalyst in acidic solutions, pH adjusting agent (adjustment of pH at 3-4) and lastly added hydrogen peroxide slowly. Effluent of the Fenton reactor (Oxidation tank) is fed into a neutralizing tank for adjusting the pH (adjustment of pH at 9), then the stream followed by a flocculation tank and a solid-liquid separation tank for removing the precipitate. A schematic representation of the Fenton oxidation treatment has been shown in Figure 8 [21].

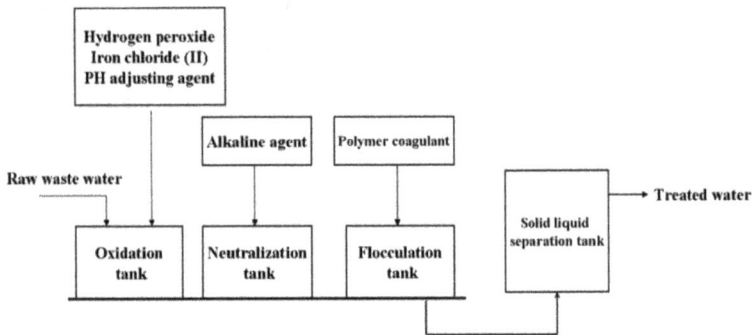

Figure 8. Typical reactor used for fenton oxidation.

3.1.2.3. Typical reactor used for ozonation

Ozone transfer efficiency should be maximized by increasing the interfacial area of contact (reducing the bubble size by using small size ozone diffusers such as porous disks, porous glass diffusers, ceramic membranes) and increasing the contact time between the gas and the water (increase the depths in the contactor, optimum being 3.7 to 5.5 m) [21].

3.1.3 .Application

3.1.3.1. Hydrogen peroxide

Hydrogen peroxide has been used in the industrial effluent treatment for detoxification of cyanide, nitrite and hypochlorite, for the destruction of phenol aromatics, formaldehyde, removal of sulfite, thiosulfate and sulfide compounds.

However, the application of hydrogen peroxide alone for wastewater treatment applications present major problems such as very low rates for applications involving complex materials, stability of H_2O_2 and mass transfer limitations. Hence, use of hydrogen peroxide alone does not seem to be a recommendable option for industrial wastewater treatment.

WHPCO process has been proposed for a variety of agro-food and industrial effluents: removal of dyestuffs from textile, treat sewage sludge, purify wastewater from pharmaceutical and chemical production, dumping site, or from cellulose production and pre-treat water streams from food-processing industries (olive oil mills, distilleries, sugar refineries, coffee production, tanneries, etc.) [22].

3.1.3.2. Fenton

Fenton process can significantly remove recalcitrant and toxic organic compounds, and increase the biodegradability of organic compounds. Leachate quality in terms of organic content, odor, and color can be greatly improved following Fenton treatment. Fenton's reagent has been used quite effectively for the treatment and pre-treatment of leachate from composting of different wastes. Reported COD removal efficiencies range from 45% to 85%, and reported final BOD_5/COD ratio can be increased from less than 0.10 initially to values ranging from 0.14 to more than 0.60, depending on leachate characteristic and dosages of Fenton reagents. Color and odor in leachate can also be reduced considerably. The decolorization efficiency is as high as 92% in Fenton treatment of a mature leachate[23]. The optimal conditions for Fenton reaction were found at a ratio $[Fe^{2+}]$/[COD] equal to 0.1. Both leachates were significantly oxidized under these conditions in terms of COD removal 77-75% and BOD_5 removal 90-98%. Fenton's reagent was found to oxidize preferably biodegradable organic matter of leachate [24]. Pirkanniemi et al. (2007) [25] tested the Fenton's oxidation to degrade complexing agents such as N-bis[2-(1,2-dicarboxyethoxy) ethyl)] glysine (BCA5), N-bis[2-(1,2-dicarboxyethoxy)ethyl]aspartic acid (BCA6) and EDTA from bleaching wastewater. It was reported that an almost complete removal of EDTA was attained at its concentration of 76 mM.

3.1.3.3. Ozonation

Ozone can be used for treatment of effluents from various industries relating to pulp and paper production (bleaching and secondary effluents), Shale oil processing, production and usage of pesticides, dye manufacture, textile dyeing, production of antioxidants for rubber, pharmaceutical production etc.

Beltrán et al. (2006) [26] reported that ozonation alone improved the removal of succinic acid up to 65% at pH 7 with an initial concentration of 339 mM. Decolorization of dye Methylene Blue can be achieved by ozonation. The COD of basic dyestuff wastewater was reduced to 64.96% and decolorization was observed under basic conditions (pH 12), complete Methylene Blue degradation occurring in 12 min. The decolorization time decrease linearly with the increase in ozone concentration. For example, increasing ozone concentration from 4.21 g/m^3 to 24.03 g/m^3 in the gas phase reduces the decolorization time of 400 mg/L dye concentration by about 88.43% [27].

3.2. Chemical oxidation technologies under high temperature and pressure

3.2.1. Classification and principle

3.2.1.1. Wet air oxidation (WAO)

WAO is based on the oxidizing properties of air's oxygen. Typical conditions for wet oxidation range from 180 °C and 2 MPa to 315 °C and 15 MPa. Residence times may range from 15 to 120 min, and the chemical oxygen demand (COD) and total organic carbon (TOC) removal may typically be about 75-90%. Insoluble organic matter is converted to simpler soluble organic compounds without emissions of NOx, SO_2, HCl, dioxins, furans, fly ash, etc.

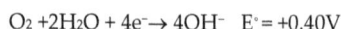

$$O_2 + 4H^+ + 4e^- \rightarrow 2H_2O \qquad E^\circ = +1.23V$$

$$O_2 + 2H_2O + 4e^- \rightarrow 4OH^- \quad E^\circ = +0.40V$$

3.2.1.2. Catalytic wet air oxidation (CWAO)

Organic pollutant is impossible to obtain a complete mineralization of the waste stream by WAO, since some low molecular weight oxygenated compounds (especially acetic and propionic acids, methanol, ethanol, and acetaldehyde) are resistant to oxidation. Organic nitrogen compounds are easily transformed into ammonia, which is also very stable in WAO conditions. Therefore, WAO is a pre-treatment of liquid wastes which requires additional treatment. The use of catalysts (WACO) allows to use milder reaction conditions but especially to promote conversion of the reaction intermediates (for example, acetic acid and ammonia) which are very difficult to convert in the absence of catalysts, as mentioned above.

Though it varies with type of wastewater, the operating cost of CWAO is about half that of non-catalytic WAO due to milder operating conditions and shorter residence time. Although the homogenous catalysts, e.g. dissolved copper salts, are effective, an additional separation step is required to remove or recover the metal ions from the treated effluent due to their toxicity, and accordingly increases operational costs. Thus, the development of active heterogeneous catalysts has received a great attention because a separation step is not necessary. Various solid catalysts including noble metals, metal oxides, and mixed oxides have been widely studied for the CWAO of aqueous pollutants. To further decrease the reaction temperature and pressure, intensive oxidants are added and form Wet Peroxide Oxidation (WPO), WHPCO is belonging to WPO.

3.2.2. Reactors

3.2.2.1. WAO reactors

The experimental set up consisted mainly of a reactor and a condenser. It was equipped with suitable measuring devices, such as thermocouple, rotameter and pressure gauge. The material of construction for reactor is titanium. The top of the reactor is connected to a reflux condenser with a stainless steel flange. The reactor was equipped with a heating jacket and a gas sparger. The gas (air or oxygen) entered the reactor through the titanium sparger. Air or oxygen bubbled out through the sparger at high speed and thus ensured proper agitation [28].

3.2.2.2. CWAO reactors

Homogeneous catalysts for CWAO are usually transition metal cations, such as Cu and Fe ions. Industrial homogenous CWAO processes have been developed such as the Ciba-

Geigy/Garnit process working at high temperature (300 °C), and the LOPROX Bayer process working with oxygen below 200 °C in the presence of iron ions. Common two-phase reactor types used in homogeneous CWO include bubble columns, jet-agitated reactors, and mechanically stirred reactor vessels.

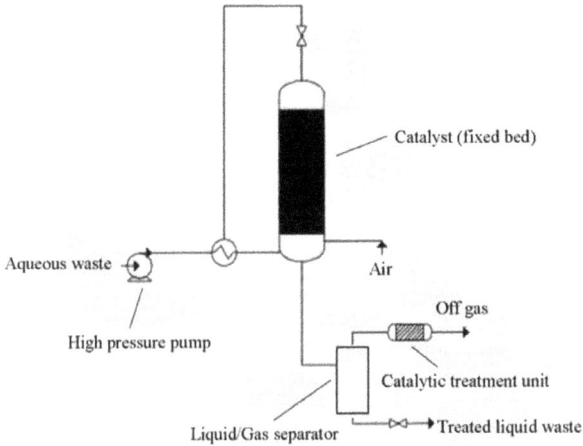

Figure 9. Typical reactor used for WACO process.

Figure 9 reports a simplified flow diagram of a WACO process which consists mainly of a high-pressure pump, an air or oxygen compressor, a heat-exchanger, a high-pressure (fixed bed) reactor and a downstream separator. The simplest reactor design is usually a cocurrent vertical bubble column with a height-todiameter ratio in the range of 5-20. A catalytic unit for the treatment of the off-gas is also typically necessary [25].

3.2.3. Application

WAO is not used as a complete treatment method, but only as a pretreatment step where the wastewater is rendered to nontoxic materials and the COD is reduced for the final treatment. For integrated WAO-biological treatment process, more detailed studies concerning the WAO pretreatment step are necessary for the design of a rational and efficient integrated process. The WAO process has been subjected to numerous investigations by researchers in the past decades as a pretreatment step before the biological treatment [29-32].

Pretreatment of Afyon (Turkey) alcaloide factory wastewater, a typical high strength industrial wastewater (COD= 26.65 kg/m^3; BOD$_5$= 3.95 kg/m^3) was carried out by WAO process. Experimental results indicated that over 26% COD removal of the wastewater could be achieved in 2.0 h of reaction time at 150°C; 0.65 MPa and with an airflow rate of 1.57×10^5 m^3/s. BOD$_5$/COD ratio is increased from 0.15 to 0.4. The experimental data also revealed that the pressure and temperature effects on the COD removal were important. The COD removal was observed to increase with an increase in both pressure and temperature. Maximum COD removal was obtained at around pH 7.0 [28].

As considering for using the CWAO process to treat Afyon (Turkey) alcaloide factory wastewater, results indicate that the presence of catalyst increases the COD removal. The COD removal for 2.0 h reaction time increased from 25.7% without catalyst to 33.2% with 0.25 kg/m^3 catalyst. While the BOD$_5$/COD ratio is increased to over 0.4 [28].

CWAO makes a promising technology for the treatment of refractory organic pollutants (phenolic compounds, carboxylic acids, N-containing compounds) in industrial wastewaters, such as Olive oil mill wastewater, Kraft bleaching plant effluents, Coke plant wastewater, Textile wastewater, Alcohol-distillery wastewater, Landfill leachate, Pulp and paper bleaching liquor, Heavily organic halogen polluted industrial wastewater and so on [33].

4. Adsorption technology

4.1. Principle of adsorption technology

Adsorption offers a cleaner technology, free from sludge handling problems and produces a high quality effluent. Over the last few decades, adsorption has gained importance as an effective purification and separation technique used in water and wastewater treatment. Adsorption is the process by which a solid adsorbent can attach a component dissolved in water to its surface and form an attachment via physical or chemical bonds, thus removing the component from the fluid phase. Adsorption is used extensively in industrial processes for many purposes of separation and purification. The removal of metals, coloured and colourless organic pollutants from industrial wastewater are considered an important application of adsorption processes using suitable adsorbents.

Adsorption is nearly always an exothermic process. We can distinguish between 2 types of adsorption process depending on which of these two force types plays the bigger role in the process. Adsorption processes can be classified as either physical adsorption (van der Waals adsorption) or chemisorption (activated adsorption) depending on the type of forces between the adsorbate and the adsorbent.

Physical adsorption occurs quickly and may be mono-molecular (unimolecular) layer or monolayer, or 2, 3 or more layers thick (multi-molecular). As physical adsorption takes place, it begins as a monolayer. It can then become multi-layer, and then, if the pores are close to the size of the molecules, more adsorption occurs until the pores are filled with adsorbate. Accordingly, the maximum capacity of a porous adsorbent can be more related to the pore volume than to the surface area.

Chemisorption involves the formation of chemical bonds between the adsorbate and adsorbent is a monolayer, often with a release of heat much larger than the heat of condensation. Chemisorption from a gas generally takes place only at temperatures greater than 300 °C, and may be slow and irreversible. Most commercial adsorbents rely on physical adsorption; while catalysis relies on chemisorption.

4.2. Development of adsorption materials

4.2.1. Activated carbon

Activated carbon is by far the most common adsorbent used in wastewater treatment. Since, during adsorption, the pollutant is removed by accumulation at the interface between the activated carbon (absorbent) and the wastewater (liquid phase) the adsorbing capacity of activated carbon is always associated with very high surface area per unit volume. Activated carbon can be manufactured from carbonaceous material, including coal (bituminous, subbituminous, and lignite), peat, wood, or nutshells (i.e., coconut). The manufacturing process consists of two phases, carbonization and activation. The carbonization process includes drying and then heating to separate by-products, including tars and other hydrocarbons, from the raw material, as well as to drive off any gases generated. The carbonization process is completed by heating the material at 400–600°C in an oxygen-deficient atmosphere that cannot support combustion. Powdered activated carbon is made up of crushed or ground carbon particles, 95–100% of which will pass through a designated mesh sieve or sieves. Granular activated carbon can be either in the granular form or extruded. It is designated by sizes such as 8×20, 20×40, or 8×30 for liquid phase applications and 4×6, 4×8 or 4×10 for vapor phase applications.

4.2.2 .Activated alumina

Activated alumina had been used in the treatment of wastewater and its adsorption capability for the removal of both organic and inorganic compounds was found to be favoured by a specific surface area, pore structure, ionic strength and chemical inertness. It can be produced from the mixtures of amorphous and gamma alumina prepared by the dehydration of $Al(OH)_3$ under low-temperatures of 300-600°C, with surface areas in the range of 250-350 m^2/g.

Research conducted on the use of microporous alumina pillared montmorillonite (clay) and mesoporous alumina aluminium phosphate as adsorbents had shown successful removal of fluoride, arsenic, selenium, beryllium, 2,4-chlorophenol, 2,4,6- trichlorophenol, pentachlorophenol, and also pesticides such as: molinate, propazine and atrazine from waster water. The removal efficiency of the pillared clay material for the herbicide was found to be higher than that of the mesoporous aluminium phosphate due to the substitution of the alkyl lateral chains of the aluminium phosphate during the sorption of s-triazines and the increase of P/Al ratio during the adsorption of propachlor.

4.2.3. Zeolites

The drawback suffered by activated carbon due to its high regeneration cost and production cost has lead to the application of zeolites as an alternative adsorbent. Zeolites are a group of natural or synthetic hydrated aluminosilicate minerals which contain both alkaline and alkali-earth metals. It has been used as an adsorbent, molecular sieve, ion-exchangers and catalysts in the past decades, because their chemical properties and large effective surface area gives them superior adsorptive qualities. There are several types of zeolites such as MCM-22, ZSM-5, ZSM-22, BETA, and Y. Their adsorption equilibrium had been studied and showed that the synthetic zeolites have higher adsorption capacity than the natural zeolites for the removal of ink, dyes and polluted wastewater.

4.2.4. Peat

Peat and other biomass materials have been used previously in the treatment of wastewater containing heavy metals and organic compounds. Peat is a yellow to dark brown residue, which occurs during the first stage of coal formation. It is composed of partly carbonizing materials such as decayed trees and peats bogs that have accumulated in water–saturated environments and swamps. The main constituents of peat moss are lignin, humic acid and cellulose. In addition, the surface functional groups of peat include aldehydes, carboxylic acids, ketones, alcohols, ethers and phenolic hydroxides, which are all involved in the adsorption of pollutants. In addition, its polar nature is responsible for its specific adsorption potential for dissolved metals and polar organic compounds.

4.2.5. Natural materials

Natural materials that are available in large quantities, or certain waste products from industrial or agricultural operations, may have potential as inexpensive adsorbents. The abundance, availability, and low cost of agricultural byproducts make them good adsorbents for the removal of various pollutants from wastewaters. Agricultural waste biomass currently is gaining importance. In this perspective rice husk, which is an agro-based waste, has emerged as an invaluable source for the utilization in the wastewater treatment. Rice husk contains ~20% silica, and it has been reported as a good adsorbent for the removal of heavy metals, phenols, pesticides, and dyes. The adsorptive capacity of rice husk silica had been evaluated by Grisdanurak et al. [34] and its adsorption capacity for chlorinated volatile organic compounds was found to be higher than that of commercial mordenite and activated carbons. It has been utilized for solving disposal problems and also as an adsorbent in treating organic wastewaters. The adsorption potential of this biomass for adsorbing phenol from aqueous solution was found to be depended on the pH, contact time and the initial phenol concentration. This result shows that phenol was adsorbed to a lesser extent at higher pH values. Phenol forms salts, which readily ionize leaving negative charge on the phenolic group while, its present on the adsorbent prevents the removal of phenolateions. In addition, the percentage adsorption of phenol for this test also decreases as the initial phenol concentration increases. The adsorption capacity determined for this test was 0.886 mg/g for phenol and the equilibrium data was fitted successfully by the Freundlich model [35].

4.2.6. Polymeric

Polymeric adsorbents are non-functionalized organic polymers which are capable of removing organics from water. The principle is quite simple. Wastewater is passed through a column containing the polymeric adsorbent. The organic materials are retained on the resin while water and some simple salts pass through. When the resin is fully loaded, the organics are stripped from the resin with solvents or caustic. The organic material may be concentrated by orders of magnitude in some cases. The following recommendations are those being used at the present time. The regenerants used are not the only ones possible. The choice of regenerant (solvent) usually depends on the availability at the particular location.

4.3. Adsorption equipment and their applications

Granular activated carbon systems are generally composed of carbon contactors, virgin and spent carbon storage, carbon transport systems, and carbon regeneration systems. The carbon contactor consists of a lined steel column or a steel or concrete rectangular tank in which the carbon is placed to form a "filter" bed. A fixed bed downflow column contactor is often used to contact wastewater with granular activated carbon. Wastewater is applied at the top of the column, flows downward through the carbon bed, and is withdrawn at the bottom of the column. The carbon is held in place with an underdrain system at the bottom of the contactor. Provisions for backwash and surface wash of the carbon bed are required to prevent buildup of excessive headloss due to accumulation of solids and to prevent the bed surface from clogging.

There are two basic types of water filters: particulate filters and adsorptive/reactive filters. Particulate filters exclude particles by size, and adsorptive/reactive filters contain a material (medium) that either adsorbs or reacts with a contaminant in water. The principles of adsorptive activated carbon filtration are the same as those of any other adsorption material. The contaminant is attracted to and held (adsorbed) on the surface of the carbon particles. The characteristics of the carbon material (particle and pore size, surface area, surface chemistry, etc.) influence the efficiency of adsorption [36].

The characteristics of the chemical contaminant are also important. Compounds that are less water-soluble are more likely to be adsorbed to a solid. A second characteristic is the affinity that a given contaminant has with the carbon surface. This affinity depends on the charge and is higher for molecules possessing less charge. If several compounds are present in the water, strong adsorbers will attach to the carbon in greater quantity than those with weak adsorbing ability.

5. Other technologies

5.1. Solvent extraction

Solvent extraction is a common form of chemical extraction using organic solvent as the extractant. It is commonly used in combination with other technologies, such as solidification/stabilization, incineration, or soil washing, depending upon site-specific conditions. Solvent extraction also can be used as a stand alone technology in some instances. Organically bound metals can be extracted along with the target organic contaminants, thereby creating residuals with special handling requirements. Traces of solvent may remain within the treated soil matrix, so the toxicity of the solvent is an important consideration.

Solvent extraction method has many advantages, such as less investment in equipment, easy to operate and lower consumption. Moreover, the major pollutants can be effectively recycled by solvent extraction method. The extraction method is widely used in a variety of organic waste, such as phenol, organic carboxylation acids, organic phosphorus nitrogen, organic sulfonic acid, organic amine, etc. Solvent extraction has been shown to be effective in treating sediments, sludges, and soils containing primarily organic contaminants such as PCBs, VOCs, halogenated solvents, and petroleum wastes. The process has been shown to be applicable for the separation of the organic contaminants in paint wastes, synthetic rubber process wastes, coal tar wastes, drilling muds, wood-treating wastes, separation sludges, pesticide/insecticide wastes, and petroleum refinery oily wastes.

Adopting solvent extraction treatment technology for organic wastewater, the most important thing is to choose the right process flow specifically for specially appointed pollution. For the general flow, most difficult degradation of the pollutants are removed after the process of solvent extraction. Crafts residue mainly contain some pollutants which are not extractive and dissolved, and they would meet the emissions standards through the secondary treatment regeneration (such as biochemistry, chemical oxidation, etc.).

5.2. Incineration

Incineration involves the combustion of the organic (carbon-containing) solids present in wastewater solids and biosolids to form carbon dioxide and water. The temperature in the combustion zone of furnaces is typically 1023K to 1143K. The solids that remain at the end of the process are an inert material commonly known as ash. Either undigested wastewater solids or biosolids may be incinerated. The terms thermal oxidation and combustion may be used interchangeably with incineration.

Incineration takes advantage of the fuel value of wastewater treatment residual solids (referred to as sludge) and biosolids. In some cases, the energy recovered from this process has been used in heat exchangers and waste heat boilers to save on energy use at

the wastewater treatment plant. For example, in Montreal, a portion of the biosolids generated at the facility are incinerated, while the remaining portion is pelletized. Waste heat from the biosolids that are incinerated is used in the thermal dryers that produce fertilizer pellets. In Europe, there is a trend to use biosolids as a fuel source in dedicated power generation facilities. In addition, incineration results in a large reduction in volume and mass in comparison to other alternatives and options. The mass of solids in the ash that results from the inceration process is approximately 10% of that of the biosolids fed into the incinerator. This reduces the mass and volume requiring disposal.

There are two common incineration technologies for wastewater solids and biosolids: fluidized bed incinerators and multiple hearth incinerators. Fluidized bed incinerators are steel cylinders lined with refractory bricks to withstand the high operating temperatures of the unit. Multiple hearth incinerators consist of a series of refractory brick hearths, stacked vertically. A rotating shaft through the centre of the hearths supports rake arms for each hearth, thereby facilitating drying and incineration. Solids are usually fed through at the top hearth and are directed to successive inner or outer dropholes as they move down through the hearths. Most of the ash is discharged from the bottom hearth.

Over the years, incineration technologies have evolved considerably and regulations and procedures have continually been enhanced to protect human and animal health and the environment. A considerable amount of scientific study has been undertaken to support the development of the regulations, and ongoing research contributes to the continuous improvement of this practice. However, some segments of the public still have concerns that incineration may be unsafe because of perceptions related to outdated technology and to experiences with incineration of other materials such as hazardous waste, municipal solid waste and medical waste.

5.3. Photocatalysis

To date, the most widely applied photocatalyst in the research of water treatment is the Degussa P-25 TiO_2 catalyst. This catalyst is used as a standard reference for comparisons of photoactivity under different treatment conditions. The fine particles of the Degussa P-25 TiO_2 have always been applied in a slurry form. This is usually associated with a high volumetric generation rate of reactive oxygen species as proportional to the amount of surface active sites when the TiO_2 catalyst in suspension. On the contrary, the fixation of catalysts into a large inert substrate reduces the amount of catalyst active sites and also enlarges the mass transfer limitations. Immobilization of the catalysts results in increasing the operation difficulty as the photon penetration might not reach every single surface site for photonic activation. Thus, the slurry type of TiO_2 catalyst application is usually preferred. With the slurry TiO_2 system, an additional process step would need to be entailed for post-separation of the catalysts. This separation process is crucial to avoid the loss of catalyst particles and introduction of the new pollutant of contami-nation of TiO_2 in the treated water [37]. The catalyst recovery can be achieved through process hybridization with conventional sedimentation [38], cross-flow filtration [39] or various membrane filtrations [40].

Natural clays have been used intensively as the support for TiO_2 owing to their high adsorption capacity and cost-effectiveness. The use of the photocatalytic membranes has been targeted owing to the photocatalytic reaction can take place on the membrane surface and the treated water could be continuously discharged without the loss of photocatalyst particles. To broaden the photoresponse of TiO_2 catalyst for solar spectrum, various material engineering solutions have been devised, including composite photocatalysts with carbon nanotubes [41], dyed sensitizers [42], noble metals or metal ions incorporation [43], transition metals and non-metals doping [44].

5.4. Ultrasonic

High-frequency ultrasound is a mechanical wave, with a shorter wavelength, the energy concentration characteristics, its application mainly on the basis of energy major, along a straight line features of these two started. The 20th century, the early 90s, some scholars have begun to study abroad, such as the ultrasonic degradation of organic pollutants in water. Ultrasound technology is simple, efficient, non-polluting or less polluting characteristics, in recent years the development of a new type of water treatment technology. It combines advanced oxidation, pyrolysis, supercritical oxidation technology in one, and the degradation of speed, be able to water of harmful organic compounds into CO_2, H_2O, inorganic ions or organic toxic than the original readily biodegradable organic matter, and therefore in dealing with difficult Bio-degradation of organic contaminants has significant advantages.

6. Treatment processes of various industrial organic wastewaters

6.1. Coking plant

Coke, produced by the pyrolysis of natural coals, is an indispensable material for most of the metallurgical facilities. During coking, coal decomposes into gases, liquid and solid organic compounds. Coke wastewater contains high concentration of ammonia, phenols, thiocyanate, cyanide and lower amounts of other toxic compounds, such as polyaromatic hydrocarbons (PAHs), e.g. naphthalene, and heterocyclic nitrogenous compounds, e.g. quinoline. The individual concentration of the contaminants depends on the quality of coal and the properties of the coking process.

Coke wastewater handling usually consists of a series of physico-chemical treatments reducing the concentration of ammonia, cyanide, solids and other substances, followed by different biological treatments, mainly activated sludge process. The application of two or three consecutive activated sludge systems is particularly favored as readily biodegradable substrates like phenol can be removed in the first step. Phenols, which contribute to the greatest extent to the total COD in coke wastewater, are not only highly toxic and carcinogenic compounds, but also inhibit advantageous biological processes like nitrification. Under optimal circumstances, thiocyanate degradation can also be achieved in the first activated sludge step.

The influent concentrations of NH_4^+-N, phenols, COD and thiocyanate (SCN^-) in the wastewater ranged between 504 and 2340, 110 and 350, 807 and 3275 and 185 and 370 mg/L, respectively. A laboratory-scale activated sludge plant composed of a 20 L volume aerobic reactor followed by a 12 L volume settling tank and operating at 35 ⊚ was used to study the biodegradation of coke wastewater. Maximum removal efficiencies of 75%, 98% and 90%

were obtained for COD, phenols and thyocianates, respectively, without the addition of bicarbonate. The concentration of ammonia increased in the effluent due to both the formation of NH_4^+ as a result of SCN^- biodegradation and to organic nitrogen oxidation. A maximum nitrification efficiency of 71% was achieved when bicarbonate was added, the removals of COD and phenols being almost similar to those obtained in the absence of nitrification [45]. An anaerobic-anoxic-aerobic (A(1)-A(2)-O) and an anoxic-aerobic (A/O) biofilm system were used to treat coke-plant wastewater. At same or similar levels of HRT, the two systems had almost identical COD and NH_3 removals, but a different organic-N removal. Set-up of an acidogenic stage benefited for the removal of organic-N and the A(1)-A(2)-O system was more useful for total nitrogen removal than the A-O system [46].

Newly studies for treatment of coking wastewaters are listed. Chu et al. investigated coking wastewater treatment by an advanced Fenton oxidation process using iron powder and hydrogen peroxide. The results showed that higher COD and total phenol removal rates were achieved with a decrease in initial pH and an increase in H_2O_2 dosage. At an initial pH of less than 6.5 and H_2O_2 concentration of 0.3 M, COD removal reached 44-50% and approximately 95% of total phenol removal was achieved at a reaction time of 1 h. The oxygen uptake rate of the effluent measured at a reaction time of 1 h increased by approximately 65% compared to that of the raw coking wastewater. This indicated that biodegradation of the coking wastewater was significantly improved. Several organic compounds, including bifuran, quinoline, resorcinol and benzofuranol were removed completely as determined by GC-MS analysis. The advanced Fenton oxidation process is an effective pretreatment method for the removal of organic pollutants from coking wastewater. This process increases biodegradation, and may be combined with a classical biological process to achieve effluent of high quality [47].

Bioaugmented zeolite-biological aerated filters (Z-BAFs) were designed to treat coking wastewater containing high concentrations of pyridine and quinoline and to explore the bacterial community of biofilm on the zeolite surface. The investigation was carried out for 91 days of column operation and the treatment of pyridine, quinoline, total organic carbon (TOC), and ammonium was shown to be highly efficient by bioaugmentation and adsorption. This bioaugmented Z-BAF method was shown to be an alternative technology for the treatment of wastewater containing pyridine and quinoline or other N-heterocyclic aromatic compounds [48].

6.2. Textile wastewater

Dyes and pigments have been utilized for coloring in the textile industry for many years. Several types of textile dyes are available for use with various types of textile materials. Textile wastewater contains dyes damages the esthetic nature of water and reduces light penetration through the water's surface, and also the photosynthetic activity of aquatic organisms. It also contains toxic and potential carcinogenic substances. Therefore it must be adequately treated before they can discharge into receiving water bodies. There are several applied treatment methods for textile effluents, involving biological, physical or chemical methods and combinations of these. Among the different technologies that can be applied for the treatment of textile wastewaters, Coagulation-flocculation (CF) and Activated Sludge Process (ASP) are widely used as they are efficient and simple to operate. Generally, these processes can be applied alone to remove suspended colloidal particles or as pre-treatment prior to Ultrafiltration (UF), Nanofiltration (NF) or Reverse Osmosis (RO) respectively for dissolved organic substances removal, decolorization and desalination.

Biological treatment resulted in a high percent reduction in chemical oxygen demand (COD), total Kjeldahl nitrogen (TKN), and total phosphorus (TP), and in a moderate decrease in color. The process was found to be independent of the variations in the anoxic time period studied; however, an increase in solids retention time (SRT) improved COD and color removal, although it reduced the nutrient (TKN and TP) removal efficiency. Furthermore, combined treatment (biological treatment and Fenton oxidation) resulted in enhanced color reduction [49].

The treatability of textile wastewaters in a bench-scale experimental system, comprising an anaerobic biofilter, an anoxic reactor and an aerobic membrane bioreactor (MBR) was evaluated by S. Grilli et al. The MBR effluent was thereafter treated by a nanofiltration (NF) membrane. The proposed system was demonstrated to be effective in the treatment of the textile wastewater. The MBR system achieved a good COD (90-95%) removal; due to the presence of the anaerobic biofilter, also effective color removal was obtained (70%). The addition of the NF membrane allowed the further improvement in COD (50-80%), color (70-90%) and salt removal (60-70% as conductivity). In particular the NF treatment allowed the almost complete removal of the residual color and a reduction of the conductivity such as to achieve water quality suitable for reuse [50].

Typical contaminants of wool textile effluents are heavy metal complexes with azo-dyes. One of the most representative heavy metals is chromium. In aquatic environments chromium can be present as Cr(III) and/or Cr(VI), mainly depending on pH and redox conditions; the two forms behave quite differently, since Cr(III) is much less soluble and therefore less mobile than Cr(VI). The heavy metal can not be removed by activated sludge effectively. The constructed wetlands (CWs) in full-scale systems and in pilot plants evidenced good performances for several elements, including chromium. Donatella et al investigated the fate of Cr(III) and Cr(VI) in a full-scale subsurface horizontal flow constructed wetland planted. The reed bed operated as post-treatment of the effluent wastewater from an activated sludge plant serving the textile industrial district. Removals of Cr(III) and Cr(VI) was 72% and 26%, respectively. The mean Cr(VI) outlet concentration was 1.6±0.9 g/l and complied with the Italian legal limits for water reuse [51].

6.3. Food and fermentation wastewater

Food processing and fermentation industries have being experiencing a significant growth in China. Wastewater streams discharged from these industries are generally characterized with high strength organic and nutrient contents, e.g., COD 10000 mg/L, TN 600 mg/L, and tend to bring serious water environment contamination if discharged without proper treatment. The conventional treatment of this kind of high strength wastewater is anaerobic/aerobic activated sludge processes.

Recent years, considerable concern has been focused on the development of the anaerobic membrane bioreactor (AMBR), which is an anaerobic reactor coupled with a membrane filtration unit. The viability of the AMBR treating high-concentration food wastewater depended upon feedwater organic concentration, loading rate, HRT, SRT, hydraulic shearing effect and membrane properties. The HRT kept at 60 h, SRT was designed for 50 days. The effluent COD removal achieved above 90% at loading rate of 2.0 kg/m^3/d and above 80% at a loading of 2.0-4.5 kg/m^3/d. The membranes all exhibited high efficiency in removal of SS, color, COD and bacteria, reaching 499.9%, 98%, 90%, and 5 logs, respectively [52].

Wang et al. [53] applied an anoxic/aerobic membrane bioreactor (MBR) to simultaneous removals of nitrogen and carbon from food processing wastewater. The system is proposed to be applied jointly with anaerobic pre-treatment. In order to simulate the quality from anaerobic pre-treatment, raw wastewater taken from a food processing factory was fed to the system after dilution. By continuous runs under appropriate operational conditions, COD, NH$_4^+$-N and TN removal was over 94, 91 and 74%, respectively. The anoxic reactor and aerobic MBR contributed 40-63 and 29-46% to COD removal, and 31-43 and 47-64% to NH$_4^+$-N removal, respectively. The maximum volumetric COD and TN loadings as high as 3.4 kg COD/m^3/day and 1.26 kg N/m^3/day were achieved.

Food processing and fermentation wastewaters can be characterized as nontoxic because they contain few hazardous compounds, have high BOD$_5$ and much of the organic matter in them consists of simple sugars and starch. Hence, this high-carbohydrate wastewater is the most useful for industrial production of hydrogen. Food Wastewaters obtained from four different food-processing industries had COD of 9 g/L (apple processing), 21 g/L (potato processing), and 0.6 and 20 g/L (confectioners A and B). Biogas produced from all four food processing wastewaters consistently contained 60% hydrogen, with the balance as carbon dioxide. COD removals as a result of hydrogen gas production were generally in the range of 5-11%. Overall hydrogen gas conversions were 0.7-0.9 L H$_2$/L-wastewater for the apple wastewater, 0.1 L/L for Confectioner-A, 0.4-2.0 L/L for Confectioner B, and 2.1-2.8 L/L for the potato wastewater [54].

Hydrogen yields were 0.61-0.79 mol/mol for the food processing wastewater (Cereal), ranged from 1 to 2.52 mol/mol for the other samples. A maximum power density of 8177mW/m^2 (normalized to the anode surface area) was produced using the two-chambered MFC and the Cereal wastewater (diluted 10 times to 595 mg COD/L), while at the same time the final COD was reduced to lower 30 mg/L (95% removal). Although more studies are needed to improve hydrogen yields, these results suggest that it is possible to link a MFC to biohydrogen to recover energy from food processing wastewaters, providing a new method to offset wastewater treatment plant operating costs [55].

6.4. Pharmaceutical wastewater

The pharmaceutical manufacturing industry produces a wide range of products to be used as human and animal medications. Treatment of pharmaceutical wastewater is troublesome to reach the desired effluent standards due to the wide variety of the products produced in a drug manufacturing plant, thus, variable wastewater composition and fluctuations in pollutant concentrations. The substances synthesized in a pharmaceutical industry are structurally complex organic chemical that are resistant to biological degradation. Soluble COD removal efficiency is about 62% at 30⊚. Therefore, there is a need for advanced oxidation methods. As the process costs may be considered the main obstacle to their commercial application. Cost-cutting approaches have been proposed, such as combining AOP and biological treatment.

Fenton's oxidation is very effective method in the removal of many hazardous organic pollutants from wastewaters. Fenton's oxidation can also be an effective pretreatment step by transforming constituents to by-products that are more readily biodegradable and reducing overall toxicity to microorganisms in the downstream biological treatment processes.

Optimum pH was determined as 3.5 and 7.0 for the first (oxidation 30 min) and second stage (coagulation 30 min) of the Fenton process, respectively. For all chemicals, COD removal efficiency was highest when the molar ratio of H_2O_2/Fe^{2+} was 150-250. At H_2O_2/Fe^{2+} ratio of 155, 0.3M H_2O_2 and 0.002M Fe^{2+}, Fenton process provided 45-65% COD removal (influent COD 35000-40000 mg/L) [56].

Real pharmaceutical wastewater containing 775 mg dissolved organic carbon (3324 mg COD) per liter was treated by a solar photo-Fenton/biotreatment. The photo-Fenton treatment time (190 min) and H_2O_2 dose (66 mM) necessary for adequate biodegradability of the wastewater. And biological treatment was able to reduce the remaining dissolved organic carbon to less than 35 mg/L. Overall dissolved organic carbon degradation efficiency of the combined photo-Fenton and biological treatment was over 95%, of which 33% correspond to the solar photochemical process and 62% to the biological treatment [57]. Due to the high COD concentration in pharmaceutical wastewaters, anaerobic processes have been made to utilize, such as upflow anaerobic sludge blanket (UASB) reactor, anaerobic filter (AF), anaerobic continuous stirred tank reactor (CSTR) and a hybrid reactor combining UASB and AF. The COD reduction of anaerobic process treating pharmaceutical wastewater containing macrolide antibiotics was 70-75%, at a total HRT of 4 d and OLR of 1.86 kg $COD/m^3/d$ [58].

The two-phase anaerobic digestion (TPAD) system comprised a CSTR and a UASBAF reactor, working as the acidogenic and methanogenic phases, respectively. The wastewater was high in COD, varying daily between 5789 and 58,792mg/L, with a wide range of pH from 4.3 to 7.2. Almost all the COD was removed by the TPAD-MBR system, leaving a COD of around 40mg/L in the MBR effluent, at respective HRTs of 12, 55 and 5 h. The pH of the MBR effluent was found in a narrow range of 6.8-7.6, indicating that the MBR effluent can be directly discharged into natural waters. As demonstrated by an overall COD removal efficiency of more than 99% [59].

6.5. Sugar refinery wastewater

Sugar refineries generate a highly coloured effluent resulting from the regeneration of anion-exchange resins (used to decolourize sugar liquor). This effluent represents an environmental problem due to its high organic load, intense colouration and presence of phenolic compounds. The colored nature of the effluent is mainly due to (1) the presence of melanoidins, that are brown polymers formed by the Maillard amino-carbonyl reaction and (2) the presence of thermal and alkaline degradation products of sugars (e.g. caramels). Most of the organic matter present in the effluent can be reduced by conventional biological treatments but the colour is hardly removed by these treatments.

The remaining colour can lead to a reduction of sunlight penetration in rivers and streams which in turn decreases both photosynthetic activity and dissolved oxygen concentrations causing harm to aquatic life. *P. chrysosporium* can remove color and total phenols from the sugar refinery effluent. A rotating biological contactor (RBC) containing *P. chrysosporium* immobilized on polyurethane foam (PUF) disks was operated with optimized decolourization medium, in continuous mode with a retention time of 3 days. During the course of operation the color, total phenols and chemical oxygen demand were reduced by 55, 63 and 48%, respectively. Addition of glucose was obligatory and the minimum glucose concentration was found to be 5 g/L [60].

Wastewater obtained from Guangxi Nanning sugar refinery (COD 86.02 g/L) is first diluted by 100 times, then treated by adding amphiphilic flocculants (CMTMC) mg/L at pH 6.6, COD removal to reached to 95%. The wastewater color changed from fuscous brown to buff yellow. After flocculation and purification, the treated water could reach the national first level discharge standards. (GB8978-88, China) [61].

Sugar refinery wastewater containing high organic load can be used as carbon sources for hydrogen production by microorganisms. As reported pretreated sugar refinery wastewater was used for the production of hydrogen by *Rhodobacter sphaeroides* O.U.001. Hydrogen was produced at a rate of 0.001 L hydrogen/h/L culture in 20% dilution of the wastewater. To adjust the carbon concentration to 70 mM and nitrogen concentration to 2 mM, sucrose or L-malic acid was added as carbon source and sodium glutamate was added as nitrogen source to the 20% dilution of SRWW. By these adjustments, hydrogen production rate was increased to 0.005 L hydrogen/h/L culture [62].

7. The cost accounting of different organic wastewater treatment

The cost of organic wastewater treatment includes two parts: the capital expenditure and the operation expenditure. The total cost relates to the characters of the influent, the technique we selected, the characters of the effluent, the time cost during the treatment etc. In this section, the pollutants are divided into degradable and reluctant ones. Some typical wastewater was selected in each group, and the feasible methods to treat it and their cost were discussed.

7.1. The degradable organic pollutants

Wastewater with degradable organic pollutants usually comes from domestic sewage, food processing, breeding industry etc. This wastewater has high BOD, and could break down in the nature condition, given enough time. Most of the techniques could be used to treat the degradable organic pollutants, and biological methods are favorite because of their efficiency and economic properties.

Sewage is one of the most important sources of degradable organic pollutants, which contributes to 37.5% of total COD in China in 2011. Therefore, sewages are treated before discharge in order to reduce the impact of the pollutants to the environment. Several biological methods, including aerobic biodegradation, activated sludge reactor, membrane bioreactor, constructed wetland etc., have been used in the sewages treatment, and their efficiency and cost have been compared. Taking the research of Song as an example[63], in response to the characteristics of decentralized domestic sewage, several treatment technologies including biogas purification tank, constructed wetland, viewing earthworm ecological, high rate algal pond, membrane bioreactor and integrated treatment equipment were applied to the domestic sewage, and their efficiency and cost were calculated and showed in table 1.

Treatment	Load (m³/d)	Capital expenditure (10⁴Yuan/m³)	Operation expenditure (10⁴Yuan/m³)	Quality of the effluent (GB18919-2002)
Biogas purification tank	20-200	0.06-0.08	0.02-0.05	2nd grade
Constructed wetland	30-3100	0.06-0.2	0.05-0.2	1st grade B
Viewing earthworm ecological	2-12	0.7-2.0	0.5-1.2	1st grade B
High rate algal pond	-	-	-	1st grade B
Membrane bioreactor	5-100000	0.19-1.0	0.25-1.05	1st grade B
Integrated treatment equipment	20-	1.0-1.5	0.27-0.8	1st grade A

Table 1. The load, cost accounting and effluent of different treatments

According to the technologies, the biogas purification tank, constructed wetland, viewing earthworm ecological and high rate algal pond were characterized by low investment, operating cost, and convenient management. The membrane bioreactor and integrated treatment equipment had the higher operating cost, and the need for professional management, which could be used in the area with higher economic development and stricter effluent qualities.

The industrial waste water from agricultural and sideline food processing industry contain high concentration of organics and suspended substance. Food wastewater is composited of natural organic matters (such as protein, fat, sugar, starch), so they are of low toxicity and high BOD/COD value (up to 0.84). Physical (such as adsorption, air flotation), chemical (flocculation) and biological methods (aerobic biodegradation, activated sludge reactor, sequencing batch reactor, oxidation pond) could be used to remove the pollutants. Most of the physical and chemical techniques are costly and need secondary treatment, therefore, food wastewater was mainly treated by biological methods. The cost varied greatly with the characters of the influent. Longda food industry compared the load and cost of oxidation pond and sequencing batch reactor, results were shown in the table 2 [64].

Treatment	Design capacity (m³/d)	Wastewater quantity (m³/a)	Total cost (Yuan/m³)	Electricity consumption (kwh/m³)
Oxidation pond	6500	1985300	0.56	0.335
Sequencing batch reactor	4500	1114200	0.455	0.25

Table 2. The load, cost accounting of oxidation pond and sequencing batch reactor

7.2. The reluctant organic pollutants

The reluctant organic pollutants, including benzene series, pharmaceutical intermediates, pesticide etc., mainly come from paper making industry, chemical industry, printing and dyeing wastewater, mechanical manufacturing industry, and agriculture [65]. This kind of wastewater is reluctant to biodegradation either owed to its toxicity or stable structure, therefore, their disposal usually costs higher than degradable ones.

The paper making wastewater reaches 10% of total industrial water. This kind of wastewater contained high concentration and complex structure pollutants, such as lignin, cellulose, hemicellulose, monosaccharide, and could cause serious pollution. The traditional two-stage biochemical treatment has relativity low cost, but the effluent could hardly meet the discharge standard of China owing to its high COD and chroma. The advanced oxidation technique could remove the pollutants from paper making wastewater efficiently, without any secondary pollution. However, the H_2O_2 used in this method is very expensive, which affects the application and extension of this technology [66]. Flocculation is another efficient method for paper making wastewater treatment, and its COD remove rate could reach 95% at the optimal condition, and the flocculants could be reused after treatment. The cost of this technique is in the middle of the two methods mentioned above (around 1.5-2 Yuan/m³).

The printing and dyeing wastewater contains of much refractory bio-degradable organism with extremely high chrome, therefore, it is hard to be efficiently treated with biological technique [67]. Advanced oxidation could degrade the organisms and reduce the toxicity of this wastewater, but it is too expensive to be used to deal with a great amount of dyeing wastewater. The membrane separation technique could also obtain high pollutants remove rate, but the high cost of the membrane and the energy also hinder the technique from widely application. The flocculation is the most common used technique owing to its moderate price and basically satisfactory results. Partial related with the character of the wastewater, the cost of the flocculation treatment ranges from 3 yuan/m³ to 5 yuan/m³. Some researchers suggested that the combination of the flocculation technique with other techniques, such as Fenton, biological technique could reduce the cost without affecting the effluent quality.

Generally speaking, among all the techniques, biological technique costs the lowest if the pollutants are degradable. The flocculation and adsorption techniques could dispose of the wastewater at a moderate price, but the flocculant and adsorbent need secondary treatment for reuse. Membrane separation and the advanced oxidation could remove pollutants efficiently, but they are costly.

8. Conclusion

The treatment technologies for organic wastewater at present were reviewed. That a variety of technologies such as biological treatment, chemical oxidation technologies, adsorption technology and the others were introduced. At last, the cost accounting of different organic wastewater treatments was discussed.

Author details

Chunli Zheng
School of Energy and Power Engineering, Xi'an Jiaotong University, China

Ling Zhao and Zhimin Fu
College of Environment & Resources of Inner Mongolia University, China

Xiaobai Zhou
The Environmental Monitoring Center of Jiangsu Province, Nanjing, China

An Li
School of Petrochemical Engineering, Lanzhou University of Technology, China

9. References

[1] Wu WE, Ge HG, Zhang KF. Wastewater biological treatment technology. Chemical Industry Press (CIP) Publishing: BeiJing, 2003 [In Chinese].

[2] Ju KS, Parales RE (2010) Nitroaromatic Compounds, from Synthesis to Biodegradation Microbiol. Mol. Biol. R. 74: 250-272.

[3] Van den Berg M, Birnbaum L, Bosveld ATC, et al.(1998) Toxic Equivalency Factors (TEFs) for PCBs, PCDDs, PCDFs for Humans and Wildlife. Environ. Health Perspect. 106: 775-792.

[4] Sims RC, Overcash MR (1983) Fate of Polynuclear Aromatic Compounds (PNAs) in Soil- Plant Systems. Residue Reviews 88: 1-68.

[5] Pope CN (1999) Organophosphorus pesticides: Do they all have the same Mechanism of Toxicity? J. Toxicol. Env. Heal. B. 2: 161-181.

[6] Aislabie J, Lloydjones G (1995). A Review of Bacterial-Degradation of Pesticides. Aus. J. Soil Res. 33: 925-942.

[7] Leahy JG, Colwell RR (1990) Microbial-Degradation of Hydrocarbons in the Environment. Microbiol. R. 54: 305-315.

[8] Scott JP, Ollis DF (1995). Integration of Chemical and Biological Oxidation Processes For Water Treatment: Review and Recommendations. Environ. Prog. 14: 88-103

[9] Pedro JJA, Walter AI. Bioremediation and Natural Attenuation: Process Fundamentals and Mathematical Models. Copyright © 2006 John Wiley & Sons, Inc.

[10] Low EU, Chase HA, Milner MG (2000) Uncoupling of Metabolism to Reduce Biomass Production in the Activated Sludge Process. Wat. Res. 34: 3204-3212

[11] Ahmed FN, Lan CQ (2012) Treatment of Landfill Leachate Using Membrane Bioreactors: A Review. Desalination 287: 41-54.

[12] Leitinga G, Hulshoff Pol L W (1991) UASB-process design for various types of wastewaters, Water Sci. Techol. 24, 87-107.

[13] Kassab G, Halalsheh M, Klapwijk A, Fayyad M, Van Lier JB (2010) Sequential Anaerobic-Aerobic Treatment for Domestic Wastewater - A Review. Bioresour. Technol. 101: 3299-3310.

[14] Peng Y, Hou H, Wang S, Cui Y, Zhiguo Y (2008) Nitrogen and Phosphorus Removal in Pilot-Scale Anaerobic-Anoxic Oxidation Ditch System. J Environ Sci 20(4):398-403.

[15] Mook WT, Chakrabarti MH, Aroua MK et al (2012) Removal of total ammonia nitrogen (TAN), nitrate and total organic carbon (TOC) from aquaculture wastewater using electrochemical technology: A review. Desalination 285: 1-13.

[16] Busca G, Berardinelli S, Resini C (2008) Technologies for the Removal of Phenol from Fluid Streams: A short review of recent developments. J. Hazard. Mater. 160: 265-288.

[17] Herney-Ramirez J, Vicente MA, Madeira LM (2010) Heterogeneous Photo-Fenton Oxidation with Pillared Clay-based Catalysts for Wastewater Treatment: A review. Appl. Catal., B. 98: 10-26

[18] Navalon S, Alvaro M, Garcia H (2010) Heterogeneous Fenton Catalysts Based on Clays, Silicas and Zeolites, Appl. Catal., B. 99: 1-26.

[19] Sillanpää MET, Kurniawan TA, Lo W (2011) Degradation of Chelating Agents in Aqueous Solution Using Advanced Oxidation Process (AOP). Chemosphere 83: 1443-1460.

[20] Li D, Qu J (2009) The Progress of Catalytic Technologies in Water Purification: A review, J. Environ. Sci. 21: 713-719

[21] Gogate PR, Pandit AB (2004) A Review of Imperative Technologies for Wastewater Treatment I: Oxidation Technologies at Ambient Conditions. Adv. in Environ. Res. 8: 501-551.

[22] Perathoner S. Centi G (2005) Wet Hydrogen Peroxide Catalytic Oxidation (WHPCO) of Organic Waste in Agro-food and Industrial Streams. Top. Catal. 33: 1-4.

[23] Deng Y, Englehardt JD (2006) Treatment of Landfill Leachate by the Fenton Process. Water Res. 40(20): 3683-3694.

[24] Trujillo D, Font X, Sanchez A (2006) Use of Fenton Reaction for the Treatment of Leachate from Composting of Different Wastes. J. Hazard. Mater. B. 138: 201-204.

[25] Pirkanniemi K, Metsärinne S, Sillanpää M (2007) Degradation of EDTA and Novel Complexing Agents in Pulp and Paper Mill Process and Wastewaters by Fenton's Reagent. J. Hazard. Mater. 147, 556-561.

[26] Beltrán FJ, Araya JFG, Giráldez I, Masa FJ (2006) Kinetics of Activated Carbon Promoted Ozonation of Succinic Acid in Water. Ind. Eng. Chem. Res. 45: 3015-3021.

[27] Turhan K, Durukan I. Ozturkcan SA, Turgut Z (2012) Decolorization of Textile Basic Dye in Aqueous Solution By Ozone. Dyes Pigment. 92: 897-901.

[28] Ka-car Y, Alpay E, Ceylan VK (2003) Pretreatment of Afyon Alcaloide Factory's Wastewater by Wet Air Oxidation (WAO), Water Res. 37: 1170-1176.

[29] Kawabata N, Urano H (1985) Improvement of Biodegradability of Organic Compounds by Wet Oxidation. Mem. Fac. Eng. Des, Kyoto Inst. Technol. Ser. Sci. Technol. 34: 64-71.

[30] Lin SH, Chuang TS (1994) Wet Air Oxidation and Activated Sludge Treatment of Phenolic Wastewater. J. Environ. Sci. Health A. 29(3): 547-64.

[31] Lin SH, Ho SJ (1996) Treatment of Desizing Wastewater by Wet Air Oxidation. J. Environ. Sci. Health A. 31(2): 355-66.

[32] Mantzavinos D, Hellenbrand R, Metcalfe IS, Livingston AG (1996) Partial Wet Oxidation of P-coumaric Acid: Oxidation Intermediates, Reaction Pathways and Implications for Wastewater Treatment. Water Res. 30(12): 2969-2976.

[33] Kim K-H, Ihm S-K (2011) Heterogeneous Catalytic Wet Air Oxidation of Refractory Organic Pollutants in Industrial Wastewaters: A review. J. Hazard. Mater. 186: 16-34.

[34] Grisdanurak N, Chiarakorn S, Wittayakun J (2003) Utilization of Mesoporous Molecular Sieves Synthesized from Natural Source Rice Husk Silica for Chlorinated Volatile Organic Compounds (CVOCs) Adsorption. Korean J. Chem. Eng. 20: 950-955.

[35] Mahvi A H, Maleki A, Eslami, A (2004) Potential of Rice Husk and Rice Husk Ash for Phenol Removal in Aqueous Systems. Am. J. Appl. Sci. 1: 321-326

[36] Focus technology go ltd (2011) Water Treatment System (Active Carbon Filter). Zhangjiagang Beyond Machinery Co. Ltd.

[37] Yang G C C, Li C J (2007) Electrofi Ltration of Silica Nanoparticle-containing Wastewater Using Tubular Ceramic Membranes. Sep. Purif. Technol. 58: 159-165.

[38] Fernandez-Ibanez P, Blanco J, Malato S. (2003) Application of the Colloidal Stability of TiO_2 Particles for Recovery and Reuse in Solar Photocatalysis. Water Res. 37: 3180-3188.

[39] Doll T E, Frimmel F H (2005) Cross-flow Microfiltration with Periodical Back-was Hing for Photocatalytic Degradation of Pharmaceutical and Diagnostic Residues-evaluation of the Long-term Stability of the Photocatalytic Activity of TiO_2. Water Res. 39: 847-854.

[40] Zhang X, Du A J, Lee P, Sun D D, Leckie J O (2008) TiO_2 Nanowire Membrane for Concurrent Filtration and Photocatalytic Oxidation of Humicacid in Water. J. Memb. Sci. 313: 44-51.

[41] Yu Y, Yu J C, Yu J G (2005) Enhancement of Photocatalytic Activity of Mesoporous TiO_2 by Using Carbon Nanotubes. Appl. Catal. A: Gen. 289: 186-196.

[42] Vinodgopal K, Wynkoop D E, Kamat P V (1996) Environmental Photochemistry on Semiconductor Surfaces: Photosensitized Degradation of a Textile Azo Dye, Acid Orange 7, on TiO_2 Particles Using Visible Light. Environ. Sci. Technol. 30: 1660-1666.

[43] Ni M, Leung M K H, Leung D Y C, Sumathy K (2007) A Review and Recent Developments in Photocatalytic Water-splitting Using TiO$_2$ for Hydrogen Production. Renew. Sust. Energy Rev. 11: 401-425.

[44] Fujishima A, Zhang X, Tryk D A (2008) TiO$_2$ Photocatalysis and Related Surface Phenomena. Surf. Sci. Rep. 63: 515-582.

[45] Vázquez I, Rodríguez J, Marañón E, Castrillón L, Fernández Y (2006) Simultaneous Removal of Phenol, Ammonium and Thiocyanate from Coke Wastewater by Aerobic Biodegradation. J. Hazard. Mater. 137(3): 1773-1780.

[46] Li YM, Gu GW, Zhao I, Yu HQ, Qiu YL, Peng YZ (2003) Treatment of Coke-plant Wastewater by Biofilm Systems for Removal of Organic Compounds and Nitrogen. Chemosphere. 52(6): 997-1005.

[47] Chu L, Wang J, Dong J, Liu H, Sun X (2012) Treatment of Coking Wastewater by an Advanced Fenton Oxidation Process Using IronPowder and Hydrogen Peroxide. Chemosphere. 86: 409-414.

[48] Bai Y, Sun Q, Sun R, Wen D, Tang X (2011) Bioaugmentation and Adsorption Treatment of Coking Wastewater Containing Pyridine and Quinoline Using Zeolite-Biological Aerated Filters. Environ. Sci. Technol. 45: 1940-1948.

[49] Fongsatitkul P, Elefsiniotis P, Yamasmit A, Yamasmit N (2004) Use of Sequencing Batch Reactors and Fenton's Reagent to Treat a Wastewater from a Textile Industry. Biochem. Eng. J. 21(3): 213-220.

[50] Grilli S, Piscitelli D, Mattioli D, Casu S, Spagni A (2011) Textile Wastewater Treatment in a Bench-scale Anaerobic-biofilm Anoxic-aerobic Membrane Bioreactor Combined with Nanofiltration. J. Environ. Sci. Heal A-Tox. Hazard. Subst. Environ. Eng. 46(13): 1512-1518.

[51] Fibbi D, Doumett S, Lepri L, Checchini L, Gonnelli C, Coppini E, Bubba MD (2012) Distribution and Mass Balance of Hexavalent and Trivalent Chromium in a Subsurface, Horizontal Flow (SF-h) Constructed Wetland Operating as Post-treatment of Textile Wastewater for Water reuse. J. Hazard. Mater. 199-200: 209-216.

[52] He Y, Xu P, Li C, Zhang B (2005) High-concentration Food Wastewater Treatment by an Anaerobic Membrane Bioreactor. Water Res. 39: 4110-4118.

[53] Wang Y, Huang X, Yuan Q (2005) Nitrogen and Carbon Removals from Food Processing Wastewater by an Anoxic/aerobic Membrane Bioreactor. Process Biochem. 40: 1733-1739.

[54] Van Ginkel SW, Oh SE, Logan BE (2005) Biohydrogen Gas Production from Food Processing and Domestic Wastewaters. Int. J. Hydrogen Energ. 30 (15), 1535-1542.

[55] Oh SE, Logan BE (2005) Hydrogen and Electricity Production from a Food Processing Wastewater Using Fermentation and Microbial Fuel cell Technologies. Water Res. 39: 4673-4682.

[56] Tekin H, Bilkay O, Ataberk SS, Balta TH, Ceribasi IH, Sanin FD, Dilek FB, Yetis U (2006) Use of Fenton Oxidation to Improve the Biodegradability of a Pharmaceutical Wastewater. J. Hazard. Mater. B. 136: 258-265.

[57] Sirtori C, Zapata A, Oller I (2009) Decontamination Industrial Pharmaceutical Wastewater by Combining Solar Photo-Fenton and Biological Treatment. Water Res. 43: 661-668.

[58] Chelliapan S, Wilby T, Sallis PJ (2006) Performance of an Up-flow Anaerobic Stage Reactor (UASR) in the Treatment of Pharmaceutical Wastewater Containing Macrolide Antibiotics. Water Res. 40: 507-516.

[59] Z Chen, N Ren, A Wang, Z-P Zhang, Y Shi (2008) A Novel Application of TPAD–MBR System to the Pilot Treatment of Chemical Synthesis-based Pharmaceutical Wastewater. Water Res. 42: 3385-3392.

[60] Guimaraes C, Porto P, Oliveira R, Mota M (2005) Continuous Decolourization of a Sugar Refinery Wastewater in a Modified Rotating Biological Contactor with Phanerochaete Chrysosporium Immobilized on Polyurethane Foam Disks. Process Biochem. 40(2): 535-540.

[61] Li S, Zhou P, Yao P (2010) Preparation of O-Carboxymethyl-N-Trimethyl Chitosan Chloride and Flocculation of the Wastewater in Sugar Refinery. J. Appl. Polym. Sci. 116: 2742-2748.

[62] Yetis M, GuÈ nduÈz U, Eroglu I (2000) Photoproduction of Hydrogen from Sugar Refinery Wastewater by *Rhodobacter sphaeroides* O.U. 001, Int. J. Hydrogen Energ. 25: 1035-1041.

[63] Song XK, Shen YL, Jiao N (2012) Analysis on Decentralized Domestic Sewage Treatment Technologies . Environmental Science and Technology. 25(3): 68-71. (In Chinese)

[64] Yan QL (2008) Treatment of agricultural and sideline products processing wastewater with the SBR technique. Ocean university of china: 3-12.

[65] Patterson JW (2008) Industrial wastewater treatment technology, Second Edition. Butterworth Publishers, Stoneham, MA. USA.

[66] Yang DM, Wang B (2010) Application of advanced oxidation processes in papermaking wastewater treatment. China pulp and paper. 29(7): 69-73. (In Chinese)

[67] Yu QY (2011) Advances in the treatment of printing and dyeing wastewater. Industrial Safety and Environmental Protection. 37(8): 41-43.(In Chinese)

Advanced Monitoring Techniques

Chapter 13

Bioassays with Plants in the Monitoring of Water Quality

Agnes Barbério

Additional information is available at the end of the chapter

1. Introduction

Currently, the concern with hydric resources has a global reach as regards the availability of this finite natural resource. The scarcity of drinking water is alarming, and many factors contribute with this scenario; the water bodies pollution occurs indiscriminately throughout the world, there is an enormous lack of control and management policies regarding the basic sanitation and proper effluents treatment. Thus, the awareness on the rational use of water, effective monitoring and treatment policies, establishing strict goals for water quality control, are more and more urgent, to ensure the survival of all forms of life.

The quality of the water springs is correlated with the human activities, since these use chemicals to reach social and economic goals and, with the absence of an ecologically correct management, human and industrial waste is dumped in these environments. Many substances with mutagenic potential can be found in food, pharmaceutical drugs, and pesticides and in the industrial and domestic effluents complexes; and some of these substances can cause harmful changes, inherited in the genetic material.

Due to the immense range of substances that compose the complex mixtures in the aquatic ecosystems, it is becoming increasingly more difficult and more expensive to make systematic and analytical analyses in these water bodies. A relatively cheap, easy and fast alternative would be the biomonitoring, because, through this method, an assessment would be made in extensive areas and, in case of positive results, chemical analyses would be made in specifically impacted locations. In this case, time and money could be spared.

The concern that the chemical agents introduced in the environment lead to possible genetic changes in the organisms was one of the main reasons that caused the development of methods to assess the genotoxicity through chemical substances.

Among the bioassays developed for detection of mutagenicity, genotoxicity, cytotoxicity and clastogenicity due to environmental pollutants, plant systems have proven to be sensitive, cheap, and effective. Plant bioassays, which are mostly sensitive for the detection of genotoxicity, may provide a warning of environmental hazards in the water [19]. The *Allium cepa* test has been widely used for monitoring potentially cytotoxic and genotoxic effects promoted by pollutants in water, air, and soil. This test has high sensitivity and an easy and rapid execution.

2. Water pollution

Water is essential for life. The amount of fresh water on earth is limited, and its quality is under constant pressure. The quality of water sources is correlated with human activities, since they use chemicals to achieve social and economic goals, and the lack of a correct environmentally management, there is discharge human and industrial waste in these environments. Preserving the quality of fresh water is important for the drinking water supply, food production and recreational water use. Water quality can be compromised by the presence of infectious agents, toxic chemicals and radiological hazards [1]. Water scarcity affects one in three people on every continent of the globe. The situation is getting worse as needs for water rise along with population growth, urbanization and increases in household and industrial uses. Almost one fifth of the world's population (about 1.2 billion people) live in areas where the water is physically scarce. Water scarcity forces people to rely on unsafe sources of drinking water. More than 10% of people worldwide consume foods irrigated by wastewater that can contain chemicals or disease-causing organisms. 1.1 billion people practised open defecation in 2010 [2].

In 2010, 783 million people still relied on unimproved drinking water sources; 2.5 billion people still lacked access to improved sanitation facilities. The MDG drinking water target, which calls for halving the proportion of the population without sustainable access to safe drinking water between 1990 and 2015, was met in 2010, five years ahead of schedule. While this a tremendous achievement, continued efforts are needed. As of 2010, 783 million people still rely on unimproved water sources (surface water from lakes, rivers, dams, or unprotected dug wells or springs) for their drinking, cooking, bathing and other domestic activities. The proportion of the world's population with access to improved drinking water sources increased from 76% to 89% globally between 1990 and 2010. While coverage is above 90% in Latin America and the Caribbean, Northern Africa and large parts of Asia, it is only 61 per cent in sub-Saharan Africa. There are also disparities between urban and rural coverage, where an estimated 96% of the urban population globally used an improved water supply source in 2010, compared to 81% of the rural population [3].

The surveillance and water quality control requires a systematic programme of surveys, which may include auditing, analysis, sanitary inspection and institutional and community aspects. It should cover the whole of the drinking water system, including sources and activities in the catchment, transmission infrastructure, treatment plants, storage reservoirs and distribution systems (whether piped or unpiped). Surveillance of drinking water quality can be defined as "the continuous and vigilant public health assessment and review of the safety and acceptability of drinking water supplies" [4-5].

One of the main sources of pollution of fresh water ecosystems refers to the discharge of industrial and urban wastes. Untreated sewage is considered an important cause to the deterioration of water quality in developing countries. Usually, this complex substance mixture the soil and the water surface of rivers and reservoirs destined for public supply. Environmental pollutants have caused a decrease in water quality, inducing harmful effects in organisms which are in direct or indirect contact with them. The effect of exposure of organisms to any potentially toxic substance depends on its concentration as well as exposure time. One of the effects frequently observed by environmental agents is the chemical alteration of the DNA affecting vital processes, like DNA duplication and transcription, gene regulation and cell division, and leading cells to pathologic processes and/or cell death. The emergence of some genetic diseases and the development of some types of cancer (e.g. leukemia, lymphomas, liver cancer, etc) are promoted by pollutants found in the water, often in drinking water.

Genotoxic agents cause primary DNA lesions (e.g. formation of DNA-adducts, oxidation of bases, base-dimerization, or cross-links), which are either repaired or otherwise lead to irreversible alterations of the DNA or cause cell death. Mutations occur on both, the gene level (gene mutations) and on the chromosomal level (structural and numerical aberrations) [6]. The pollutants can promote breaks in the genetic material are known as clastogenic and those who undertake to chromosome disjunction during division are aneugenic. Micronuclei and bridges are examples of results clastogenic, while c-mitosis and stickness, aneugenics.

3. Plant bioassays

The concern that the chemical agents introduced in the environment might lead to possible genetic changes in the organisms was one of the key reasons that promoted the development of methods to assess the genotoxicity by chemicals. An international collaborative study on the use of bioassays with plants for biomonitoring of genotoxicity of environmental pollutants was initiated and organized by global bodies such as the International Program Chemical Safety (IPCS), the United Nations Environmental Program (UNEP) and World Health Organization (WHO). Since then, the aim of the International Program on Plant Bioassays (IPPB) was to put in practice the applicability of bioassays: micronucleus test and chromosomal abnormalities and/or mitotic anomalies in the *Allium/Vicia*; micronucleus test and stem hairs mutations in *Tradescantia*. Thus, many laboratories in different cities throughout the world have been using these bioassays to assess the genotoxicity in air, water and soil. Sentinel bioassays are biological indictors of environmental contamination with the concept of detection and prevention of disease [7]. For assessing chromosome damage, the most frequently used plant species is *Vicia faba* which has large chromosomes amenable to the study of chromosome aberrations in somatic cells (1) during mitotic divisions and (2) as micronuclei in root tip cells following the mitotic divisions. *Vicia faba* was used in radiation experiments as early as 1913. *Allium cepa* and *Hordeum vulgare* are the next most commonly used species for studying chromosome aberrations although *Crepis capillaris* has been used by many Russian investigators. A few plants such as *Hordeum vulgare*, *Lycopersicon esculentum*, *Pisum sativum* and *Zea mays* can be used for the study of both gene mutations and chromosome aberrations. The crucifer *Arabidopis thaliana* is only used for mutation studies as the chromosomes are very small and the total genome contains only about 70000 kb in contrast to over a million kilobases in most other plants [7].

Among the higher plants used as test organisms for the detection of genotoxics substances in the environment are the following: *Tradescantia, Arabidopis thaliana, Hordeum vulgare, Pisum sativum, Crepis capillaris, Vicia faba, Zea mays, Allium cepa*, among others.

Tradescantia, commonly called spiderwort, is a herbaceous plant which has almost world-wide adaptation and can grow around the year in the field of subtropical regions of the world or in green-houses everywhere. Its relative small plant size (most species are less than 50 cm in height) and six pairs of relatively large chromosomes in its somatic cells made it a favorable experimental material for cytogenetic studies [8]. Using chromosome damage as the indicator of the carcinogenic properties of environmental agents, the Trad-MCN bioassay is a quick and efficient tool for screening carcinogens in gaseous, liquid and solid forms. Test results can be obtained within 24-48 h after the exposure either on site or in the laboratory, thus, the Trad-MCN can be used in a global scale to detect carcinogens as a preventive measure of cancer [9]. The *Tradescantia* assays currently utilize a diploid hybrid clone 4430 heterozygous for flower color with blue dominant and pink recessive that is highly sensitive to chemical mutagens. *Tradescantia* has two assay systems: (1) the *Tradescantia* stamen hair assay which can be used to detect airborne, soil, and aqueous mutants and (2) the *Tradescantia* micronucleus assay (Figure 1) for the detection of chromosomal aberrations from airborne, soil, and aqueous clastogens. The *Tradescantia* stamen hair bioassay has been an important research tool in the field of genetic toxicology for several decades. This has been exploited as a somatic mutation test in the fields of radiobiology, chemical mutagenesis, and ambient air monitoring [7]. For water pollutants from runoffs of industrial sites or power plants, monitoring can be conducted either by growing plants in nearby fields or taking water samples to be tested in the laboratories [8].

When studying the quality of the water in a reservoir in Illinois, [9] collected water samples from the reservoir for two years and assessed the frequency of micronuclei through the *Tradescantia* micronucleus (Trad-MCN) test, and found an average of 12-14 MCN/100 tetrads, showing the samples mutagenicity. Drinking water from the tap was tested in parallel with lake water, and its mutagenicity tended to follow with the mutagenicity of the lake water. The Trad-MCN bioassay was used to determine the clastogenicity of wastewater samples collected from the Arena canal which contains effluent industrial in the city of Queretaro, Mexico; micronucleus frequencies of all the exposed groups at the Conalep site, a predominantly industrial area, were higher than that of the laboratory control groups throughout the 2 year period [10]. The genotoxicity of untreated and treated sewage from two municipal wastewater treatment plants in the municipality of Porto Alegre, in the southern Brazilian state of Rio Grande do Sul, was evaluated over a one year period using the *Tradescantia pallida* var. purpurea (Trad-MCN) bioassay. Inflorescences of *T. pallida* var. *purpurea* were exposed to sewage samples in February (summer), April (autumn), July (winter) and October (spring) 2009, and the micronuclei (MCN) frequencies were estimated in each period. The results indicated that the short-term Trad-MCN genotoxicity assay may be useful for regular monitoring of municipal wastewater treatment plants [11].

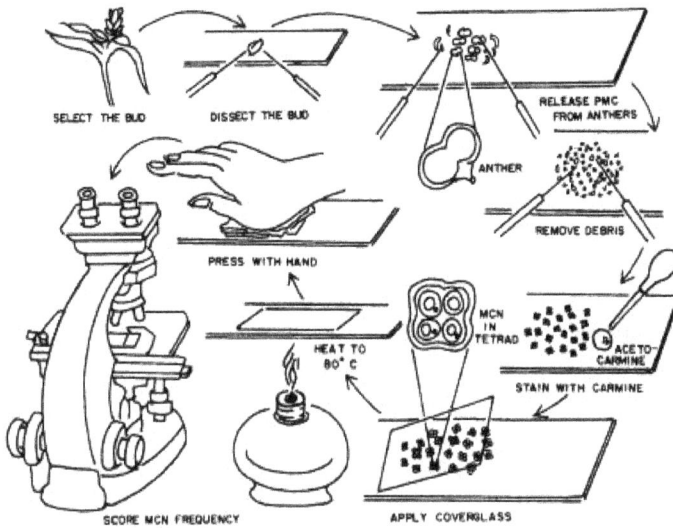

Figure 1. *Tradescantia* micronucleus assay - TRAD-MCN [8].

Vicia root tip mitotic and pollen mother cell meiotic tests are two major kinds of cytogenetic tests for environmental mutagens. Mitotic tests to determine the frequencies of chromosome or chromatid aberrations and/or sister chromatid exchange from root-tip meristematic cells can be used. Treatment of root tip meristem can be done by allowing the newly germinated roots to absorb the chemical mutagens from a water solution. Pollen mother cells can be treated by spraying the solution or pipetting the liquid over the flower buds. After an appropriate recovery time, the samples are fixed and stained, and the slides are prepared for metaphase or anaphase figures for scoring aberration frequencies. Slides for meiotic tests are prepared for metaphase I and/or Anaphase I stages for scoring chromosome aberration frequencies. Results of both cytogenetic tests should be expressed in terms of number of breaks per cell or per 100 cells. The *Vicia* root tip mitotic test is reliable, efficient, and relatively inexpensive [12]. A collaborative study involving laboratories in six countries was initiated under the sponsorship of the International Programme on Chemical Safety (IPCS) to determine the sensitivity, efficiency and reliability of the *Vicia faba* root tip meristem chromosomal aberration assay using a standardized protocol. The conclusions from this study suggest that the *Vicia faba* chromosomal aberration bioassay is an efficient and reliable short-term bioassay for the rapid screening of chemicals for clastogenicity [13]. *Vicia* root micronucleus assay was used to determine the clastogenicity of water samples from Xiaoqing River that passes through Jinan City. Positive results were obtained from eight water collecting sites. This indicates that the water in most areas of this river was polluted with industrial waste and municipal sewage. Results of this study proves that biomonitoring with *Vicia* root micronucleus test is an efficient way to assess the water quality of this river [14].

Zea mays (maize), a member of the Poaceae, is the third most important crop plant in the world. Maize is the oldest plant to have a fully established gene map with the basic genome consisting of 10 chromosomes, is an excellent plant for testing for mutations. The maize bioassay is a particularly favorable experimental assay for the study of chromosome aberrations that may be scored in both mitotic and meiotic cells and pollen. The sister chromatid exchanges have been induced in root tips of maize. Plants of maize have also been used to detect urban air particulate matter, mutagenicity from lake water, municipal water, bottom sediment of a water reservoir, municipal sewage sludge, waste water and soil contaminated with coal fly ash [15].

In this context, toxicity and genotoxicity biological tests are mandatory for the evaluation of reactions of living organisms to environmental pollution, as well as for the identification of potential synergetic effects of several pollutants [16].

One of the principal objectives of using chromosomes as a monitoring system is to determine whether or not a particular chemical is a clastogen – that is, capable of breaking chromosomes. If chemical is a clastogen, then this would permit exchanges with subsequent cytological or genetic damage. At the same time it has been recognized that turbagens (chemicals which cause mitotic disturbance) while not necessarily affecting DNA directly, may result in chromosome segregation errors, and therefore, should not be considered genetically insignificant [17]. Pollutants with mutagenic and cytotoxic potentials produce effects such as DNA fragmentation, induction of chromosome aberrations, inhibition of cellular division, and arrest of the cellular cycle, that can be cytologically detected [18]. Plant roots are generally useful tools in biological tests, because they are the first structures to be exposed to chemical variations in the water and soil [19]. Many types of assays for evaluation of cytotoxicity and genotoxicity employing microorganisms and mammalian cells have been used for monitoring complex environmental samples such as river water. Plant assays and the *Allium cepa* test, particularly, have some advantages over microbial and mammalian cell tests because they are highly sensitive to many environmental pollutants, including heavy metals, and are also useful for monitoring the potential synergistic effects of mixtures of pollutants, including hydrophylic and lipophylic chemicals. Furthermore, test plants can be directly exposed to complex mixtures or environmental samples either in the laboratory or in situ. Because of the large size of their chromosomes, some higher plants are suitable for cytological analysis and the responses so obtained are highly correlated with those seen in other biological systems, thus making plant tests also good candidates for evaluating the genotoxicity of environmental samples.

A. cepa has been indicated as an efficient bioindicator in genotoxicity testing, due to cellular proliferation kinetics, the rapid growth of its roots, large numbers of cells in division, high tolerance to different cultivation conditions, year-round availability, easy management and reduced number of large chromosomes ($2 n = 16$) [20-21]. The *Allium cepa* test has been used since the 1930s [22] and subsequently standardized by [20] in other studies (1985). Since then, many studies adopt this methodology and the results confirm its efficiency in detecting the effects caused by toxic substances found in the environment. A literature review by [23] shows that 148 chemicals had the clastogenic potential when tested through the *Allium* test. Among these substances, 76% had positive behavior for the induction of chromosome aberrations. The author then suggested the introduction of the *Allium* test to the tests routinely used for the detection of chromosomal damage induced by chemicals substances. Considering only the analysis of aberrations in anaphase and telophase cells, the *A. cepa* test is the simplest and considerably one of the most reliable among the investigation methods.

3.1. *Allium cepa* test '*in focus*'

Among the several classes of contaminants detected by the sensitivity of the *Allium* test are: heavy metals, domestic and industrial sewage, landfill extracts and samples of water from rivers and lakes, whose solutions encompass a complex mixture of substances of different compositions, thus attesting the sensitivity and efficacy of this bioassay. Some heavy metals, such as iron, nickel and chrome, are among the contaminants commonly found in the waters of rivers and lakes, act as toxic agents, and can influence the mitotic rates, induce chromosomal aberrations and the formation of micronuclei in tests conducted with *Allium cepa*. Certain metals in aqueous solution can cross the cellular membrane or enter through phagocytosis or pinocytosis and can cause damages to the DNA molecule structure [25].

The first studies conducted in *Allium cepa* roots were performed by [23] when investigating the toxic action of colchicine. Since then, numerous works have used this protocol to estimate the environmental danger in several situations.

[26] when assessing samples of domestic sewage from the residual water from a municipal water treatment station, observed a dramatic reduction in the mitotic index, so that it was impossible to count chromosome aberration; but the dilution is a factor that can influence these parameters. Similar results were also observed by [27] and [28] when estimating the genotoxicity in pharmaceutical/hospital effluents samples and [29] when assessing water samples from a lake environmentally affected with heavy metals. Other works also observed the environmental impact of the presence of heavy metals in rivers, lakes and soil aqueous extracts [30-34]. In addition to the domestic sewage [26,35-37], industrial effluents are also historically impacting: [38] submitted seeds of *A. cepa* to the effluent from a textile industry and observed a mutagenic effect, the type of cell damage may be transmitted to subsequent generations, possibly affecting the organism as a whole, as well as the local biota exposed to the effluent discharge. If the damage results in cell death, the development of the organism may be affected, which could also lead to its death. Chromosome aberration assay was carried out in *A. cepa* meristematic cells exposed to the Guaecá river waters, located in the city of São Sebastião, SP, Brazil, which had its waters impacted by an oil pipeline leak. Analyses of the aberration types showed clastogenic and aneugenic effects for the roots exposed to the polluted waters from Guaecá river, besides the induction of cell death. Probably all the observed effects were induced by the petroleum hydrocarbons derived from the oil leakage [39]. In the work of [40], the general toxicity (root growth inhibition and malformation) and genotoxicity (induction of chromosome aberrations in root cells) of an oil field wastewater in the Nigeria have been investigated by the *Allium* test. The wastewater is mitodepressive and increased significantly the frequency of chromosome aberrations in root cells (sticky chromosomes, c-mitosis, spindle multipolarity, bridges and fragments). At

lower concentrations c-mitosis was the most common aberration. The suitability of the *Allium* test in genotoxicity screening is highlighted and the impact and significance of positive results on the environment and human health should be discussed.

The uses of medicinal plants have always been part of human culture. The World Health Organization estimates that up to 80% of the world's population relies on traditional medicinal system for some aspect of primary health care. However, there are few reports on the toxicological properties of most medicinal plants especially, their mutagenicity and carcinogenicity. In this study, nine medicinal plants had their mutagenic potential assessed; the extracts inhibited the root growth, and promoted a mitodepressive effect and induction of chromosomal aberrations in *A. cepa* [41]. In other studies, plant extracts are assessed for the toxicity of their potential medicines: [42] detected a cytotoxic and genotoxic action when assessing extracts of *Inula viscosa*. [43] assessed the toxic effects of five medicinal plants, all the tested extracts were observed to have mitodepressive effects on cell division and induced mitotic spindle disturbance in *Allium cepa*.

3.2. Evaluated parameters in *Allium cepa* test

Among the plants commonly used as indicators for studies of potential toxicity of river water, *Allium cepa* L. constitutes a convenient system for the analysis of anatomical (root growth, deformity, twist) and microscopic parameters (chromosome abnormalities, altered mitotic index (MI), and micronucleus (MN) formation) [44]. Root growth and mitotic index are parameters of the cytotoxicity; micronucleus and chromosome abnormalities parameters of genotoxicity. The mitotic index is a parameter that allows to estimate the frequency of cells in cell division, reflects cell proliferation and is regarded as an important parameter when determining the rate of plant root growth. Therefore, a low mitotic index (mitodepressive effect) characterize a few dividing cells [45-46] and is more frequently reported in the literature. This parameter predicts the root growth behavior, which is conducted by the frequency of cellular division in the root tissue. This index has been clearly correlated to the root length in the *A. cepa* test, as for instance decreasing in value with increasing concentrations of toxic metals like chromium and cadmium [46].

Chromosomal aberration types detected may depend on DNA lesions and the cell cycle phase that is considered [46]. The mitotic anomalies often detected by the *Allium* test are: chromosome stickiness, multipolar anaphase, anaphasic bridge, c–mitosis or c–metaphase, and micronuclei (Figure 2).

The induction of micronuclei in root meristems of *A. cepa* or in any cell of any other organism is the manifestation of chromosome damage and disturbance of the mitotic process. The micronucleus is formed by the development of a new membrane around the chromatin matter that failed to move to either pole during the anaphase of mitosis. Such chromatin matter arises either from anomalous disjunction of chromosomes due to spindle abnormalities or the breakage of chromosomes resulting in formation of acentric fragments, dicentric chromosomes and chromatin bridges. Thus, the induction of micronuclei may suggest that the environmental pollutant is either a spindle inhibitor or a clastogen. Induction of micronucleus formation, the outcome of chromosome breaks/fragments or spindle poisoning which induces an anomalous disjunction of chromosome at anaphase, has usually been considered a genotoxic indicator. On the other hand, chemicals substances can induce micronuclei promoting changes in the achromatic fuse (aneugenic effect). Micronuclei arise when the lost genetic material is involved by a nuclear envelope, independent of nuclear envelope of the principal nucleus [48]. When the interphase cell is exposed to environmental pollutants, damage is produced either in whole chromosomes - in G1, or damage in individual chromatids - S and G2 phases. Anomalies typically found in S phase are gaps and chromatid breaks [49]. Moreover, during mitosis, all types of AC may be found. Chromosome breaks, bridges and lagging chromosomes may result from sticks and are considered to induce aneuploid and polyploid cells. The stick form can be related to the depolymerization of DNA, the dissolution of nuclear proteins or with an increase in chromosome condensation.

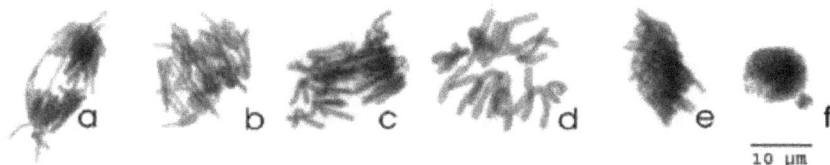

Figure 2. Mitotic anomalies and micronucleus observed in root meristematic cells of *Allium cepa* exposed to water samples of the Paraiba do Sul river - Brazil: a. chromosome bridges; b. not identified; c. multipolar anaphase; d. c-mitose; e. stickiness; f. micronuclei [47].

The use of *Allium cepa* as a test system was introduced by [23], when the effects of colchicine were investigated. This author defined it as an inactivation of segments achromatic spindle fibers by condensed chromosomes randomly dispersed throughout the cytoplasm cell. The formation of c-mitotic inhibitors relates to fibers of the spindle. Calmodulin, involved in regulating the concentration of calcium in the cell was specifically localized in the zone achromatic of mitotic cells, suggesting its involvement in the movement of chromosomes by controlling the polymerization and depolymerization of microtubules [50]. The storage irregular calcium in the cell inhibits the polymerization of microtubules, causing the formation of c-mitosis [51]. The c-mitosis may be defined, as well as the permanence of the equatorial plate anaphase chromosomes rather than move to their respective poles [21]. C-mitosis results from inactivation of the mitotic spindle followed by a random scattering of the condensed chromosomes [46].

The chromosome stickiness arises from improper folding of the chromosome fiber into single chromatids and chromosomes. As a result there is an intermingling of the fibers, and the chromosomes become attached to each other by means of subchromatid bridges. Chromosome stickiness has generally been inflicted by highly toxic agents, usually of an irreversible type, and probably leads cells to death [18,46]. Chromosome breaks - Any descriptive classification must include a category for the break or discontinuity, a simple severance of the chromosome or chromatid to give an acentric fragment, and which is not clearly associated with any exchange process [52]. Usually, the mitotic anomalie is the most frequently observed in studies conducted with *A. cepa*.

The chromosome bridges can involve one or more chromosomes. Bridges associated with stickiness promote a usually irreversible toxic effects.

The chromosome aberration terminology used in cytogenetic experiments generally comprehends chromosome changes designated as numerical (euploidy and aneuploidy) and structural (deletions, inversions, duplications, and translocations) aberrations. However, in experiments involving the *A. cepa* test, other parameters have also been considered to be important tools to inform on chromosome abnormalities as induced by cytotoxic or genotoxic agents. In this respect, we consider the term mitotic anomalies more appropriate for chromosomal changes observed in the *Allium* test [46].

3.3. The *Allium cepa* protocol

In the *Allium cepa* test protocol (based in Invittox Protocol n. 08 by [53]) for complex mixtures (as is the case of rivers and lakes waters), after having their ring of the root primordia carefully cleaned, the onion bulbs are exposed to clean water (it may be good quality tap water, Milli-Q water, distilled water, Hoagland solution, or mineral water, etc) for 48 hours in order to allow for the rooting of the bulb. After this period, bulbs were exposed to treatment for 24 hours: solution or water which needs to be tested, a negative control (no effect on the cell samples; the same solution for rooting is usually used) and a positive control (a drug used to promote the formation of abnormalities). A 20 L bucket was used for collecting the samples. The water was collected at bridges over the River passing alongside the mentioned cities (Figure 3). A mixture was composed from water sampled from the margins and the middle of the River, and transferred to 40 L containers.

All root tips are fixed in absolute ethanol-glacial acetic acid 3:1 (volume/volume) for 5 minutes and subjected to the Feulgen reaction en bloc. The acid hydrolysis pertinent to the Feulgen reaction is carried out in 4 M HCl at 24 °C for 75 minutes. Each stained root was squashed between slide and coverslip, and the squashes frozen in liquid nitrogen for the coverslip removal, and air dried. The preparations were then counterstained with Fast Green at pH 2.7 [54], rinsed in distilled water, air dried, cleared in xylene, and mounted in Canada balsam (Figure 3).

Although the *Allium cepa* test has been widely used to identify potentially cytotoxic and genotoxic pollutants in aquatic environments, variable non-standardized choices have been made regarding the number of plant bulbs and roots analyzed. We propose numbers for bulbs and roots per bulb when comparing the frequencies of micronuclei and mitotic anomalies with this test. When comparing quantitative biological data, one of the most frequent issues is the selection of an adequate number of samples. This decision is influenced by several factors, including the purpose of the study, the size of the population, the risk of using a sample that differs markedly from the population, and the sample error allowed. Determining the sample size is therefore of great importance; exceeding the necessary sample size may prove detrimental in terms of time and human and financial resources. However, too small of a sample size may lead to inconsistencies. As sample sizes increase, their variability tends to decrease, leading to a better hypothesis testing, a higher statistical power, and smaller confidence intervals. Nevertheless, there has not been a standardization of the ideal sample size of bulbs and their roots to be analyzed with the *A. cepa* test. The sampling of ten bulbs and five roots per bulb was adequate for comparative studies to evaluate the potential damage inflicted by pollutants in aquatic environments. Furthermore, even one bulb and one root per bulb was sufficient in discerning this damage, thereby shortening the time required to attain a statistically confident comparative evaluation. However, to allow for the use of statistical programs based on the evaluation of average values, and to avoid criticism based on genetic variability, we propose that three bulbs and three roots per bulb be considered as standard sample sizes for the *A. cepa* test [46].

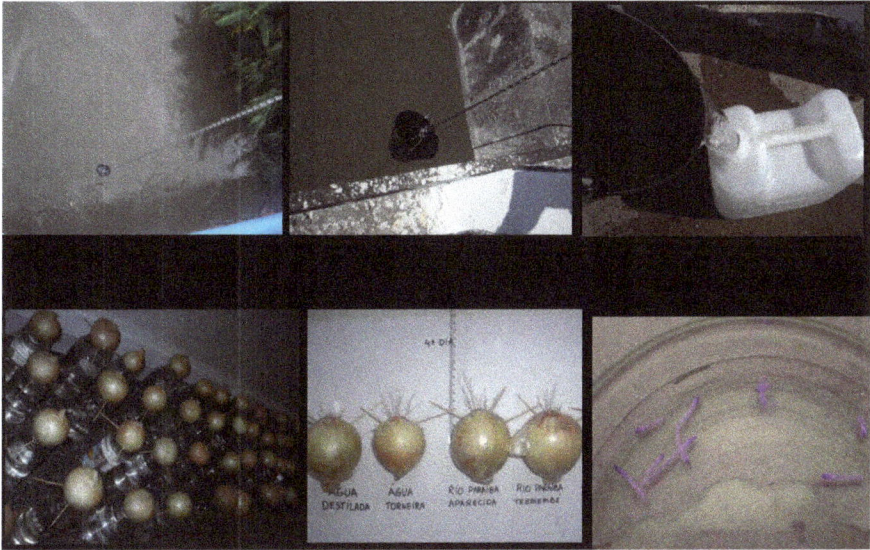

Figure 3. *Allium cepa* bioassay: a 20 L bucket was used for collecting the samples and transferred to 40 L containers (up); bulbs were exposed to treatment for 24 hours; growth root and root stained with Schiff (down).

4. Biomonitoring in the Vale do Paraíba region, São Paulo state, Brazil

The São Paulo state is the most developed area of Brazil, with a high level of industrialization, expanding urbanization and high demographic growth rates. Consequently, there is a trend towards the worsening of some situations, as regards the deficiency in several aspects, such as infrastructure (difficulties to accommodate a population that grows every year), basic sanitation, lack of hydric resources before the demand and damage to their quality, mainly resulting from the direct dumping of urban and industrial sewage, untreated or inadequately treated, in rivers, lakes and reservoirs, among others. The main source of water pollution in the state of São Paulo is the dumping of domestic and industrial effluents, as well as the diffuse load of urban and agricultural.

The Paraíba do Sul river basin is located in the country's key economic area, which highlights for the diversity of its industrial park, especially the aeronautical and automotive industries, paper and cellulose, chemical, mechanical, electrical and electronic, and extrativist industries, in addition to technological research centers. In agriculture, the predominance is of cultivations destined for animal production (cattle), extensive areas with the cultivation of eucalyptus, as well as the presence of the cultivations of rice, beans and corn. The basin draining area is about 55 500 km², encompassing the states of São Paulo, Minas Gerais and Rio de Janeiro [47].

The Paraíba do Sul river (Figure 4) is one of the components of the Paraíba do Sul basin, of importance to the southeast region of Brazil, as it serves for urban supply, irrigation, generation of energy, and assimilation of urban, industrial and agricultural discharges in the mentioned region. For this reason, since 2005, samples of water of this important river, collected in cities along its course, have been periodically evaluated. Anatomo-morphologic parameters (root growth), mitotic indices (MI), cell division phase indices (PI), frequency of micronuclei (MN) and chromosome anomalies (CA) are investigated.

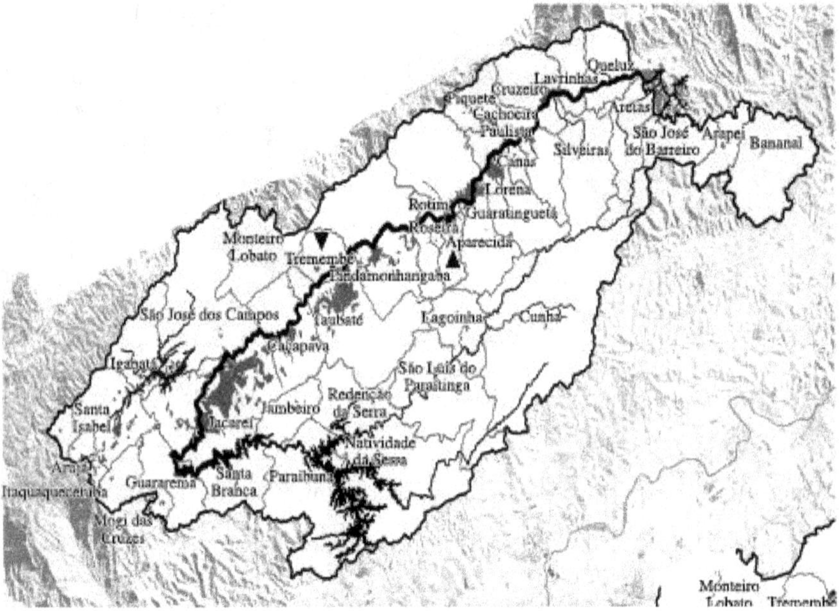

Figure 4. Indication of the cities (Tremembé and Aparecida) 'triangles' at which the water was sampled from the Paraíba do Sul river [19].

The Environmental Sanitation and Technology Company (CETESB) performs the monitoring of the water bodies in the state of São Paulo, and has been trying to contribute with the pollution control and water quality recovery actions in the rivers and reservoirs in the state of São Paulo developed by the city, state and federal governments. In this monitoring, CETESB use 50 water quality variables (physical, chemical, hydrobiological, microbiological and ecotoxicology) [17]. In the last 20 years, 1007 surface water samples were analyzed through the Ames test at CETESB, and 137 (14%) of those showed mutagenic activity. The reason for this contamination was, usually, the discharge of industrial effluents [55-56]. Other ecotoxicological tests, such as those conducted with *Ceriodaphnia dubia*, have shown, according to the CETESB 2007 Report, chronic toxicity in 21% of samples collected in the Paraíba do Sul River.

Considering the data involving root length growth and especially MI values, a cytotoxic potential is suggested for the water of the river Paraíba do Sul at Tremembé and Aparecida, in 2005. On the other hand, since in this year the MN frequency was not affected by the river water treatments, genotoxicity is not assumed for the river water sampled at the mentioned sites. By considering the frequencies of MN, CA and PI, a cytotoxic and genotoxic effect is supposed for *A. cepa* treated with the water collected from the river Paraíba do Sul at sites of Tremembé and Aparecida, in 2007. The *Allium* test has proved to be a sensitive tool for detecting cytotoxicity and genotoxicity effects promoted by the Paraíba do Sul water in the mentioned sites. Present findings reinforce the sensitivity of the *Allium cepa* test, especially concerning the MI evaluation, for monitoring river waters, thus serving as a tool for the early warning of the presence of cytotoxins in the hydric environment. We consider that this test could even be recommended for the prescreening of cytotoxicity in wastewaters [17,19]. In this study, MI and the MN frequency were determined by examination of 2,000 cells per slide, using 3-5 root tips from each bulb. The number of bulbs used depended on the availability of roots produced. The slides were examined from right to left, up end down, and the first 2,000 cells were scored for MI and MN frequency. Data on root length, MI, and the MN frequency were compared using ANOVA and the Tukey test. Variance heterogeneity was tested. Data on root length regarding the Tremembé site and the negative control in the year of 2005 were compared by a Welch test (t-test for different variances). The analyses were done using Statistica and Statsdirect software [19].

Still in the year 2007, the genotoxic and citotoxic potential in water samples from the basin of Tapanhon river, Pindamonhangaba, São Paulo, using the *Allium* test was evaluated. Water samples of the Tapanhon river and its affluents (the Primeira Água and Segunda Água Streams, the Galega Brook) as well as a sample from the Borba Spring have been collected for negative control. Six bulbs of *Allium cepa* have been submitted to each of the samples for 24 hours. As such, 6000 cells have been collected for evaluation of the mitotic index (MI) and micronuclei (MNC) as well as 300 cells in the metaphase and anaphase phases, to evaluate chromosome aberrations (CA). No MNC has been found in the samples tested. The MI values have shown to be elevated in all samples tested when compared to the control with statistical relevance, however, only for the samples from the Primeira Água Stream and the Galega Brook (p<0,05). Chromosome aberrations have been observed in all samples (p<0,05). The averages of the parameters assessed were compared through ANOVA (0.05) of a factor, and the heteroscedasticity premise was assessed through the Levene test. The averages obtained were compared to one another through the Tukey test. The analyses were done using Statistica and Statsdirect software [25].

In 2008 a new study was carried out to characterize the mutagenic potential of the pollutants in the water of the Paraíba do Sul river, in Tremembé city, São Paulo State, Brazil, analyzing chromosomal changes in the meristematic cells of *Allium cepa*, in the summer (April) and winter (August). The bulbs were exposed for 72 h to the treatments: water from river, Hoagland solution (negative control) and 15 µg/L from MMS – methyl methanesulfonate (positive control). In each treatment, three bulbs were exposed and for each bulb, five slides were prepared. For mitotic index (MI) and micronucleus (MN) frequency rate, a total of 2,000 cells per root/slide were analyzed and 100 cells for the chromosome aberrations (CA). In April, the pollutants induced a high mutagenic potential in the meristematic cells of *Allium cepa*, the frequency rate of MN, stickiness and CA from non identified type were greater than the negative control. In August, the only significant change found was the chromosome bridges. There was no significant change for MI. These results bioindicator, therefore, it is important to keep biomonitoring and adopting effluents control measures.

Macroscopic Parameters (root growth) were evaluated in the roots of 5 to 10 bulbs of *Allium cepa* submitted to samples of water of the Paraíba do Sul river collected in the city of Tremembé in August 2009 and 2010. Tap water was used as negative control. After 48 hours the roots of each bulb were measured in 72 hours (1st day) e 168 hours (5th day). There was no difference in the growth of roots in 2009, however, there was a difference in 2010, as well as between the years of 2009 and 2010 for the samples of the river. Several factors may have influenced this toxicity. August 2010 was the driest month in the state of São Paulo, with only 11 mm of rainfall registered. The low rainfall index may have promoted less dispersion of substances along the river. In the same period, the ecotoxicological result carried out by the CETESB using *Ceriodaphnia dubia* was chronic, what shows the presence of toxic substances present in water which inhibited the reproduction of this organism. The same toxicity was observed in our study for the decrease of the index of cellular division in the meristematic root of *A. cepa*.

Positive results, detected by the analysis of some parameters in bioassays with superior plants, indicate the presence of genotoxic and/or citotoxic substances in the environment, demonstrating a direct or indirect potential risk for living beings in contact with it.

Water is an essential resource to sustain life. As governments and community organizations make it a priority to deliver adequate supplies of quality water to people, individuals can help by learning how to conserve and protect the resource in their daily lives. In conclusion, for being an efficient and low cost, and easy to be executed tool, the *Allium* test is recommended for use in order to magnify the system of biomonitoring carried out by the inspection agencies.

5. Conclusion

The physical-chemical monitoring process is expensive, once it demands time, equipment, solutions and specialized labor. The biomonitoring techniques do not replace the assessments of the different parameters assessed through the conventional methods, but have great usefulness for a first diagnosis of the monitored and non-monitored areas, in order to effectively contribute in the pollution control process. The bioassays that use plants are effective, fast, easy to apply to any area, have low cost, do not need sophisticated equipment and can be performed in partnership with teaching institutions, by providing specialized labor in the environmental area. In this sense, it is suggested to the official pollution control bodies that they implement the use of bioassays with plants as a screening before the expensive physical-chemical analyses, in order to spare time and money; and that partnerships with private initiative and teaching institutions can build on the relationship of the world with the quality of life throughout the terrestrial biome as regards the pollution control and environmental preservation.

Author details

Agnes Barbério
Institute of Bioscience, University of Taubaté, Brazil

6. References

[1] World Health Organization – WHO (2012a) Health topics: water. Available: http://www.who.int/topics/water/en/. Accessed 2012 Apr 10.

[2] World Health Organization – WHO (2012b) Water scarcity. Available: http://www.who.int/features/factfiles/water/water_facts/en/index1.html. Accessed 2012 Apr 11.

[3] World Health Organization – WHO (2012c) Global Health Observatory: Water and sanitation. Available: http://www.who.int/gho/mdg/environmental_sustainability/en/. Accessed 2012 Apr 10.

[4] WHO (1976) Surveillance of drinking-water quality. Geneva, World Health Organization. Available: http://whqlibdoc.who.int/monograph/WHO_MONO_63.pdf. Accessed 2012 Feb 01.

[5] World Health Organization – WHO (2011) Guidelines for drinking-water quality. WHO Library Cataloguing-in-Publication Data. Available: http://www.who.int/water_sanitation_health/publications/2011/dwq_guidelines/en/. Accessed 2012 Feb 01.

[6] Majer BJ, Grummt T, Uhl M, Knasmueller S (2005) Use of plant bioassays for the detection of genotoxins in the aquatic environment. Acta Hydrochim. Hydrobiol. 33: 45-55.

[7] Grant WF (1994) The present status of higher plant bioassays for the detection of environmental mutagens. Mutat. Res. 310: 175-185.

[8] Ma TH (1981) *Tradescantia* micronucleus bioassay and pollen tube chromatid aberration test for in situ monitoring and mutagen screening. Environ. Health Persp. 37: 85-90.

[9] Ma TH (2001) *Tradescantia* micronucleo bioassay for detection of carcinogens. Folia Histochem. Cytobiol. 39(Suppl. 2): 54-5.

[10] Ma TH, Anderson VA, Harris MM, Neas RE, Lee TS (1985) Mutagenicity of drinking water detected by the *Tradescantia* micronucleus test. Can. J. Genet. Cytol. 27(2): 143-150.

[11] Ruiz EF, Rabago VM, Lecona SU, Perez AB, Ma TH (1992) *Tradescantia* micronucleus (Trad-MCN) bioassay on clastogenicity of wastewater and in situ monitoring. Mutat. Res. 270(1): 45-51.

[12] Thewes MR, Junior DE, Droste A (2011) Genotoxicity biomonitoring of sewage in two municipal wastewater treatment plants using the *Tradescantia pallida var. purpure*a bioassay. Gen. Mol. Biol. 34(4): 689-693.

[13] Ma TH (1982) *Vicia* cytogenetic tests for environmental mutagens: A report of the U.S. environmental protection agency Gene-Tox program. Mutat. Res. 99(3): 257-271.

[14] Kanaya N, Gill BS, Grover IS, Murin A, Osiecka R, Sandhu SS, Andersson HC (1994) Vicia faba chromosomal aberration assay. Mutat. Res. 310(2): 231-247.

[15] Miao M, Fu R, Yang D, Zheng L (1999) *Vicia* root micronucleus assay on the clastogenicity of water samples from the Xiaoqing River in Shandong Province of the People's Republic of China. Mutat. Res. 426(2): 143-5.

[16] Grant WF, Owens ET (2006) *Zea mays* assays of chemical/radiation genotoxicity for the study of environmental mutagens. Mutat. Res. 613: 17-64.

[17] Barbério A (2009) Efeitos citotóxicos e genotóxicos no meristema radicular de *Allium cepa* exposta à água do rio Paraíba do Sul – estado de São Paulo – regiões de Tremembé e Aparecida. PhD thesis, Unicamp, Campinas.

[18] Grant WF (1978) Chromosome aberrations in plants as a monitoring system. Environ. Health Perspect. 99: 273-291.

[19] Barbério A, Barros L, Voltolini JC, Mello MLS (2009) Evaluation of the cytotoxic and genotoxic potential of water from the Brazilian river Paraíba do Sul with the *Allium cepa* test. Braz. J. Biol. 69(3): 837-842.

[20] Fiskesjö G (1988) The *Allium* test – an alternative in environmental studies: the relative toxicity of metal ions. Mutat. Res. 197(2): 243-260.

[21] Fiskesjö G (1985) The *Allium* test a standard in environmental monitoring. Hereditas 102(1): 99-112.

[22] Ma TH, Xu Z, Xu C, McConnell H, Rabago EV, Arreola GA, Zhang H (1995) The improved *Allium/Vicia* root tip micronucleus assay for clastogenicity of environmental pollutants. Mutat. Res. 334(2): 185-195.

[23] Levan A (1938) Effect of colchicines on root mitosis in *Allium*. Hereditas 24(1): 471-486.

[24] Grant WF (1982) Chromosome aberration assays in *Allium*: A report of the U. S. Environmental Protection Agency Gene-Tox Program. Mutat. Res. 99(3): 273-291.

[25] Amaral AM, Barbério A, Voltolini JC, Barros L (2007) Avaliação preliminar da citotoxicidade e genotoxicidade da água na bacia do rio Tapanhon (SP-Brasil) através do teste *Allium* (*Allium cepa*). Rev. Bras. Toxicol. 20(1-2): 65-71.

[26] Grisolia CK, Oliveira ABB, Bonfim H (2005) Genotoxicity evaluation of domestic sewage in a municipal wastewater treatment plant. Gen. Mol. Biol. 28(2): 334 -338.

[27] Bakare AA, Okunola AA, Adetunji OA, Jenmi HB (2009) Genotoxicity assessment of a pharmaceutical effluent using four bioassays. Gen. Mol. Biol. 32(2): 373-38.

[28] Muzio MPH, Mendelson A, Magdaleno A, Tornello C, Balbis N, Moretton J (2006) Evaluation of genotoxicity and toxicity of Buenos Aires city hospital wastewater samples. J. Braz. Soc. Ecotoxicol. 1(1): 1-6.

[29] Barbosa JS, Cabral TM, Ferreira DN, Agnez-Lima LF, Batistuzzo De Medeiros SR (2010) Genotoxicity assessment in aquatic environment impacted by the presence of heavy metals. Ecotoxicol. Environ. Saf. 73: 320-325.

[30] Cotelle S, Masfaraud JF, Férard JF (1999) Assessment of the genotoxicity of contaminated soil with the *Allium/Vicia* micronucleus and the *Tradescantia* micronucleus assays. Mutat. Res. 426: 167-171.

[31] Evseeva TI, Geras'kin SA, Shuktomova II (2003) Genotoxicity and toxicity assay of water sampled from a radium production industry storage cell territory by means of *Allium* test. J. Environ. Radioact. 68: 235-248.

[32] Palacio SM, Espinoza-Quiñones FR, Galante RM, Zenatti, DC, Seolatto AA, Lorenz EK, Zacarkin CE, Rossi N, Rizzutto MA, Tabacniks MH (2005) Correlation between heavy metal ions (Copper, Zinc, Lead) concentrations and root length of *Allium cepa* L. in polluted river water. Braz. Arch. Biol. Technol. 48: 191-196.

[33] Bortolotto T, Bertoldo JB, Silveira FZ, Defaveri TM, Silvano J, Pich CT (2009) Evaluation of the toxic and genotoxic potential of landfill leachates using bioassays. Environ. Toxicol. Pharmacol. 28: 288-293.

[34] Radić S, Stipaničev D, Vujčić V, Rajčić MM, Širac S, Pevalek-Kozlina B (2010) The evaluation of surface and wastewater genotoxicity using the *Allium cepa* test. Sci. Total Environ. 408: 1228-1233.

[35] Cabrera GL, Rodriguez DMG (1999) Genotoxicity of leachates from a landfill using three bioassays. Mutat. Res. 426: 207-210.

[36] Cabrera GL, Rodriguez DMG, Maruri AB (1999) Genotoxicity of the extracts from the compost of the organic and the total municipal garbage using three plant bioassays. Mutat. Res. 426: 201-206.

[37] Monarca S, Feretti D, Collivignarelli C, Guzzella L, Zerbini I, Bertanza G, Pedrazzani R (2000) The influence of different disinfectants on mutagenicity and toxicity of urban wastewater. Wat. Res. 34(17): 4261-4269.

[38] Caritá R, Marin-Morales MA (2008) Induction of chromosome aberration in the *Allium cepa* test system caused by the exposure of seeds to industrial effluents contaminated with azo dyes. Chemosphere 72: 722-725.

[39] Leme DM, De Angelis DF, Marin-Morales MA (2008) Action mechanisms of petroleum hydrocarbons present in waters impacted by an oil spill on the genetic material of *Allium cepa* root cells. Aquat. Toxicol. 88: 214-219.

[40] Odeigah O, Nurudeen O, Amund OO (1997) Genotoxicity of oil field wastewater in Nigeria. Hereditas 126:161-167.

[41] Akintonwa A, Awodele O, Afolayan G, Coker HAB (2009) Mutagenic screening of some commonly used medicinal plants in Nigeria. J. Ethnopharmacol. 125: 461-470.

[42] Çelik TA, Aslantürk OS (2010) Evaluation of cytotoxicity and genotoxicity of *Inula viscosa* leaf extracts with *Allium* test. J. Biom. Biotechnol. doi:10.1155/2010/189252

[43] Akinboro A, Bakare AA (2007) Cytotoxic and genotoxic effects of aqueous extracts of five medicinal plants on *Allium cepa* Linn. J. Ethnopharmacol. 112: 470-475.

[44] Egito LCM, Medeiros MG, De Medeiros SRB, Agnez-Lima LF (2007) Cytotoxic and genotoxic potential of surface water from the Pitimbu river, northeastern/RN Brazil. Genet. Mol. Biol. 30(2): 435-441.

[45] Fiskesjö G (1993) The *Allium* test in wastewater monitoring. Environ. Toxicol. Water. Qual. 8: 291-298.

[46] Barbério A, Voltolini JC, Mello MLS (2011) Standardization of bulb and root numbers for the *Allium cepa* test. Ecotoxicol. 20(4): 927-935.

[47] Oliveira LM Oliveira LM, Voltolini JC, Barbério A (2011) Potencial mutagênico dos poluentes na água do rio Paraíba do Sul em Tremembé, SP, Brasil, utilizando o teste *Allium cepa*. Ambi-Agua 6(1): 90-103.

[48] Grover IS, Kaur S (1999) Genotoxicity of wastewater samples from sewage and industrial effluent detected by the *Allium* root anaphase aberration and micronucleus assays. Mutat. Res. 426(2): 183-188.

[49] Natarajan AT, Boei JJ, Darroudi F, Van Diemen PC, Dulout F, Hande MP, Ramalho AT (1996) Current cytogenetics methods for detecting exposure and effects of mutagens and carcinogens. Environ. Health Perspect. 104(Suppl.3): 445-448.

[50] Li JX, Sun DY (1991) A study on CaM distribution in cells of living things. Chin. J. Cell Biol. 13(1): 1-6.

[51] Liu D, Jiang W. Li M (1992) Effects of trivalent and hexavalent chromium on root growth and cell division of *Allium cepa*. Hereditas 117: 23-29.

[52] Savage JRK. (1975) Classification and relationships of induced chromosomal structural changes. J. Med. Gen. 12: 103-122.

[53] Fiskesjö G (1989) Invittox Protocol nº 8 - *Allium* test. Nottingham: Russel and Burch House.

[54] Mello MLS, Vidal BC Práticas de biologia celular. São Paulo: Edgard Blücher/Funcamp, 1980. p. 57-58.

[55] Umbuzeiro G de A, Roubicek DA, Sanchez PS, Sato MI (2001) The Salmonella mutagenicity assay in a surface water quality monitoring program based on a 20-year survey. Mutat. Res. 491(1-2): 119-126.

[56] Oliveira DP, Carneiro PA, Rech CM, Zanoni MV, Claxton LD, Umbuzeiro GA (2006) Mutagenic compounds generated from the chlorination of disperse Azo-Dyes and their presence in drinking water. Environ. Sci. Technol. 40(21): 6682-6689.

Chapter 14

Eutrophication: Status, Trends and Restoration Strategies for Palic Lake

Vera Raicevic, Mile Bozic, Zeljka Rudic, Blazo Lalevic, Dragan Kikovic and Ljubinko Jovanovic

Additional information is available at the end of the chapter

1. Introduction

Eutrophication is the general term used by aquatic scientists to describe the suite of symptoms that a lake exhibits in response to fertilization with nutrients [1]. Common symptoms include dense algal blooms causing high turbidity and increasing anoxia in the deeper parts of lakes from the decay of sedimenting plant material. The anoxia can in turn cause fish kills in midsummer [2].

Aging process is a principal characteristic of each lake. This process is generaly very slow in geological sense, but the aging of shallow lakes (like Palic lake is) is much faster. Eutrophication is not necessarily harmful or bad; it's a Greek word, which means "well nourished" or "good food." However, eutrophication can be artificially accelerated, eventually leading to the suffering of its inhabitants, as the nutrients input increases far beyond the lake's natural capacity.

Eutrophication of lakes and reservoirs is a degradation process originating from the introduction of nutrients from agricultural run-off and untreated industrial and urban discharges [3]. Accelerated eutrophication of lakes and reservoirs experienced during last century in most parts of the world represents a serious degradation of water quality, not only in the developing countries [4-6], but also in developed countries [7-10].

Surface waters are not isolated from their environment, neighter from human activities. Therefore, the inorganic and organic matter continuously feed these waters and consequently accelerate the process of natural eutrophication. Waters coming from treatment plants in Subotica town, have been loading this lake for decades, attributing to high nitrogen and phosphorus loadings of Palic Lake together with numerous diffuse pollutants from surrounding villages and agricultural surfaces. Agriculture has been identified as one of the primary contributors to non-point source nutrient losses in the USA [11].

Over the past decades, numerous restoration projects have been carried out to control and reduce the negative effects of eutrophication and to improve water quality.

Engineering approaches for improving water quality have focused mainly on control of external nutrient loading, sediment degrading and prevention of the release of phosphorus from sediments through chemical treatments [12]. However, lakes often exhibit a delayed response of a reduction in external nutrition. That situation is likely caused by the release of phosphorus from a pool that has accumulated in the sediment during the period of high nutrient loading [13-14].

Research also suggests that restoring macrophytes could provide long-term improvement in water quality [15]. Thus, in recent years, there has been an increasing focus on the importance of constructed wetlands in ecosystem restoration [16-18]. Possessing multi-functions such as water storage, flood detention, water purification, nutrients transformation, and ecosystem biodiversity, wetlands have been recognized as an important part of aquatic ecosystems [19].

The aim of this review is to compare the water quality parameters of Palic Lake in past 20 years with its quality in 2010, as well as to perceive the possibility to apply ecoremediation technologies in remediation of Palic Lake.

1.1. Lake characteristics, nutrient input and vegetation

The Palic Lake is a shallow Pannonian lake, created million years ago, during creation of pits and dunes by wind erosion. The lake was filled mostly by atmospheric precipitation. It is situated 8 km from Subotica, near the town of Palic and covers an area of 3.8 km².

This area is characterized by continental climate, with a severe winters, hot summers and irregular distribution of precipitation. The average air temperatures are 10.8°C (Table 1), air humidity is 69% and air pressure is 1007mb. Annual number of rainy days is 105 and the average perennial mean precipitation values are 561.1mm, while the number of days with snow cover is 59 per year. This area is also characterized by strong winds (wind speed more than 6 Bf, i.e. 34 km/h), during 104 days per year [20].

jan	feb	mar	apr	may	jun	jul	aug
-0,8	1,2	5,9	11,1	16,8	19,9	21,6	20,9
sep	oct	nov	dec	min	med	max	
16,5	11,1	5,2	0,7	9,6	10,8	12,7	

Table 1. Mountly mean temperatures (°C)

Highest precipitation output [20] is during late spring and beginning of summer (June), and the lowest is in winter time – January, February and March (Table 2).

jan	feb	mar	apr	may	jun	jul
32,5	32,3	33	44,9	53,1	75,3	62,3
aug	sep	oct	nov	dec	year	
59,2	44,1	36,8	43,2	44,4	561,1	

Table 2. Mountly mean precipitation (mm)

This lake is separated into 4 sectors (Fig. 1). Sector 2, which covers the area of 81 ha, has been made to improve the quality of inflowing water from wastewater treatment plant. Sector 4, the biggest part of the lake, is designed for recreational purposes. It coveres around 372 ha and the initial depth of lake was 2 m. Nowdays, some parts of the lake are covered with thick sediment layer, causing variations in lake's depth.

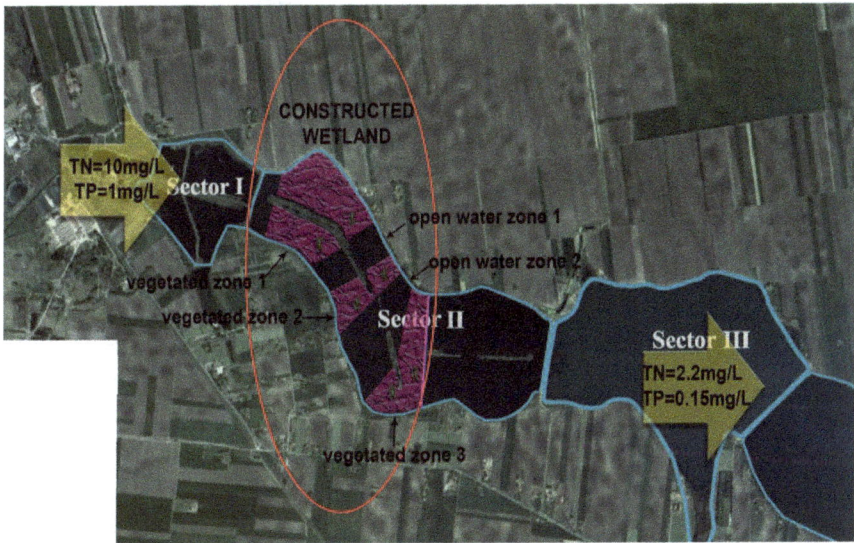

Figure 1. Sectors in Palic Lake (red circle - target place for restoration)

Central section for wastewater treatment of Subotica town lies in the depression of the far west end of Palic Lake. Water from this treatment unit inflows into the lake. Wastewater treatment plant (WWTP) of the city of Subotica was built in 1977, and since than the capacity of the plant was enlarged (in 1983). In 1997 it has been revealed that the amount of influent was 80% higher than the planned one. This exessive wastewater has predominantly originating from precipitation that clogged and blocked the operation of treatment facility unit. In the period of extremely high precipitation, the excess water was transported into lagoons and further on, as untreated water, into the lake. In 1985 the capacity of treatment unit was enlarged, but due to numerous omissions, the so called "treated" wastewaters have been discharged into the lake. The analyzed data show that the lake was loaded with more than 4t of total phosphorus in August 1970.

The new treatment section for wastewaters has been released in 2009, but it wasn't technicaly accepted until 2011. The quality of treated wastewater, discharged into the Sector 1 of Palic Lake, meets the objectives of Directive 91/271/EEC, according to which the following parameters are being controlled: BOD_5 25 mg/l; COD_5 125 mg/l; suspended solids 35 mg/l; total N 10 mg/l; total P 1 mg/l.

The capacity of new WWTP is 36.000m³, while during rainy days it gets enlarged up to 72.000m³. The new treatment facility has shown to be very efficient in achieving good effluent quality regarding all the analyzed parameters. The amount of discharged water from new WWTP into the lake is bigger, but its quality is undoutably better (Figure 2). This will certainly influence the processes in the lake.

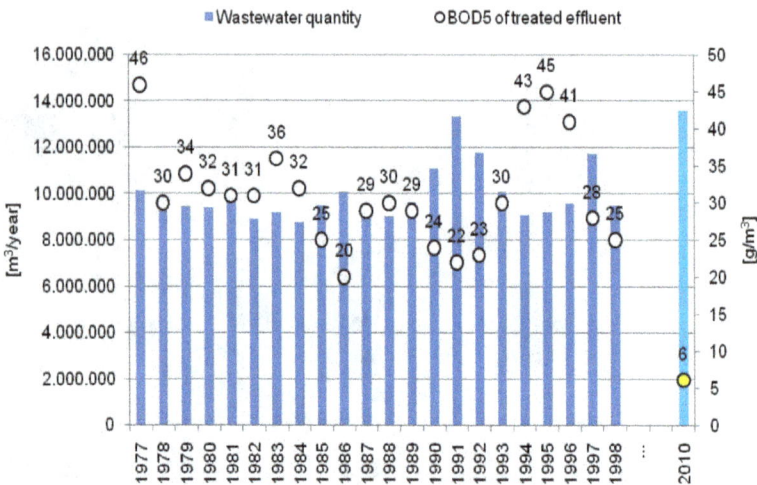

Figure 2. Wastwater amount and effluent quality (BOD_5) during period from 1977-2010

Beside the effluent from the treatment facility being the greatest source of nutrients in Palic Lake (although the treatment of domestic and industrial wastewaters has positive effect), there is contamination from numerous pollutants additionally destroying the biological balance of the lake. There are cultivated areas near the lake with intensive farming, which assumes the application of mineral fertilizers and pesticides; as well as livestock farm without appropriate collection and treatment of wastewaters. There are also resident weekend facilities and rural settlments nearby with catchpits. All the surrounding wastewaters should be collected and transfered to the wastewater treatment plant, before the remediation and cleaning of Palic lake sludge.

In recent years there have been some changes in turistic and recreational characteristics of lake's environment, leading to the deterioration of indigenous vegetation of coastal zone. But land reclamation for establishment of arable agricultural land influenced these changes as well.

There are conditions for survival of community *Ceratophylletum demersi* in 3rd sector, famous for its submerged stands, the indicators of deep and cold waters in warm and eutrophic aquatic ecosystems. The domination of these dense populations of *Ceratophylletum demersi* community, is an indicator of significant nutritional values of this area. The characteristic growth of the community Lemnetum gibbae on one locality of Palic lake in reed cover and around reed vegetation, indicates the presence of shallow, warm and trophogenic backwater.

Marsh vegetation is a dominant flora here with reed cover, presented by the association Scipro-Phragmitetum W, Koch, and tipical stands (subass: Phragmitetosum Schmale 1939) and stand of timothy grass, i.e. subass: typhetosum (angustifoliae-latifoliae) Soo 1937. Reed and timothy grass are present in the form of dense populations by the coastal edge of the island in the second sector of the lake. They are less frequent in third sector, while in the first sector of the lake they form narrow region around the lagoons and on the lagoon dikes [21].

In geographical aspect aquatic marsh plants haven't got great significance, because they are widely spread in most cases. But their significance lies in the mentainance of ecosystem's balance, and they also represent natural habitat for a large number of birds from the surrounding area. Beside this, semiaquatic coastal vegetation has also the antierosion role. In certain parts of sector 2 in the coastal zone the stand of high yields of order Magnocaricetaria PIGN1953 is fragmentaly developed.

The northern loess coast of Palic lake represents one of the rare settlements of the presereved acient steppe. One part of this rare habitat is permanently exposed to rockslide and total devastation, due to high water levels, which disables the growth of coastal reed belt. Because of formation of pathways for local fishermans and peasants, very near to the coast of the lake, the narrow belt of steppe vegetation is under intensive invasion of weed, especialy row weed [21]. Beside these populations, there are some pioniring sand stone species like *Bromus squarrosus* i *Centaurea arenaria*. There is only partial presence of steppe elements like *Festuca rupicola, Agropyron cristatum* subssp. pectinatum f. puberulum are, with dense populations of *Allium scorodoprasum* subssp. Waldsteinii. The coast of Palic lake is inhabited by exceptional trees of *Salix alba* and rare individual trees of *Populus alba*. Very invasive species *Eleagnus angustifolia* is grown all around this coastal area. On the islands of the sector 2 there is a stand of *Populus alba*, from which the invasive belt of *Acer negundo* is being spread. Shrubby vegetation covers the high coast of the lagoons in sector 1, which also enters the reed cover of the lake. There are also numerous individual plants like *Sambucus nigra, Lonicera caprifolium, Rosa canina* and *Spiraea media*. Herb stratum is very well developed, and being followed by the trees of white poplar and white willow.

2. Status and trends: Phosphorus and nitrogen loadings and BOD5

Lakes are not only a significant source of precious water but also provide valuable habitats to the biological world. The major impact of eutrophication, due to overloading with nitrogen and phosphorus, are changes in structure and functioning of lake's ecosystem, reduction of biodiversity and reduction of fishery and tourism. Impairment of water quality due to eutrophication can lead to a series of problems and result in loss of ecological integrity, sustainability and safe use of aquatic ecosystems [22]. Phosphorus and nitrogen are major nutrients causing eutrophication.

Phosphorus is usually the main nutrient responsible for eutrophication of freshwater. The results from each lake sectors (X - axes) of average annual phosphorus concentrations have shown to have linear regression (Figure 3). When separately analyzing each single year, the results are regression curves of similar incline angle. Almost all of them are translatory moved in regard to the resulting average for the studied period. These results show that the Palic Lake regulates the content of phosphorus as well as the existence of self-cleaning ability, but also the fact that it is being loaded with excessive amount of phosphorus.

Determination of total phosphorus by ammonium molybdenum and ascorbic acid, have shown the decrease of its total content in 2010 comparing to previous years.

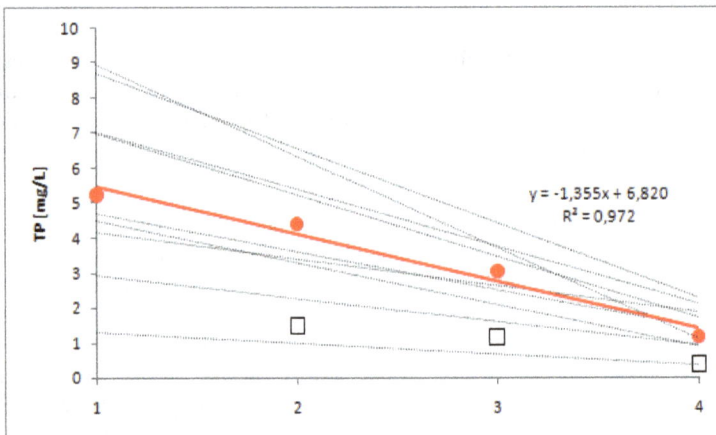

Figure 3. Total phosphorus concentration trends in sectors of Palic lake (mg/l) during period from 1978 – 2010

Nitrogen reaches aquatic ecosystems through direct binding with living organisms, atmospheric precipitation and nitrogen rich inflowing waters. The previous research conducted during period from 1977-1998 show that nitrogen loadings were mainly originating from wastewater treatment facility, with discharges of 591,8 t of nitrogen into sector 1 [23].

The highest amount of nitrogen in Palic Lake is in sector 1, while its content decreases in sector 4. This trend is also noticeable in the year of 2010. These results show the efficiency of lake in reducing total nitrogen content.

Although it should be taken into account that certain portion of generated nitrogen during production of organic matter is being built in live organisms. Hence, after the mineralization of dead organisms it is being returned into the matter cycle in the lake.

The red line (circle) indicates trend in decreasing nitrogen concentrations of each examined year (Figure 4). Thin point lines represent yearly trends (1978 - 2010), and squares indicate the check measurement conducted in 2010.

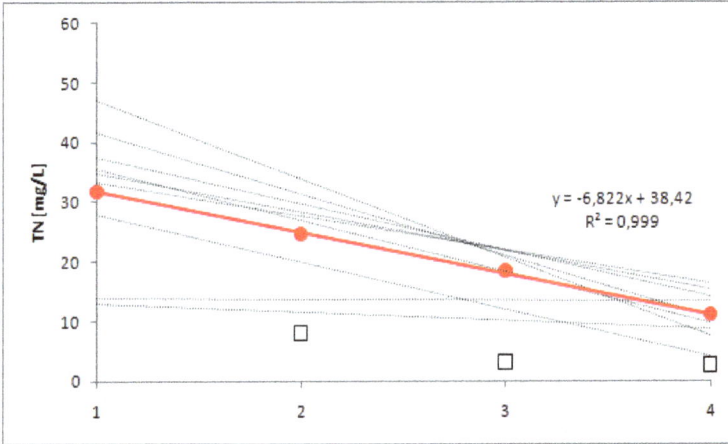

Figure 4. Total nitrogen concentration (mg/l) trends in sectors of Palic lake during period 1978 - 2010

BOD values expressing organic matter loading are rather high for surface water and emphasize the high level of pollution.

In the period before the year 2000, BOD₅ values have been descending from sector 1 to sector 4 (X – axes), but from the beginning of year 2003 these values have been decreasing from sector 1 to sector 3, and then rapidly ascended in sector 4 (Figure 5).

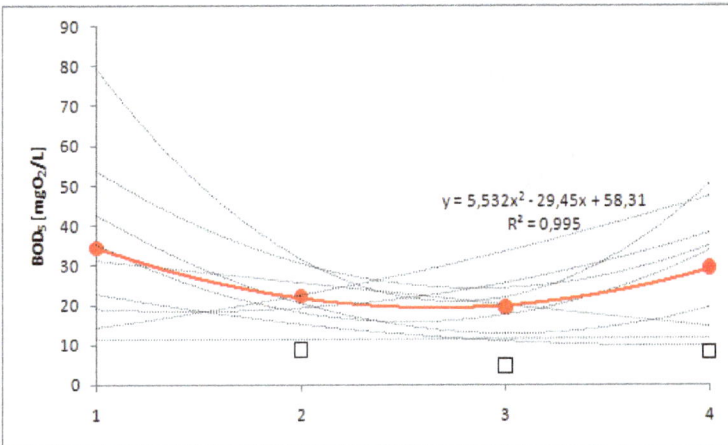

Figure 5. BOD (mg/l) trends in sectors of Palic lake during period 1978 - 2010

These high BOD$_5$ values are the consequence of biomass hyper production in sector 4. In comparison to measurements from previous 20 years and measurements from the year 2010, it is noticeable to have the decrease of BOD$_5$ in all sectors of Palic Lake.

3. Restoration of the lake

The following review deals in detail with the quality of water and sediment in sector 2, since this sector is designed to additionally clean the wastewater.

3.1. Methods for determination of water and sediment quality in sector 2 of Palic Lake

Water sampling has been conducted by 3 liters Friedinger bottle, and the undamaged sediment has been taken by "core sampler" method (Eijeklkamp). Two different sediment layers are clearly defined in Palic lake by color and consistency: the black (oily) one, 20-25cm thick and the grey layer which is in contact with water, being 40-45cm thick. All samples were transferred to laboratory during 4h from the ending of sampling. The analyses of easy variable parameters and microbial activity of samples were performed immediately after their receiving in laboratory.

As mentioned previously, the content of total phosphorus has been determined by spectrophotometer using ammonium molybdenum and ascorbic acid. The same method, with N,N-dimethyl-p-phenylendiamine, was used to determine H$_2$S. Electrochemical methods helped in defining the oxygen saturation, BOD$_5$, electro conductivity, pH values in the field and laboratory. Chemical oxygen demand (COD) was determined by oxidation of organic matters with KMnO$_4$ and K$_2$Cr$_2$O$_7$.

Method of thermo catalytic destruction has been used to set out the total nitrogen. Ion chromatography method was used for determination of NH$_4^+$, SO$_4^2$, NO$_3^-$, NO$_2^-$, and Cl.

Total organic carbon was determined by SRPS ISO 8245.1994 method, total sulphides by EPA 9030B method, and carbonates and bicarbonates by SPRS EN 13137:2005 method.

Trace metals as Pb, Cd, Cr, Cu, Ni, Fe, and Mn have been determined by ICP-OES technique after acidic digestion with concentrated nitrogen acid and H$_2$O$_2$. Total arsenic was defined by AAS-atomic absorption spectrophotometry/hibrid technique.

The grain-size distribution was determined by using the pipette method and was classified into clay, silt, and sand fractions.

MPN method was used to determine aerobic mesophylic bacteria, fecal and total coliform bacteria, sulphite reducing clostridia and fecal streptococci.

3.2. Water and sediment quality in sector 2 of Palic Lake

Sector 2 of Palic Lake is characterized by water with high phosphorus and nitrogen content as well as high pH values (Table 3).

In sampling period the water was covered with foam which indicates presence the surface active materials in water. This is due to Serbia still used anionic detergent with phosphate which than caused acceleration eutriphication in shallow water. The content of total organic carbon (TOC) is high which indicates the intensive organic output. High concentration of chloride and sulfate are consequence of theirs high content in treated waters from WWTP. The concentration of ammonium ions are moderate high, but nitrite nitrogen had extremely high. Five day (BOD_5) was 2.2 (Table 3) and is due presence of easily biodegradable organic material the, but in sector 4 the sums was 8 [24].

PARAMETERS	VALUES	PARAMETERS	VALUES
pH	8.6	COD - $K_2Cr_2O_7$ (mg/l)	30
Cl^- (mg/l)	163.7	As (mg/l)	0.045
SO_4^{2-} (mg/l)	434.5	Cu (mg/l)	0.006
NH_4^+ (mg/l)	1.24	Zn (mg/l)	0.0027
NO_2^- (mg/l)	0.902	Fe (mg/l)	0.065
BOD_5 (mg/l)	2.2	Cr^{6+} (mg/l)	< 0.05
NO_3^- (mg/l)	5.7	Cr total (mg/l)	< 0.002
total N (%)	7.93	Cd (mg/l)	< 0.0008
Total PO_4 (mg/l)	1.46	Mn (mg/l)	0.024
TOC (mg/l)	9.81	Ni (mg/l)	0.006
H_2S (mg/l)	< 0.02	Pb (mg/l)	< 0.005
COD - $KMnO_4$ (mg/l)	7	Hg (mg/l)	< 0.0005

Table 3. Chemical analyses of water in sector 2

From the results of the content of heavy and toxic metals in water it can be noticed that heavy and toxic metals do not affect significantly the water lake Palic (Table 5, Figure 5). We should also bear in mind that the water in the lake has high pH values which adversely affects the solubility and availability of metals in water.

The accelerated eutrophication in this sector resulted in formation of significant sediment deposit. This sediment is alkaline by its nature with high percentage of inorganic content and nutrients (Table 4 and 5), the same as the water is.

Phosphorus is a major nutrient for aquatic ecology, and its excess supply can lead to eutrophication. When the external loading of P increases, the sediments act like pool and absorb it. However, after the external loading is reduced, the sediments now as a source would release the adsorbed P back into the water. The characteristics of sediments, environmental factors, as well as the concentrations of P in the overlying water, will affect the transfer direction of phosphate on the interface of the sediment–water [25-26].

The supply of nutrients can directly and indirectly limit the metabolic activity of heterotrophic microorganisms. For example, there is evidence for direct positive effects of N and P on bacterial growth [27] and accordingly, the total bacterial biomass is strongly correlated with concentrations of total phosphorus in fresh-water and marine ecosystems.

The deeper parts of the sediment layer have higher concentrations of heavy metals as a result of historical pollution of the lake. The concentration of As and Ni exceed the defined Holland [28] and Canadian [29] limits, as well as the concentration of Cr which is also higher than the recommended limits, while Cu concentration is just a bit above the maximum threshold (Figure 6).

PARAMETERS	SECTOR II OPEN WATER ZONE 1*	SECTOR II OPEN WATER ZONE 2*
pH	8.45	8.71
moisture (%)	51.64	63.48
dry matter (%)	48.39	36.52
organic part (%)	20.12	22.66
inorganic part (%)	79.88	77.34
total P (mg/kg)	7010	24300
TOC (mg/kg)	44000	65900
SO_4^{2-} (mg/kg)	132	69
Cl^- (mg/kg)	75	158
S^{2-} (mg/kg)	594	263
$CO_3^{2-} + HCO_3^-$ (mg/kg)	37540	31650
total N (mg/kg)	4100	6800
aerobic mesophylic bacteria	4000	10000
fecal coliform bacteria	0	0
total coliform bacteria	0	0
fecal streptococci	negative	negative
sulphite reducing clostridia	30000	10000
other isolated microorganisms	saprophytic cocci, *Bacillus* sp.	saprophytic cocci, *Bacillus* sp.

* location of open water zones are given in figure 1

Table 4. Chemical and microbiological analysis of sediment in sector 2

layer depth	water in sample (%)	sand			silt	clay
		> 0.2 mm	0.2-0.02 mm	total		
0-20 cm	43.19	13.44	56.86	70.30	21.90	7.80
20-40 cm	44.38	43.50	38.50	82.00	12.80	5.20
40-60 cm	65.72	41.05	45.75	86.80	9.00	4.20

Table 5. Mechanical analysis of sediment from sector 2

Fecal coliforms and fecal streptococcus haven't been detected in the water (Table 4). It should be noted that in some sectors had detected the presence of fecal coliforms [24]. Periodically in the tourist part of the lake, due to the presence of pathogenic bacteria and reduced quality of water it is necessary to prohibit swimming. Continuously monitoring the presence of potentially pathogenic bacteria in water, as warned in previous investigations [30], the diseases related to water are the main cause of morbidity and mortality worldwide.

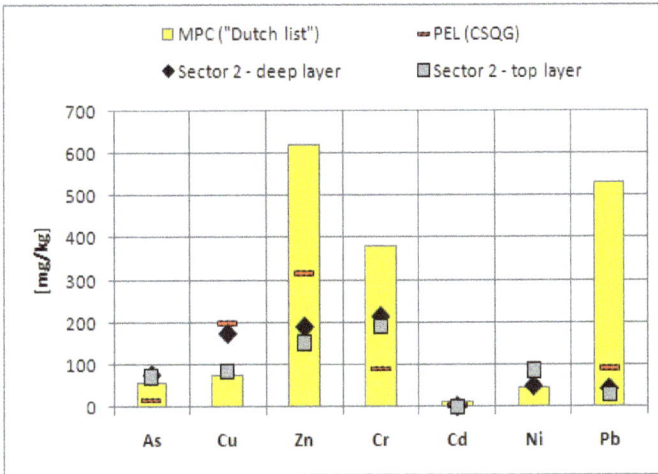

Figure 6. The concentration of As and Ni in Palic Lake (sector 2) compared to Holland and Canadian limits

4. Problem and targets

General problem of Palic Lake is enormous amount of deposited sediment in each sector of the lake (1.9 mil m³), which is a result of accelerated eutrophication. In the last decades, the wastewater discharge (treated and untreated) mainly increased the nutrient load to the Lake. Human negligence resulted in deterioration of biodiversity and ecological imbalance, i.e. general perturbation of the environment. A restoration of ecological balance and improvement of lake water quality can be achieved by implementation of technical solutions based on ecoremediation principles.

Attempts to manage lake eutrophication have most frequently involved controls of nitrogen (N) and/or phosphorus (P) loads from both diffuse sources [31] and point sources as well as internal loads from lake bed sediments [32].

Hence, essential problems of Palic Lake are inadequate water quality that recharges the lake and abundance of nutrient rich sediment. Construction of wetland in sector 2 would resolve both problems, improvement of water quality and sediment problem. In this manner sector 2 will now have the purpose of additional water treatment. Wetland technology, a design based on natural principles, has the aim of water treatment and water quality improvement using plants, microorganisms, soil and sediment. The additional important role of constructed wetlands is preservation of biodiversity, but also recreation, education, flood control, etc.

Constructed wetland is efficient and reliable technology if properly designed, created and specifically managed. Proper maintenance is also important for their efficiency. Constructed wetlands can remove most of pollutants that are carried by urban, atmospheric and industrial wastewater. Wetland design stimulates a decrease of biological oxygen demand (BOD_5), suspended matter, nutrients (nitrogen, phosphorous), metals (chromium, cadmium, manganese, zinc) and toxic organic pollutants.

Key elements and design criteria are defined in regard to overview and verification of current state [33], systematization and analyses of existing technical documentation and field investigations (survey, geomechanical investigations of soil and sediment, laboratory analyses of physical and chemical characteristics of water and sediment).

1. Sediment:
    - sectors 2-4 of the Palic Lake are covered by sediment layer (total volume of 1900160 m^3) whose thickness varies in the range 0.3 to 1.5 m; solely in sector 4 there is 1311356 m^3 of deposited sediment;
    - A portion of inorganic matter in sediment is approximately 80%. Inorganic part consists of 73 % sand (both coarse and fine sand, fraction size 2.0-0.2 mm);
    - sediment in sector 3 and 4 is overloaded with plant nutrients – nitrogen and phosphorous, but it is practically free of heavy and toxic metals (concentrations are within acceptable limits, compared to both Dutch and Canadian sediment quality guidelines);
2. Wastewater treatment plant of the city of Subotica (WWTP)
    - new WWTP started in year 2009;
    - a capacity of new WWTP is 36000 m^3/day, up to 72000 m^3/rainy day;
    - a quality of effluent that flows in the sector 1 is defined according to the Directive 91/271/EEC:

- BOD_5	25 mg/L
- COD	125 mg/L
- suspended matter	35 mg/L
- total N	10 mg/L
- total P	1 mg/L

    - Achieved level of effluent quality in 2011 contributes to lake water quality, without its impairment.
3. Required water quality for sector 4 of the Palic Lake
    - required water quality for the Palic Lake, according to Regulation for water classification of the Republic of Serbia, is a class 2a, which means the satisfaction of following standards:

- suspended matter	30 mg/L
- dry matter	1000 mg/L
- pH value	6.8 – 8.5
- BOD_5	4 mg/L

        - most probable number of coliform in 100 ml of water is max 6000
        - without smell, colour and visible waste.

- concerning that the effluent is the main source for lake recharge and that annual volume of discharged effluent overrides lake's volume, the quality of the effluent has to satisfy more strict limits than those set by European Union and Serbian legislative. Acceptable values of maximum tolerable risk are:
  - total N   2.2 mg/L
  - total P   0.15 mg/L
- previously listed parameters have to be met at the entrance of the sector 4, to decrease nutrient input and subsequently a risk of accelerated eutrophication.

5. Design elements and technical solution

Basic parameters for design and sizing of future wetland are water quality that is discharged in the lake and requested lake water quality. Since the new wastewater treatment plant started, effluent quality improved significantly, although it still represents nutrient burden for the lake (Table 6).

PARAMETERS	2009	2010	2011
BOD$_5$ (mgO$_2$/l)	8.1	6.2	4.4
COD (mgO$_2$/l)	36.5	35.2	41.1
Suspended matter (mg/l)	28.5	14.3	9.5
Total N (mg/L)	16.6	12.8	11.5
Total P (mg/L)	2.1	1.4	1.0

Table 6. Effluent quality (annual average)

The required area for future constructed wetland was calculated using input parameters, effluent quality and demanded remedy level. In addition, sizing was done according to efficiency of particular aquatic plants. Calculation results showed that the minimum area for wetland construction in sector 2 is 50 ha. Nutrient loading from diffuse sources (agricultural fields) is negligible in comparison to WWTP, so it is excluded from calculation.

Hydrology of constructed wetland considerably affects treatment efficiency. Hydraulic retention time must be sufficient for biological purification, but not for too long, due to risk of creating of anaerobic conditions. Retention time in constructed wetland in sector 2 is planned to be at least 3 days (recharge 36000 m^3/day without precipitation), up to 5 days (recharge 72000 m^3/rainy day).

Hydrological analysis considered influence of the quantity of water that recharges the lake, weather conditions, water loss and evapotranspiration. The efficiency of natural system like this one could easily be decreased due to severe precipitation, increased water flow or shortage of retention time.

Biological community of constructed wetland consists of both plants and microbial population (Table 7). Wide range of different species could be used for planting, which depends on climate and environment. Indigenous plants, well adapted to existing conditions are mostly used for this purpose.

Plant parts	Role
The roots and underground parts	Habitat for microbial population Filtration and adsorption Sediment stabilization Nutrient storage
Plant parts above the ground	Reduction of sunlight and limitation of algae growth Moderation of wind, i.e. gas exchange between water and atmosphere Gas transfer to submerged parts of plant Nutrient storage

Table 7. Roles of plant parts in constructed wetlands

Plant species	pH optimum	Distance (m)	Rooting depth (m)	Nitrogen content (%)	Phosphorous content (%)	Dry matter yield (t/ha)
Emerged plants						
Typha spp.	4-10	0.60	0.3-0.5	14	2	30
Scirpus spp.	4-9	0.30	0.6	18	2	20
Phragmites spp.	2-8	0.60	0.4	20	2	40
Juncus spp.	5-7,5	0.15	0.3	15	2	50
Carex spp.	5-7,5	0.15	0.2	1	0.5	5
Submerged plants						
Potamogeton spp.	6-10	0.3		2-5	0.1-1	3
P. myriophillum spp.	6-10	0.3		2-5	0.1-1	9
Ceratophillum spp.	6-10	0.3		2-5	0.1-1	10
Floating plants						
Lemna spp.				6	2	20 t/year

Table 8. Basic ecophysiological traits of proposed plants for future constructed wetland

In studied case, selected plants fulfil following conditions: they are tolerant to high nutrient content and they live in continuous aquatic conditions, i.e. plants that periodically need dry conditions are avoided. Furthermore, it is best to use perennial plants, as well as plants with bigger rate of biomass increase. Plants that have slower growth rate should be grown with bigger density. The best practice is to grow many different plant species, especially the ones that represent important habitat to animals. Monoculture should be avoided, because it can lead to quicker spread of diseases and be attacked by insects. According to data given from other authors [34], the Indigenous plants that are appropriate for future wetland creation in sector 2 of the Palic Lake are chosen together with basic ecophysiological traits and given in table 8.

The purpose of future constructed wetland inflicts a choice of the wetland type. Concerning input parameters and natural processes that occur in wetlands, a sequential model has been chosen. Generally, the basic drawback of this technology is occupation of large area; however in this case this is not a limiting factor.

Designed solution considers hydraulic dredging of deposited sediment in sector 4, and deposition through long pipeline in sector 2. Dredged sediment will be used as a substrate for constructed wetland. Future wetland will consist of indigenous plants. Plant structure is chosen to achieve requested water quality and to remediate sediment/substrate.

Water from the sector 1 inflows the sector 2 through the culvert and overflows the area of 53.5 ha where macrophyte zones interchange with open water zones. Container with biofilters is designed to be placed in the culvert for initial water treatment. Three vegetated zones with emerged macrophytes are designed to cover 30 ha totally. They should be established on substrate formed by dredged sediment, deposited to the elevation 102.0-102.2 m.a.s.l., to provide necessary depth for plants 0.4-0.6 m.

In the open water zone, minimum water depth is 1.4 m. Natural processes of reaeration are facilitated by means of submerged plants existence. Floating wetlands are also planned in this zone. The purpose of floating wetlands is decrease of nutrient and pollution loading naturally. They affect the increase of existing aquatic vegetation and intensify existing interaction between water, plants, microorganisms and atmosphere.

In this manner total area of constructed wetland will have a part in water treatment, which is joint work of all zones, as depicted in Figure 7. All processes that are important for achievement of recommended water quality are increased. Some modifications that encompass introduction of biofilters and floating wetland were also done, to increase efficiency of constructed wetland.

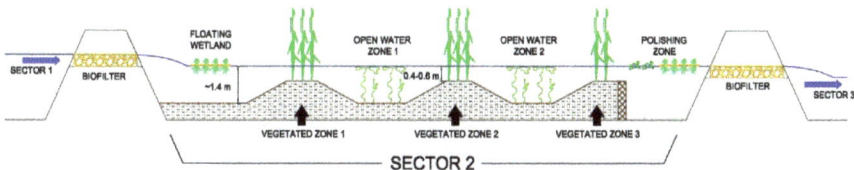

Figure 7. Scheme of the sequential model of wetland in sector 2 (longitudinal section)

Sequential model of constructed wetland makes the most of chemical, biological and physical processes that will further assure improved water quality.

6. Vegetated, open water and "polishing" zone

In the vegetated zone it is necessary to maintain the water depth up to 0.6 m. Retention time in vegetated zones is 3-5 days, which is optimal time for the development of sedimentation and biological processes for nutrient transformation. Several species will be planted in constructed wetland, with plants successively distributed. It is planned to grow emerged plants that are very common in studied area, e.g. *Typha spp., Phragmites communis, Juncus spp., Scirpus spp., Carex spp.*

Suggested plants have branched roots, "airy" stems and leaves. Roots take nutrients from the sediment and transport them in airy parts of the plant. After the wilting of the plant, nutrients are released in the water or in the pore water of the sediment. The largest amount of released nutrients is to be used by epiphytic microflora. Therefore, necessary maintenance measure is timely plant removal from the constructed wetland.

Plants that are considered for future wetland are the following:

1. *Typha spp.* is commonly found in different ecological conditions, which makes it ideal for constructed wetlands. It produces large biomass and after the period of three months it makes very dense vegetation.
2. *Juncus spp.* is a perennial plant that can be found in wet areas, mostly along shores. It is adapted to a wide range of environmental pH. It produces a large biomass during vegetation period.
3. *Phragmites communis* is a perennial plant with large perennial rhizome, the most utilized plant, very efficient in oxygen supply.
4. *Carex spp.* and *Scirpus spp.* are very suitable for planting in constructed wetlands that are located in colder climates. They are effective during periods of low temperatures (around 0°C). During these conditions their efficiency is 20-30% higher in comparison to other plants, which makes their usage in this case compulsory.

Constructed wetland will be planted with quoted plants after sediment stabilization. Plants can be grown from seed, by division, using seedlings or adult plants. Soil in the natural wet habitat in the vicinity can be used as a seed source, because it contains seeds of numerous indigenous species that are well adapted to local conditions. Moreover, indigenous plants have bigger survival rate.

Rhizomes and tubers can also be used for reproduction. They are collected in late autumn, at the end of the growing season, or in early spring, at the beginning of the growing season. Rhizomes can be stored in wet peat or sand until planting. Whole roots with some soil can be taken, by reason of inherent microbial population that are important for future wetland formation. Local plants, as well as local plant nursery are recommended, as the most confident source of planting material. Plant nurseries on the spot are good choice, since they are reliable and their cost is low. For that reason, a part of constructed wetland will be used as a nursery (approx. 5 ha). Also, it is recommended to plant by hand, using regular gardening tools and material. The largest share of seedlings is planted in spring; a process of planting should last between March and July.

In sector 2 large areas are already overgrown with common reed. Around every adult plant there are around 10 sprouts. Design of the constructed wetland anticipates getting half adult plants (with sprouts) from 1 m² of reed bed. Therefore, for 600.000 plants that is necessary for 30 ha of vegetated zones (designed plant distribution is 2 plants/m²), required size of plant nursery is 4 ha. Reed in so called nurseries will propagate, thus empty places in reed beds will be fulfilled by the end of the year.

Constructed wetland becomes efficient in 2-3 years, as plants completely develop and reach high effectiveness in water treatment. In this manner natural potential of selected indigenous plants is used. Plant production is continuous work, since it is necessary to replace damaged or wilted plants for the duration of wetland.

Maintenance of wetland through regular plant removal is obligatory. Early harvest before nutrient translocation and multiple harvests are significant for nutrient removal and also for extension of wetland endurance. Plant removal/harvest can be done manually or using machinery. Produced biomass can be used for composting, biofilters, energy production or as building material.

Submerged macrophytes alleviate reaeration processes, thus level of dissolved oxygen rises. Oxygen is necessary for oxidation of carbon compounds and reduction of BOD, and further for nitrification of ammonium and nitrates. These reactions are fundamental for nitrogen and phosphorous reduction. On the other hand, this zone has a particulate role in suppression of coliform population.

Open water zone can be planted with submerged macrophytes that are rooted in lake sediment. Photosynthetic parts of submerged vegetation are in the water body, or float on the water surface. Submerged plants remove ammonia from the water indirectly. Plants use carbon dioxide as a carbon source, consequently increasing pH and ammonia diffusion in the atmosphere.

Existing species in open water zone are: *Ceratophyllum, Elodea, Potamogeton,* etc. Selected species are widespread and adapted to growing even in saline conditions. They propagate fast using rhizomes and have high rate of biomass production. Establishing plants and maintenance in open water zone ought to be done in the similar way as in vegetated zone.

Polishing zone is covered with vegetation and it allows water denitrification process. Nitrification occurs in the water column in presence of aerobic conditions, but denitrification is limited to the sediment area. Floating and decorative plants are dominant in this zone.

Green parts of plants, where photosynthesis takes place, are on water surface, or immediately above it, while roots are in water column, extracting nutrients. Roots of floating plants are great medium for filtration process, adsorption of suspended matter, and furthermore for bacteria growth.

Lemna sp. (Duckweed) grows best in the water whose temperature is around 27°C, consequently doubling the area that covers every 4 days. It consists mainly of metabolically active cells, but it has a low content of structural fibres. Duckweed has a significant role in wastewater treatment, due to nitrogen transformation and competition with algae. In only 2-3 weeks, a quality of wastewater is improved, regarding both organic matter content and dissolved oxygen content. Duckweed contains proteins, fat, nitrogen, phosphorous, which makes it good supplement for livestock nutrition. Redistribution of these plants requires special attention in order to accomplish complete cover of polishing zone, but also timely harvesting. Due to the chemical composition, harvested crop can serve as animal food or a fertilizer (compost).

7. Biofilters

Biological filters are made of shredded herbal material, i.e. remains. Biofilters are supposed to be put in culverts between sectors: one located between sectors 1 and 2 and another between sectors 2 and 3. Before entering subsequent sector water flows through fine herbal material, leaving suspended matter, nitrogen, phosphorous and metals in it. Their efficiency depends on plant species whose remains are used: wheat, barley, rye, as straw degrades slowly and supply low peroxide concentration. Besides straw, shredded reed remains can be used.

Duration of biofilter effectiveness is at least 6 months. Biofilters can be recharged more frequently, which depends on a degree of nutrient enrichment of water and on time of year. After utilization, biofilter material can be composted and empty containers are refilled with new shredded herbal material. During winter when wetland activity is reduced, the significance of biofilters rises. The floating reed bed together with submerged barley straw in a eutrophic lowland reservoir was used to reduce phosphorous and nitrate as well as to limit algal growth [35].

8. Floating wetlands

Floating emergent macrophyte treatment wetlands (FTWs) are a novel treatment concept that employ rooted, emergent macrophytes growing in a floating mat on the surface of the water rather than rooted in the sediments [36]. This makes CFWs extremely suited for treatment of event-driven waterflows such as storm water or combined sewer overflow water [37].

Floating wetlands increase the effectiveness of indigenous vegetation. Design of this type of wetland strengthens interaction between environment and living organisms. Plant root system, the most active part that is placed in water, have significant active surface for nutrient assimilation. Roots are also important for the development of bacterial populations that take part in nutrient transformation. Wetland capacity is increased by means of functionality of living organisms and providing habitat for fauna. Growing plants in this manner allows strict control of plant growth and for that reason usage of non indigenous plants.

Floating wetlands are more effective in comparison to other types. Large amount of nutrients is assimilated and eliminated from treated water, due to the removal of whole plant from the water, not just above ground parts of plant.

Because the plants are not rooted in soils in the base of the wetland, they are forced to acquire their nutrition directly from the water column, which may enhance rates of nutrient and element uptake into biomass. Their buoyancy enables them to tolerate wide fluctuations in water depth. This provides potential to enhance treatment performance by increasing the water depth retained during flow events to extend the detention time of storm waters in the wetland [38].

Green salad, clover, alfalfa, mustards, sunflower are some of plant species that can be used for floating wetland creation besides indigenous plants. If used plants are not marsh plants, water below should be aerated. Usage of indigenous plants is common. Planting material that encompasses whole plants and rhizomes is gained locally. Planting is to be done manually on floating beds, made of woods, plastic or styrofoam.

Position of floating wetlands is optional: they can be attached to the dam between sectors or to the shore. After achievement of maximum yield of biomass, floating wetland will be pooled on the shore and plants will be replaced. Removal of whole plants from the water is very important. During growing season, control of plant growth and replacement of damaged plants has to be done on a regular basis. Timely removal or harvest of plants is one of the most important issues regarding every wetland type.

9. Nutrient removal

Collected data that depict the capacity of new WWTP are used as input for further calculation. Generally, WWTP discharges 36000 m^3 of effluent loaded with 1 mg/L of total phosphorous (TP). Calculated phosphorous loading is 36 kg TP/day, or 13 t TP/year. Accordingly, calculated nitrogen loading is 360 kg TN/day, or 130 t TN/year.

Constructed wetland design has been conducted according to nutrient loading that originates from the WWTP of the city of Subotica. Moreover, current quality and nutrient enrichment of the lake water has been considered. The lake has considerable autopurification capacity that even in current conditions influences decrease of available phosphorous.

Main mechanisms for phosphorous and nitrogen removal are adsorption, chemical precipitation and assimilation by macrophytes. Plants absorb significant amount of inorganic phosphorous, but after wilting nutrients are released in the water again. Floating wetlands (1 ha) will remove approximately 780 kg TP/year and 6.6 t TN/year. Emerged vegetation (30 ha) assimilate 19 t TP/year and 160 t TN/year (Table 9). Submerged plants assimilate lesser amounts of nutrients: 0.7 t TN/year and 0.5 t TP/year. Floating plants incorporate approximately 1.2 t TN/year and 0.4 t TP/year, which depends on size of area that cover [34].

Since, phosphorus loading is estimated to be the most important factor of lake eutrophication, more detailed calculation relates phosphorous mass balance (Table 9). Inputs of phosphorous are from the WWTP and deposited sediment. Release of phosphorous from sediment depends on environmental conditions, i.e. oxygen content, environmental pH value. Benthic algae and submerged plants intercept sediment phosphorous, i.e. they have a role similar to some products for phosphorous inactivation

[39]. Water column is in contact with active layer of sediment, whose thickness varies from a millimetre to few centimetres. If anaerobic conditions prevail in superficial sediment layer, phosphorous becomes mobile in pore water of active layer that is assumed to be 2 cm. Examined phosphorous loading is derived using concentrations of total phosphorous, thus calculated values represent sediment potential.

Plant species and planned proportion	Area (ha)	Dry matter yield (t/ha)	Dry matter yield (t)	Content TN (%)	Quantity TN (t)	Content TP (%)	Quantity TP (t)
Typha spp., 30%	9,0	30	270	14	37,8	2	5,4
Scirpus spp., 15%	4,5	20	90	18	16,2	2	1,8
Phragmites spp., 30%	9,0	40	360	20	72,0	2	7,2
Juncus spp., 15%	4,5	50	225	15	33,8	2	4,5
Carex spp., 10%	3,0	5	15	1	0,2	0,1	0,001
TOTAL:	30,0		960		160,0		19,0

Table 9. Calculation of biomass yield and assimilated quantities of nutrients by emerged plants

Output is a sum of phosphorous that is consumed by plants in vegetated zones of designed wetland and plants in floating wetlands; while one portion is represented as autopurification capacity (Table 10). Here, a portion that is affiliated by submerged plants and free floating vegetation is neglected, due to the fact that it is hard to control this kind of vegetation and predict their effects reliably.

INPUT [t]		OUTPUT [t]	
effluent from WWTP (designed regime)	14	designed wetland (vegetated zones)	19
sediment active layer	27	floating wetlands (per hectare)	0.78
		autopurification capacity in current environment (16%)	2.24
Total	41	Total	22.02
Difference	+18.98		

Table 10. Mass balance of total phosphorous

Input values represent phosphorous reserves and potential for bioavailable phosphorous forms, while output values represent used bioavailable phosphorous which is further incorporated in plant body. Positive difference confirms that it is not possible to remediate water and whole amount of (active) sediment in one year. Roughly, designed wetland is capable of additional water treatment and remediation of nearly one third of sediment active layer. Accordingly, in addition to water treatment, designed wetland will gradually remediate deposited sediment as well.

Macrophytes will assimilate considerable part of phosphorous from lake water, but some will precipitate in sectors 2 and 3, considering high water pH value. Together with proved lake autopurification, it is possible to accomplish requested water quality in sector 4, in regard to phosphorous content. Eventually, nutrient removal can be improved by increasing number of floating wetlands, if necessary. Nevertheless, the aesthetic value of floating wetlands gives them additional advantages.

The design of constructed wetland (Figure 8) depends on many factors but the plant species and microbial populations directly affect the efficiency in water treatment. Starting from the quality and quantity, especially treated water coming from the WWTP into the lake, and in order to obtain the required water quality for the sector 4 we proposed a model of constructed wetland that will maximize the potential of plants and microorganisms.

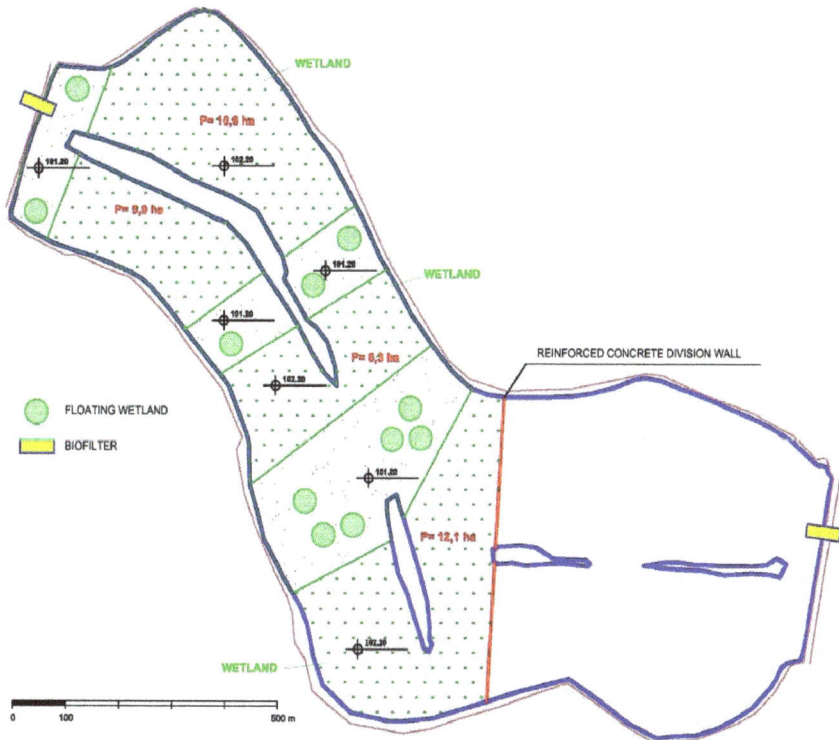

Figure 8. Wetland model in sector 2

We should also bear in mind that decomposition of macrophytes in wetland may also release large amounts of nutrients [40], and thus could reduce wetland treatment efficiency [41]. Plant decomposition plays an important role in nutrient cycling in aquatic ecosystems [43-43]. Harvesting is considered a direct and effective way to solve the re-pollution problem. Using a numerical model was suggested that harvesting the above-ground macrophyte biomass at the end of the growing season could significantly reduce phosphorus release in the decomposition process of macrophytes in shallow lakes [42]. In previous researches was reported that about 20.1% of nitrogen and 57.0% of phosphorus over nutrient removal was contributed by harvesting of emergent plants [44].

The plants produced within constructed wetlands or on floating islands can be harvested and subsequently used as animal feeds, or even human food, or be processed into biogas, bio-fertilizer and bio-materials. This may justify the practical application of the technology using the potential economic returns [45-46].

10. Conclusions

Over decades the high concentration of nutrients, especially phosphorus, (which arrived in the lake, primarily from treated or untreated wastewater), contributed to the disruption of ecological balance in the lake and the formation of thick layers of sediment. Although constructed wetland building up in recent years throughout the world, in Serbia this is a relatively new concept.

Considering the input parameters of water quality, and current water quality, the proposed model has a few activity zones in sector 2 (plants-water-plants-water...). In this way model achieves the maximum use of all physical, chemical and biological processes that are important for improvement of water quality. In order to increase efficiency a few modification are made as well as applying biofilters (after sector 1 and after sector 2) and floating wetland.

Model used the plants that are tolerant to high levels of nutrients and predominantly indigenous species whose grow in vicinity of lake.

All this will contribute to the cost and environmentally friendly and effective technologies to improve water quality, increase biodiversity and ecological balance in the lake Palic.

Author details

Vera Raicevic* and Blazo Lalevic
*University of Belgrade, Faculty of Agriculture,
Department for Microbial Ecology, Zemun-Belgrade, Serbia*

Mile Bozic and Zeljka Rudic
Institute for the Development of Water Resouorces "Jaroslav Cerni", Belgrade, Serbia

Dragan Kikovic
Faculty of Natural Science, Kosovska Mitrovica, Serbia

* Corresponding Author

Ljubinko Jovanovic
Educons University, Sremska Kamenica, Serbia

Acknowledgement

This research was partially supported by Ministry of Education and Science of Republic Serbia (grant number TR31080).

11. References

[1] Hutchinson GE (1973) Eutrophication. Am. Sci. 61:269-279.

[2] Schindler DW, Hecky RE, Findlay DL, Stainton MP, Parker BR, Paterson MJ, Beaty KG, Lyng M, Kasian SEM (2008) Eutrophication of lakes cannot be controlled by reducing nitrogen input. Results of 37-year whole-ecosystem experiment. P. natl. acad. sci usa 105(32): 11254-11258.

[3] Qin BQ (2009) Lake eutrophication: Control countermeasures and recycling exploitation. Ecol. eng. 35(11): 1569-1573.

[4] Teixera MG, Costa MC, de Carvalho VL, Pereira MS, Hage E (1993) Gastroenteritis Epidemic in the Area of the Itaparica Dam. vol. 27. Bull. pan am. health Organ. 27(3): 244-253.

[5] Albay M, Akcaalan R, Tufekci H, Metcalf JS, Beattie KA, Codd GA (2003) Depth profile of cyanobacterial hepatotoxins (microcystins) in three turkish freshwater lakes. Hydrobiologia 505: 89-95.

[6] Pyo D, Jin J (2007) Production and degradation of cyanobacterial toxin in water reservoir, Lake Soyang Bull. korean chem. soc. 28: 800-804.

[7] Kouzminov A, Ruck J, Wood SA (1996) Wood New Zealand risk management approach for toxic cyanobacteria in drinking water. Aust. nz j. publ. hcal. 31: 275-281.

[8] Saker ML, Thomas AD, Norton JH (1999) Cattle mortality attributed to the toxic cynaobacterium *Cylindrospermopsis raciborskii* in an outback region of North Queensland. Environ. toxicol. 14: 179-182.

[9] Murphy T, Lawson A, Nalewajko C, Murkin H, Ross L, Oguma K, McIntyre T (2000) Algal toxins-initiators of avian botulism? Environ. toxicol. 15: 558-567.

[10] Alonso-Andicoberry C, Garcia-Villada L, Lopez-Rodas V, Costas E (2002) Catastrophic mortality of flamingos in a Spanish national park caused by cyanobacteria. Vet. record 151: 706-707.

[11] USEPA (1990) National Water Quality Inventory. Report to Congress. U.S. Gov. Print. Office, Washington, DC.

[12] Chen KN, Bao CH, Zhou WP (2009) Ecological restoration in eutrophic Lake Wuli: A large enclosure experiment. Ecol. eng. 35(11): 1646-1655.

[13] Phillips G, Jackson R, Bennet C, Chilvers A (1994) The importance of sediment phosphorus release in the restoration of very shallowlakes (The Norfolk Broads, England) and implications for biomanipulation. Hydrobiologia 275/276: 445-456.

[14] Søndergaard M, Jeppesen E, Jensen JP, Lauridsen T (2000) Lake restoration in Denmark. Lakes reservoirs: res. manage. 5: 151-159.

[15] Coveney MF, Stites DL, Lowe EF, Battoe LE, Conrow R (2002) Nutrient removal from eutrophic lake water by wetland filtration. Ecol. eng. 19: 141-159.

[16] Zedler JB, Kercher S (2005) Wetland resources: status, trends, ecosystem services, and restorability. Ann. rev. environ. resour. 30: 39-74.

[17] Verhoeven TA, Arheimer B, Yin C, Hefting MM (2006) Regional and global concerns over wetlands and water quality. Trends in ecol. evol. 21(2): 96-103.

[18] Mitsch WJ, Gosselink JG (2007) Wetlands. (4th ed.) New York: John Wiley & Sons, Inc. 582 p.

[19] Hu L, Hu, W, Deng J, Li Q, Gao F, Zhu J, Han T (2010) Nutrient removal in wetlands with different macrophyte structures in eastern Lake Taihu, China. Ecol. eng. 36: 1725-1732.

[20] Republic hydrometeorological service of Serbia (2006) Meteorological yearbook. Belgrade, 220 p.

[21] Provincial secretariat for plant protection of Republic of Serbia (2011) Park of nature „Palic" – proposal for putting under protection as protected area 3rd category. Study of protection. 234 p.

[22] National Research Council (2000) Clean Coastal Waters: Understanding and Reducing the Effects of Nutrient Pollution. National Academy Press. Washington.

[23] Selesi D (2000). Water of the Palic lake from 1781 to 1999. JP "Palic-Ludas".

[24] Raicevic V, Bozic M, Rudic Z, Lalevic B, Kikovic D (2011) The evolution of the eutrophication of the Palic Lake (Serbia). Afr. j. biotech. 10(10): 1736-1744.

[25] Kim LH, Choi E, Stenstrom MK (2003) Sediment characteristics, phosphorus types and phosphorus release rates between river and lake sediments. Chemosphere 50(1): 53-61.

[26] Dunne EJ, Clark MW, Mitchel J (2010) Soil phosphorus flux from emergent marsh wetlands and surrounding grazed pasture uplands. Ecol. eng. 36: 1392-1400.

[27] Farjalla VF, Esteves FA, Bozelli LR, Roland F (2002) Nutrient limitation of bacterial production in clear water Amazonian ecosystems. Hydrobiologia 489: 197-205.

[28] Warmer H, van Dokkum R (2002) Water pollution control in the Nederlands. Policy and practice 2001. RIZA report 2002.009.

[29] Canadian Environmental quality guidelines (2003) Summary table. Canadian Council of Ministers of the environment.

[30] Who (2003) Emerging Issues in Water and Infectious Disease. World Health Organization.

[31] Jeppesen E, Søndergaard M, Kronvang B, Jensen JP, Svendsen LM, Lauridsen TL (1999) Lake and catchment management in Denmark. Hydrobiologia 395/396: 419-432.

[32] Cooke GD, Welch EB, Peterson SA, Nichols SA (2005) Restoration and Management of Lakes and Reservoirs. Taylor&Francis/CRC Press. 616 p.

[33] Vymazal J, Kropfelova L (2008) Wastewater Treatment in Constructed Wetlands with Horizontal Sub-Surface Flow. In: Alloway BJ, Trevors JT, editors. Dordrecht: Springer. pp. 1-566.

[34] Crites RW, Middlebrooks EJ, Reed SC (2006) Free water Surface constructed wetlands. In: Natural wastewater treatment system. Boca Raton: CRC press, Taylor & Francis. pp. 259-334.

[35] Garbett P (2005) An investigation into the application of floating reed bed and barley straw techniques for the remediation of eutrophic waters. Water environ. j. 19: 174-180.

[36] Fonder N, Headley T (2010) Systematic nomenclature and reporting for treatment wetlands In: Vymazal J, editors. Water and Nutrient Management in Natural and Constructed Wetlands. Dordrecht: Springer. pp. 191-220.

[37] Van de Moortel AMK, Laing GD, Pauw ND, Tack FMG (2012) The role of the litter compartment in a constructed floatingwetland. Ecol. eng. 39: 71-80.

[38] Tanner CC, Headle TR (2011) Components of floating emergent macrophyte treatment wetlands influencing removal of stormwater pollutants, Ecol. eng. 37(3): 474-486.

[39] Meis S, Spears B (2009) An assessment of water and sediment conditions in Loch Flemington (May - August 2009). Centre for Ecology and Hydrology. Edinburgh.

[40] Shilla D, Asaeda T, Fujino T, Sanderson B (2006) Decomposition of dominant submerged macrophytes: implications for nutrient release in Myall Lake, NSW, Australia. Wetlands ecol. manage. 14: 427-433.

[41] Chimney MJ, Pietro KC (2006) Decomposition of macrophyte litter in a subtropical constructed wetland in south Florida (USA). Ecol. eng. 27: 301-321.

[42] Asaeda T, Trung VK, Manatunge J (2000) Manatunge Modeling the effects of macrophyte growth and decomposition on the nutrient budget in Shallow Lakes. Aquat. bot. 68: 217-237.

[43] Xie D, Yu B (2004) Ren Effects of nitrogen and phosphorus availability on the decomposition of aquatic plants. Aquat. bot. 80: 29-37.

[44] Li L, Li Y, Biswas DK, Nian Y, Jiang G (2008) Potential of constructed wetlands in treating the eutrophic water: evidence from Taihu Lake of China. Bioresource technol. 99: 1656-1663.

[45] Li M, Wu YJ, Yu ZL, Sheng GP, Yu HQ (2007) Nitrogen removal from eutrophic water by floating-bed-grown water spinach (Ipomoea aquatica Forsk.) with ion implantation. Water res. 41: 3152-3158.

[46] Li XN, Song HL, Li W, Lu XW, Nishimura O (2010) An integrated ecological floating-bed employing plant, freshwater clam and biofilm carrier for purification of eutrophic water. Ecol. eng. 36: 382-390.

Chapter 15

Relationship of Algae to Water Pollution and Waste Water Treatment

Bulent Sen, Mehmet Tahir Alp, Feray Sonmez,
Mehmet Ali Turan Kocer and Ozgur Canpolat

Additional information is available at the end of the chapter

1. Introduction

Pollution of surface water has become one of the most important environmental problems. Two types of large and long-lasting pollution threats can be recognized at the global level: on the one hand, organic pollution leading to high organic content in aquatic ecosystems and, in the long term, to eutrophication. It is a well-known fact that polluted water can reduce water quality thus restricting use of water bodies for many purposes.

Organic pollution occurs when large quantities of organic compounds from many sources are released into the receiving running waters, lakes and also seas. Organic pollutants originate from domestic sewage (raw or treated), or urban run off, industrial effluents and farm water. Organic pollution could negatively affect the water quality in many ways. During the decomposition process of organic water dissolved oxygen in the water may be used up greater rate than it can be replenished thus, giving rise to oxygen depletion which causes severe consequences on the aquatic biota. Organic effluents also frequently contain large quantities of suspended solid which reduce the light available to photosynthetic organisms mainly algae. In addition organic wastes from people and animals may also rich in disease causing (pathogenic) organisms [1,2,3].

2. Algae and water pollution

Algae are the main the primary producers in all kinds of water bodies and they are involved in water pollution in a number of significant ways. Firstly, enrichments of the algal nutrients in water through organic effluents may selectively stimulate the growth of algal species producing massive surface growths or 'blooms' that in turn reduce the water quality and affect its use. However, certain algae flourished in water polluted with organic wastes play an important part in "self-purification of water bodies". Some pollution algae may frequently are toxic to fish and also mankind and animals using polluted water. In fact, algae can play significant part of food chain of aquatic life, thus whatever alters the number and kinds of algae strongly affects all organisms in the chain including fish.

Algae are also known to be causes of tastes and odors in water [4]. In fact, a large number of algae are associated with tastes and odors that vary in type. Certain diatoms, blue-green algae and coloured flagellates (particularly Chrysophyta and Euglenophyta) are the best known algae to pose such problems in water supplies, but green algae may also be involved. Some algae produce an aromatic odor resembling to that of particular flowers or vegetables. In addition, a spicy, a fishy odor and a grassy odor can also be produced by odor algae [5,6].

3. Algae as bioindicators

Bioindicator organisms can be used to identify and qualify the effects of pollutants on the environment. Bioindicators can tell us about the cumulative effects of different pollutants in the ecosystem and about how long a problem may persist. Although indicator organisms can be any biological species that defines a trait or characteristics of the environment, algae are known to be good indicators of pollution of many types for the following reasons.

- algae have wide temporal and spatial distribution.
- many algal species are avaliable all the year.
- response quickly to the charges in the environment due to pollution.
- Algae are diverse group of organisms found in large quantities.
- easier to detect and sample.
- The presence of some algae are well correlated with particular type of pollution particularly to organic pollution

Algae of many kinds are really good indicators of water quality and many lakes are characterized based on their dominant phytoplankton groups. Many desmids are known to be present in oligotrophic waters whilst a few species frequently occurs in eutrophic bodies of water [7]. Similarly, many blue-green algae occurs in nutrient-poor waters, while some grow well in organically polluted waters [8]. The ecosystem approach to water quality assessment also include diatom species and accociations used as indicators of organic pollution. Five algal species were selected as indicators of the degree of pollution in rivers in England. *Stigeoclonium tenue* is present at the down stream margin of the heavily polluted part of a river, *Nitzschia palea and Gomphonema parvulum* always appear to be dominant in the mild pollution zone whilst *Cocconeis* and *Chamaesiphon* are reported to occur in unpolluted parts of the stream or in repurified zone [9]. *Navicula accomoda* is stressed to be a good indicator of sewage/organic pollution as the species comfortably occur in the most heavily polluted zones in which other species can not occur. The same hold true for species and varietes of *Gomphonema* [10] which is commonly found in highly organically polluted water. *Amphora ovalis* and *Gyrosigma attenuatum* are also introduced as good examples of diatoms to be affected by high organic content of water [11].

A list of more than 850 algal taxa was published based on the reports of considerable number of authors. According to this list, many algal genera have species that grow well in water containing a high concentration of organic wastes. Green algae *Chlamydomonas*,

Euglena, diatoms, *Navicula, Synedra* and blue- green algae *Oscillatoria* and *Phormidium* are emphasized to tolerate organic pollution [12]. At species level, *Euglena viridis* (Euglenophyta), *Nitzschia palea* (Bacillariophyta), *Oscillatoria limosa, O.tenuis, O.princeps* and *Phormidium uncinatum* (Cyanophyta) are reported to be present than any other species in

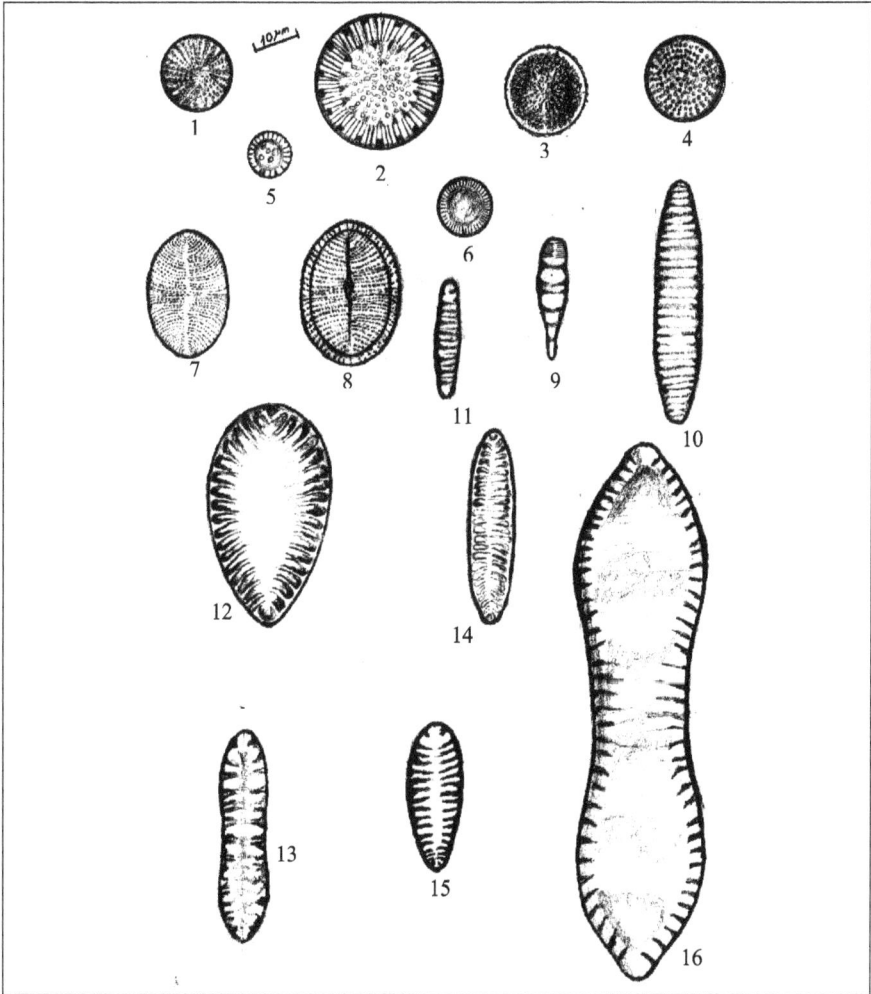

Plate 1. 1.Stephanodiscus hantzschii Grunow 2.Cyclotella comta (Ehrenberg) Kützing 3.Thalassiosira weissflogii (Grunow) G.Fryxel & Hasle 4.Aulacoseira distans (Ehrenberg) Simonsen 5.Cyclotella ocellata Pantocsek 6. C. kützingiana Thwaites 7.Cocconeis pediculus Ehrenberg 8.C.placentula Ehrenberg 9. Meridion circulare (Greville) C. Agardh 10. Diatoma vulgaris var. lineare Grunow 11. D. tenuis C.Agardh 12. Surirella ovalis Brébisson 13. S. ovata var. apiculata W. Smith 14. S. linearis W. Smith 15. S. minuta Brébisson 16. Cymatopleura solea (Brébisson) W. Smith [14].

organically polluted waters [13]. Some diatom taxa in a stream polluted with the waste water of a slaugher house are given in Plate 1-4 [14].

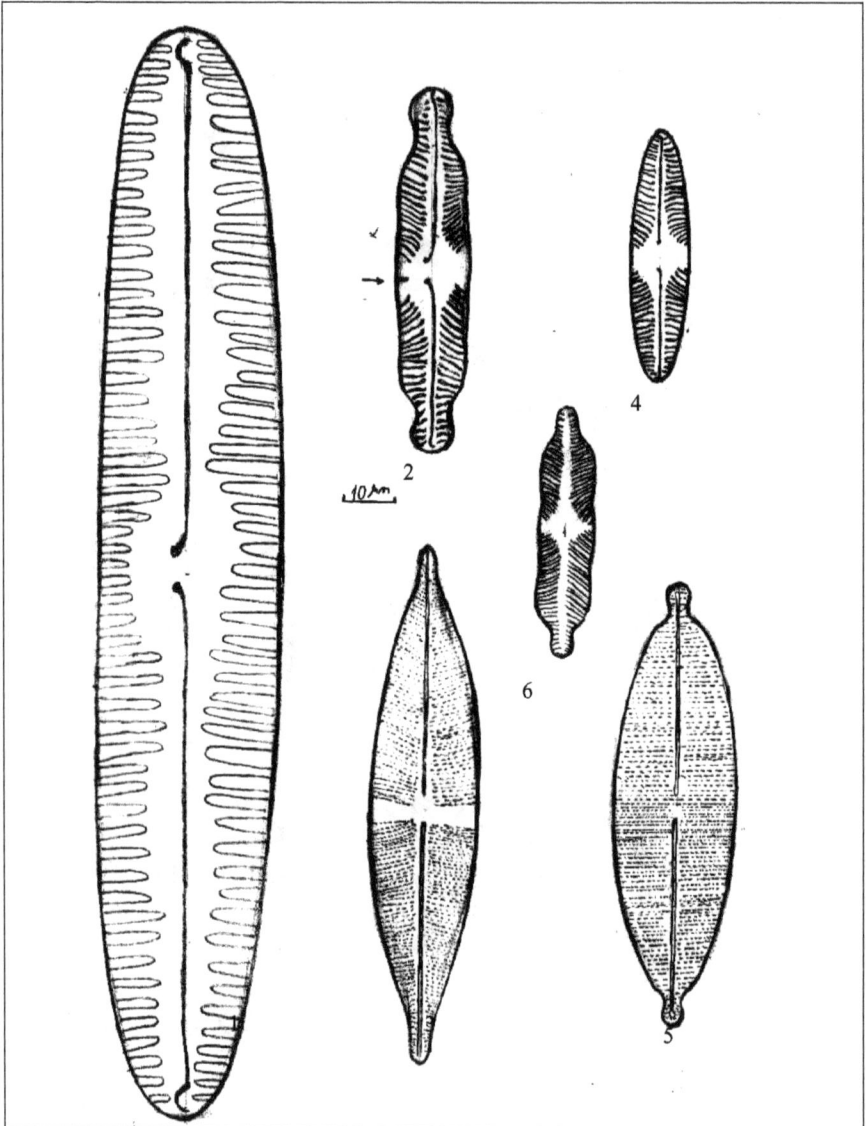

Plate 2. 1. *Pinnularia viridis* (Nitzsch) Ehrenberg 2. *P. biceps* W. Gregory 3. *Stauroneis phoenicenteron* (Nitzsch) Ehrenberg 4. *Pinnularia brebissonii* (Kützing) Rabenhorst 5. *Craticula ambigua* (Ehrenberg) D. G. Mann 6. *Pinnularia mesolepta* (Ehrenberg) W. Smith [14].

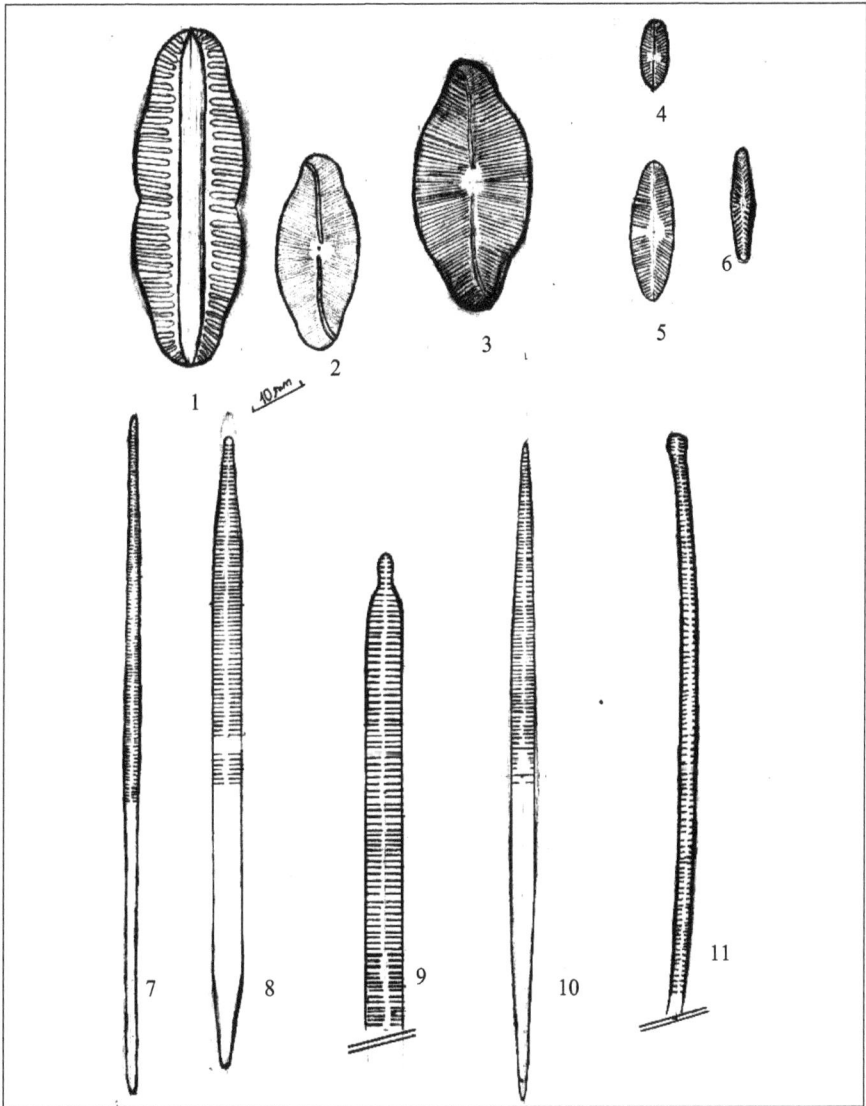

Plate 3. 1. *Rhopalodia gibba* (Ehrenberg) Otto Müller 2-3. *Eucocconeis flexella* (Kützing) Meister 4. *Achnanthidium minutisimum* (Kützing) Czarnecki 5. *Eucocconeis quadratarea* (Østrup) Lange-Bertalot 6. *Achnanthes marginulata* Grunow 7. *Ulnaria delicatissima* var. *angustissima* (Grunow) M. Aboal & P. C. Silva 8. *U. acus* (Kützing) M. Aboal 9. *U. amphirhyncus* (Ehrenberg) Compère & Bukhtiyarova 10. *U. danica* (Kützing) Compère & Bukhtiyarova 11. *U. biceps* (Kützing) P. Compère [14].

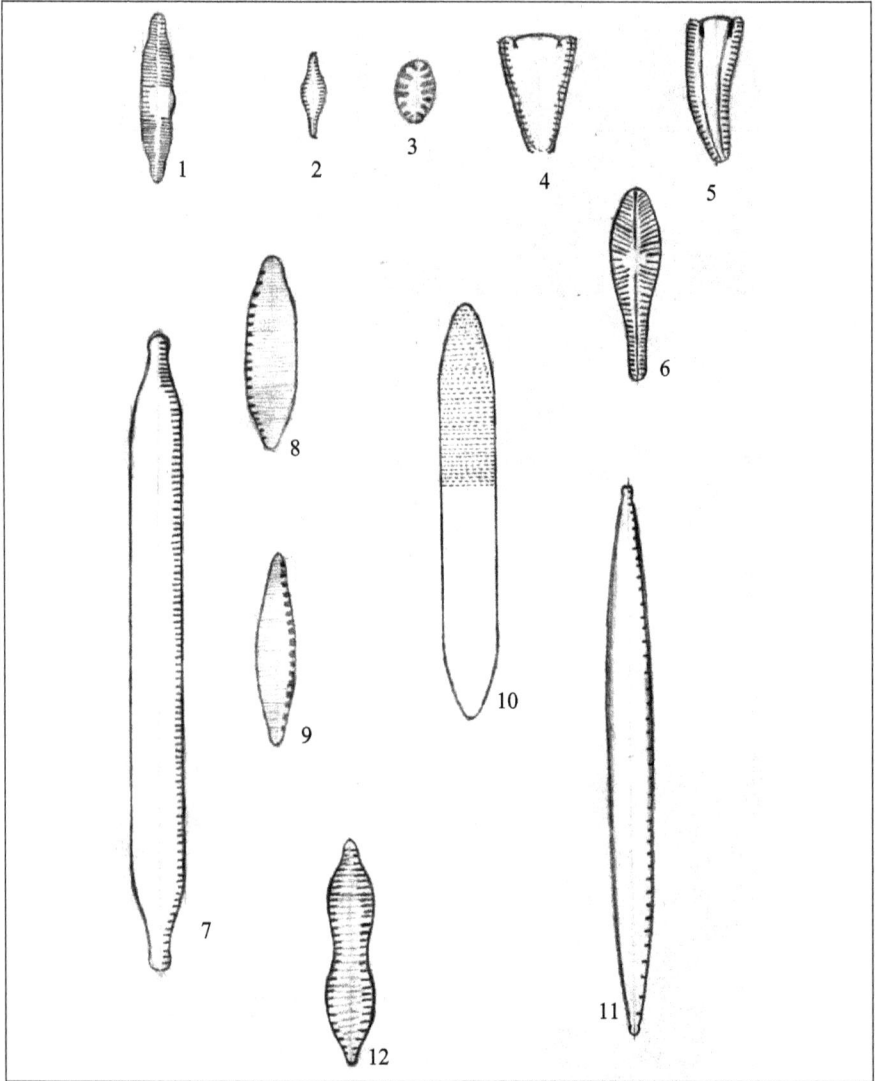

Plate 4. 1.*Fragilaria capucina* var. *vaucheriae* (Kützing) Lange-Bertalot 2.*Pseudostaurosira brevistrita*
(Grunow) D.M.Williams & Round 3.*Staurosirella pinnata* (Ehrenberg) D.M.Williams & Round
4.*Gomphonema truncatum* Ehrenberg 5.*Rhoicosphenia abbreviata* (C. Agardh) Lange-Bertalot 6.*Gomphonema
olivaceum* (Hornemann) Brébisson 7.*Nitzschia sublinearis* Hustedt 8.N. *umbonata* (Ehrenberg) Lange-
Bertalot 9. *N. hantzschiana* Rabenhorst 10. *Tryblionella angustata* W. Smith 11. *Nitzschia linearis* (C.
Agardh) W. Smith12. *N. constricta* (Kützing) Ralfs. [14].

Algae are also good indicators of clean water since many species occur insistently and predominately in the clean water zone of the streams. However it is more satisfactory to emphasize the presence or absence of several species of clean water algae rather than of any one species to define the clean water zone. Approximately 46 taxa has been announced as representatives of the clean water algae including many diatoms, several flagellates and certain green and blue-green algae [12]. However it is emphasized that minute flagellates are better indicators of clean water than many larger algae. A few of the clean water algae are planktonic whilst many are benthic, attached to substrata at the bottom or sides of the running waters.

There are many studies by various authors emphasizing the relationships of algae to clean water. A community composed of the diatom *Cocconeis* and the blue-green alga *Chamaesiphon* is claimed to be present in the portion of the stream which has returned to normal following purification of a polluted condition [9]. Kolkwitz listed 61 diatoms, 42 green algae, 41 pigmented flagellates, 23 blue-green algae, and 5 red algae as organisms of oligosaprobic and /or unpolluted zones and Lackey found 77 species of planktonic algae in the clean water portion of a small stream, 40 of which were absent in the polluted area[15,16]. The flagellates *Chromulina rosanoffi, Mallomonas caudata,* the green algae *Ulothrix zonata* and *Microspora amoena* are also reported as oligosaprobic zone organisms [17]. Two groups of algae, Cryptophyta and Chrysophyta, are reported to be indicators of clean and/or unpolluted water as the members of these algal groups tend to occur in abundance, oppositely reacting adversely to pollution [18]. The absence of blue-green algae was also accepted an indication of clean water [19].

4. Use of algae in saprobien system

The classic scheme for the interpretation of streams ecological conditions based on the biota was first introduced by Kolkwitz and Marsson [20]. They defined five zones based on the degree of pollution and proposed the use of aquatic organisms as indicators of different pollution and/or recovery zones of rivers which were polluted with organic matter such as sewage. However more recently Werner proposed nine different zones in the saprobic system in a stream organically polluted [21]. Survey of the saprobic zones and the corresponding communities are given Table 1. Pollution zones proposed in that saprobient system were basically termed "**Coprozoic**" "**Polysaprobic**", "**Mesosaprobic**", "**Oligosaprobic**" and "**Katharobic**". Each zone was different in chemical and physical characteristics and containing characteristic species. He listed indicator species of these zones except the last one which is infact clean water.

Polysaprobic zone was characterized by almost complete absence of algae except for blue-green alga *Arthrospira (Spirulina) jenneri* and green alga *Euglena viridis*. Bacteria and Protozoa were the most common groups in this zone. The preponderance of blue-green algae (Cyanophyta) was characteristic of alfa-mesosaprobic zone while diatoms (Bacillariophyta) and green algae (Chlorophyta) were dominant organisms in beta-mesosaprobic zone. Peridiniales (Dinophyta) and Charales (Charophyta) occurred in any quantity only in the oligosaprobic zone. In the same zone, the bacterial count was low but there was a great variety of plants and animals (including fish) in considerable numbers.

5. Use of algae in wastewater treatment

Recently, algae have become significant organisms for biological purification of wastewater since they are able to accumulate plant nutrients, heavy metals, pesticides, organic and inorganic toxic substances and radioactive matters in their cells/bodies [22-25]. Biological wastewater treatment systems with micro algae have particularly gained importance in last 50 years and it is now widely accepted that algal wastewater treatment systems are as effective as conventional treatment systems. These spesific features have made algal wastewaters treatment systems an significant low-cost alternatives to complex expensive treatment systems particularly for purification of municipal wastewaters.

In addition, algae harvested from treatment ponds are widely used as nitrogen and phophorus suplement for agricultural purpose and can be subjected to fermentation in order to obtain energy from metane. Algae are also able to accumulate highly toxic substances such as selenium, zinc and arsenic in their cells and/or bodies thus eliminating such substances from aquatic enviroments. Radiation is also an important type of pollution as some water contain naturally radioactive materials, and others become radioactive through contamination. Many algae can take up and accumulate many radioactive minerals in their cells even from greater concentrations in the water [12]. MacKenthun emphasized that *Spirogyra* can accumulate radio-phosphorus by a factor 850.000 times that of water [26]. Considering all these abilities of algae to purify the polluted waters of many types, it is worth to emphasize that algal technology in wastewater treatment systems are expected to get even more common in future years.

Wastewater treatment which is applied to improve or upgrade the quality of a wastewater involves physical, chemical and biological processes in primary, secondary or tertiary stages. Primary treatment removes materials that will either float or readily settle out by gravity. It includes the physical processes of screening, commination, grit removal, and sedimentation. While the secondary treatment is usually accomplished by biological processes and removes the soluble organic matter and suspended solids left from primary treatment. Tertiary or advanced treatment is process for purification in which nitrates and phosphates, as well as fine particles are removed [27]. However initial cost as well as operating cost of wastewater treatment plant including primary, secondary or advanced stages is highly expensive [28].

It is well known that algae have an important role in self purification of organic pollution in natural waters [29]. Moreover, many studies revealed that algae remove nutrients especially nitrogen and phosphorus, heavy metals, pesticides, organic and inorganic toxins, pathogens from surrounding water by accumulating and/or using them in their cells [30,31,32,33,34,35,36,37,38]. Also, studies showed that algae may be used successfully for wastewater treatment as a result of their bioaccumulation abilities [39].

Zone I.	Coprozoic zone
	the bacterium community
	the *Bodo* community
	both communities
Zone II.	α-Polysaprobic zone
	Euglena community
	Rhodo-Thio bacterium community
	Pure Chlorobacterium community
Zone III.	β-Polysaprobic zone
	Beggiatoa community
	Thiothrix nivea mommunity
	Euglena community
Zone IV.	γ-Polysaprobic zone
	Oscillatoria chlorina community
	Sphaerotilus natans community
Zone V.	δ-Mesosaprobic zone
	Ulothrix zonata community
	Oscillatoria benthonicum community (O.brevis, O.limnosa, O.splendida with O.subtilissima, O. princeps and O.tenuis present as associate species)
	Stigeoclonium tenue community
Zone VI.	β-Mesosaprobic zone
	Cladophora fracta community
	Phormidium community
Zone VII.	γ-Mesosaprobic zone
	Rhodophyceae community (Batrachospermum moniliforme or Lemanea fluviatilis)
	Chlorophyceae community (*Cladophora glomerata or Ulothrix zonata* (clean-water type))
Zone VIII.	Oligosaprobic zone
	Chlorophyceae community (*Draparnaldia glomerata*)
	Pure Meridion circulare community
	Rhodophyceae community (Lemanea annulata, Batrahcospermum vagum or Hildenbrandia rivularis)
	Vaucheria sessiis community
	Phormidium inundatum community
Zone IX.	Katharobic zone
	Chlorophyceae community (Chlorotylium cataractum and Draparnaldia plumosa)
	Rhodophyceae community (*Hildenbrandia rivularis*)
	Lime-incrusting algal communities (*Chamaesiphon polonius* and various *Calothrix* species)
a, b, c, d, e	= as alternatives
1, 2, 3	= as differences in degree

Table 1. Aquatic communities representing various zones of pollution. Survey of the saprobic zones and the corresponding communities [21].

6. Advantages of use of algae in wastewater treatment

There are a symbiotic relation among bacteria and algae in aquatic ecosystems. Algae support to aerobic bacterial oxidation of organic matter producing oxygen via photosynthesis whilst released carbon dioxide and nutrients in aerobic oxidation use for growth of algal biomass. Considering ammonium, carbon dioxide and orthophosphate as main nutrient sources, Oswald determined that oxygen release ratio is 1.5 g O_2/1 g algal biomass [40]. Grobbelaar et al. reported to oxygen release ratio of 1.9 g O_2/1 g algal biomass [41]. Arceivala, accounting latitude, climate and atmospheric conditions, calculated that 4-6% of mean daily solar radiation reaching on treatment pond in 40°N latitude use for new biomass production and production rate of algal biomass may reach 80 kg O_2/1 ha-day [42].

Most of nitrogen in algal cell bound to proteins which compose to 45-60% of dry weight and phosphorus is essential for synthesis of nucleic acids, phospholipids and phosphate esters. Algae using nitrogen and phosphorus in growth may remove to nutrients load of wastewater from a few hours to a few days [43].

In comparison to common treatment systems, oxidation ponds supporting growth of some species may be effective of nutrient removal (Fig. 1). Increasing dissolved oxygen concentration and pH cause for phosphorus sedimentation, ammonia and hydrogen sulphur removal. High pH in algal ponds also leads to pathogen disinfection [44]. Removal efficiency of heavy metals by algae shows changes among species. In fact, studies showed that chrome by *Oscillatoria*, cadmium, copper and zinc by *Chlorella vulgaris*, lead by *Chlamydomonas* and molybdenum by *Scenedesmus chlorelloides* may remove successfully [45,46,47,48,49]. Although algae have adaptation ability to sub-lethal concentrations, accumulation of heavy metals in cells may be potentially toxic effects to the other circles of food web [50].

7. Algal-bacterial ponds

Algal-bacterial pond is water body which is designed to keep and improve of wastewater in a certain time. Although wastewater is treated in pond via physical, chemical and biological processes and/or mechanical processes like aeration, there are also ponds completely based on processes of natural conditions. Ponds, where stabilization of dissolved compounds and suspended solids is in completely aerobic conditions, are named "oxidation ponds". When stabilisation in anaerobic or facultative conditions, ponds are named "waste stabilisation ponds". Stabilisation pond systems are assessed in different types: facultative, anaerobic, aeration and maturation ponds. Common pond type which utilizes from algae is facultative stabilisation ponds. Facultative ponds are designed for purposes such as decrease of waste retention time, achieve of effective treatment or algal culture (Fig. 2). Algal photosynthesis and bacterial decomposition is principal mechanism of algal-bacterial ponds. The processes including oxidation, settling, sedimentation, adsorption, disinfection in the ponds are results of symbiotic relation between algae and bacteria populations [51].

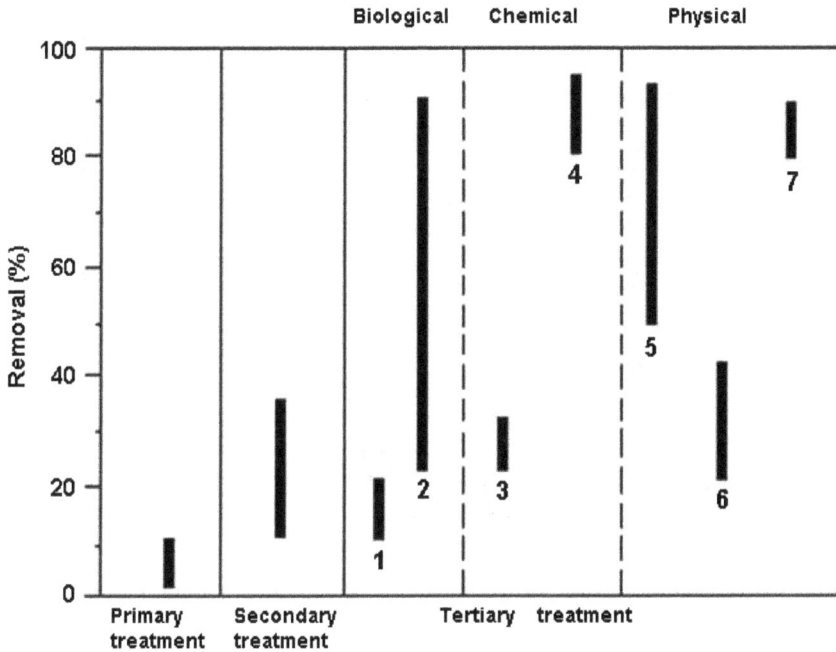

Figure 1. Removal efficiency of organic nitrogen in treatment methods [44]. 1-nitrification, 2-oxidation pond, 3-chemical coagulation, 4-chlorination, 5-ammonia removal, 6-filtration, 7-reverse osmosis

Facultative ponds (usually 2.5m in depth) are systems where effluent quality improves between 5 and 30 days depending on factors such as climate, temperature, wind and surface area [52]. There are three main zones in such ponds; two upper zones with oxygen whilst anaerobic conditions prevail in bottom. Algal photosynthesis and atmospheric diffusion are main oxygen source. Wastes are stabilized by aerobic bacteria in upper zone and by facultative bacteria in intermediate depths while degraded by anaerobic bacteria in bottom zone [53]. Zooplankton controls to excessive bacterial growth and algal blooms through grazing as well as contributing to carbon dioxide production for algal photosynthesis. Food web in a facultative pound is given in Figure 2.

Acceptable effluent quality is the most important advantage of facultative ponds though low operation and maintenance costs. However there are some disadvantages such as high land costs, odour problem in high waste loading, loss of nitrogen to atmosphere, limiting the nutrient reuse by phosphate sedimentation also limiting of irrigation potential by salinity increase during high evaporation period [54]. Although temperature largely affects retention time of wastewater, facultative ponds are widely used in different climate regimes. For example there are more than 3000 facultative ponds in Germany and France and 7000 in United States [53].

8. High rate algal ponds

Municipal wastewater treatments with high rate algal ponds were first proposed by Oswald and Golouke and thereinafter were used in many parts of the world [55,56]. High rate algal pond is usually shallow (20-50 cm) and is equipped with mechanical aeration and mixing by means of paddle wheels. High oxygen level resulting from photosynthesis and aeration allows to low retention times in these ponds. Removal rates of high rate algal ponds are almost similar to conventional treatment methods but may also be more efficient with lower retention time. In fact biochemical oxygen demand (BOD) up to 90% and more than 80% of nitrogen and phosphorus are treated in high rate algal ponds in a few days. However required time for treatment of biochemical oxygen demand up to 90% using by conventional activated sludge and bio filtration techniques, which are highly expensive secondary treatment methods, is between five and eight hours during which lower ratio of nitrogen and phosphorus may be removed. Further, construction and energy costs are highly lower and land requirement is half the required for facultative ponds [57]. It is a well-known fact that only a small amount of nitrogen and phosphorus are removed in active sludge and bio filtration techniques, In addition active sludge and bio filtration techniques require expensive chemicals and complex systems.

Cost of harvest in high rate algal ponds may be most important problem. Thus sedimentation of algae with flocculating is aimed when the wheels are stopped for harvest. In addition growth of resistant algal species to sinking such as *Chlorella, Euglena, Chlamydomonas* and *Oscillatoria* is undesired algae in the ponds. *Scenedesmus* or *Micractenium*, non-preferred species due to their cell morphology for grazing, are dominant in well mixed ponds [40]. Harvested algae may use for industrial and agricultural use as well as effluent in aquaculture (Fig. 3).

9. Advanced integrated wastewater ponds

Advanced integrated wastewater pond systems are an adaptation of waste stabilisation ponds systems based on a series of four advanced ponds: A facultative pond; a high rate algal pond; an algal settling pond and finally a maturation pond for solar disinfection and pathogen abatement. The first pond in series is a facultative pond with depth of 4 to 5 m containing a digester pit, which functions much like an anaerobic pond while surface zone remains aerobic. Effluent of the facultative pond flows to the high rate algal pond for remove to dissolved organic matter and nutrients, then to settling pond with residence time of one or two days for sedimentation of algae and suspended solids. The last unit is maturation ponds where treated water is exposed to the sun and wind leading to natural oxygenation and solar disinfection, and thus an inactivation of pathogens [58].

Wastewater Treatment and Reclamation Plant in St. Helena, California, by US Department of Energy built of formed earth rather than of reinforced concrete in the early 1960s (Fig. 4).

The total pond area needed is much larger than that needed for a conventional plant, but ponds should still cost only one-third to one-half as much to build. Another important advantage of the plant is the small amount of sludge they produced. For example, during nearly 3 decades of operation, St. Helena's wastewater treatment plant has never had to remove residue and measured less than 1 meter of residue had accumulated at the bottom of the deep digester pit [59].

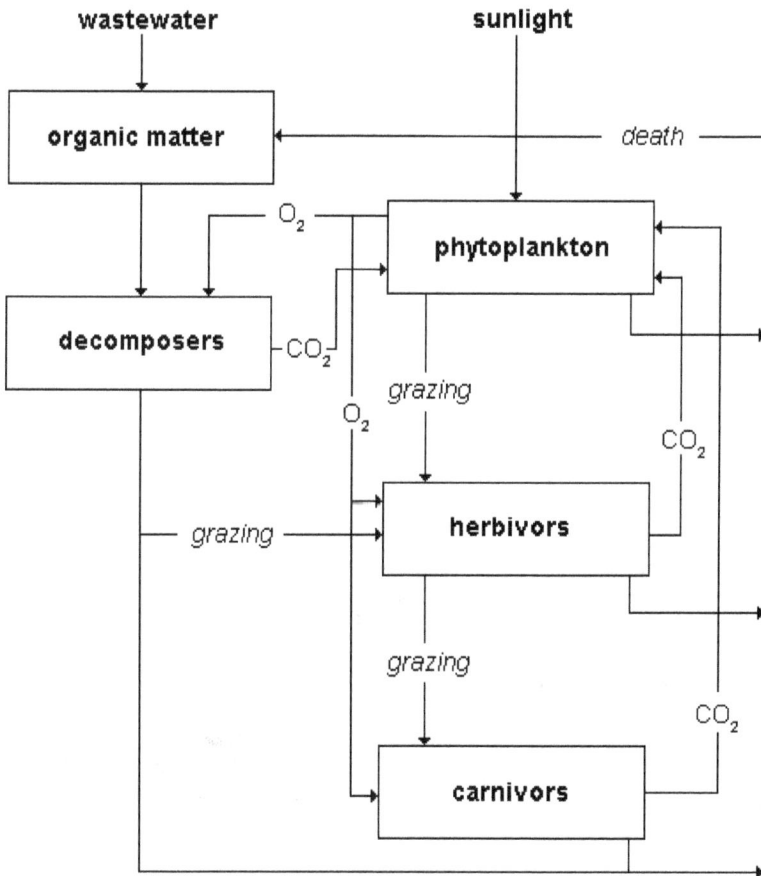

Figure 2. Food web in facultative wastewater treatment pond [51]

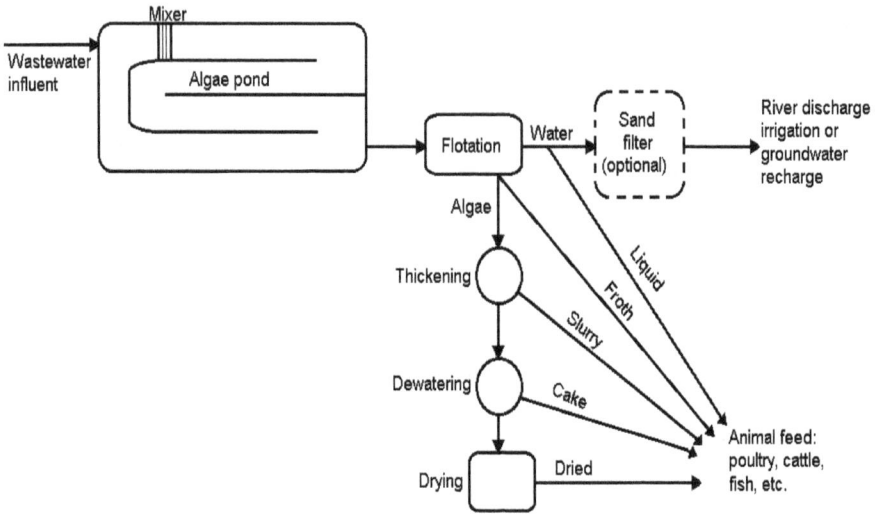

Figure 3. Flow Scheme of the Accelerated Photosynthetic Process for Waste-Water Treatment and Algal Protein Production [56]

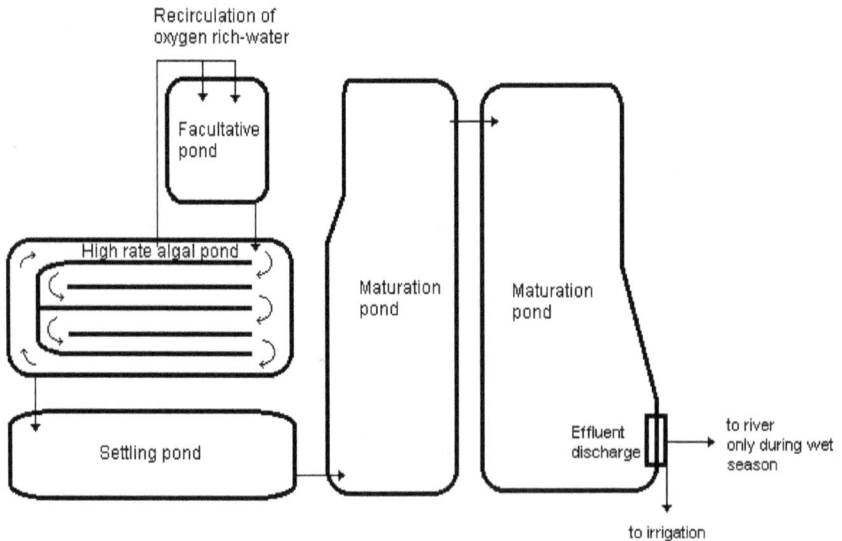

Figure 4. Diagram of St. Helena advanced integrated pond system [59]

Comparison of algae involved wastewater treatment systems is given at table 2. Many useful criteria have been used in the table for more constructive and trustable comparison of wastewater treatment systems connected to algae.

Criteria	Bacteria-Algae Pond	High rate Algae Pond	Integrated Pond
Depth	2.5 m	0.2-0.5 m (20-50 cm)	4-5 m
Salinity	increase	-	-
Retention time	low	lower (because O2 level is high)	high
Land required	high	low	low
Odor problem	occur	not occur	not occur
Loss of N to atmosphere	occur	not occur	not occur
Operation/maintenance cost	low	lower (5 folds lower)	high
PO4 sedimantation	occur	not occur	not occur
Time required for treatment	5-30 days	a few days	5-6 days
Energy requirement	low	low	high
Efficiency quality	low	high	high
BOD removal	fair	good	good
Suspended solid removal	fair	good	good
Harvesting cost		high	low

Table 2. Comparison of wastetreatment ponds in terms of various criteria

10. Conclusion

The water flows from lands into aquatic environments contribute enormous amounts of organic matters and plants nutrients to the aquatic systems which give rise to eutrophication and pollution. With increased urbanization the need for sewage treatment plants (STP) became more important. Wastewater treatment which is applied to improve or upgrade the quality of a wastewater involves physical, chemical and biological processes in primary, secondary or tertiary stages.More sewage plants are designed to remove solids (primary process), followed by a secondary process which involves either activated sludge or trickling filters to reduce the Biological Oxygen Demand (BOD). Removal of the nutrients left after secondary treatment is possible by a variety of processes, one of which involves growth and harvesting of algae from the effluents: others involve ion exchange electro chemical, electrodialysis, reserve osmosis, distillation, chemical precipitation as tertiary processes. However initial cost as well as operating cost of wastewater treatment plant including primary, secondary or advanced stages is highly expensive.

Recently, algae have become significant organisms for biological purification of wastewater since they are able to accumulate plant nutrients, heavy metals, pesticides, organic and inorganic toxic substances and radioactive matters in their cells/bodies with their bioaccumulation abilities. Particularly, biological wastewater treatment systems with micro algae have gained great importance in last 50 years and it is now widely accepted that algal wastewater treatment systems are as effective as conventional treatment systems. Removal rates of particularly high rate algal ponds are almost similar to conventional treatment methods but it is more efficient with lower retention time. With these spesific features algal wastewater treatment systems can be accepted as an significant low-cost alternatives to complex expensive treatment systems particularly for purification of municipal wastewaters.

Author details

Bulent Sen, Feray Sonmez and Ozgur Canpolat
University of Firat, Faculty of Fisheries Department of Aquatic Basic Sciences Elazığ, Turkey

Mehmet Tahir Alp*
University of Mersin, Faculty of Fisheries Department of Aquatic Basic Sciences Mersin, Turkey

Mehmet Ali Turan Kocer
Mediterranean Fisheries Research Production and Training Institue, Antalya, Turkey

11. References

[1] European Inland Fisheries Advisory Commission, Working Party on Water Quality Criteria for European freshwater fish (EIFAC), 1980. Report on combined effects on freshwater fish and other aquatic life of mixtures of toxicants in water. EIFAC Technical Paper 37, FAO, Rome.

[2] Xu, S., Nirmalakhandan, N.,. Use of QSAR models in predicting joint effects in multi-component mixtures of organic chemicals. Water Res. 1998; 32, 2391_/2399.

[3] Altenburger, R., Backhaus, T., Boedeker, W., Faust, M.,Scholze, M., Grimme, L.H. Predictability of the toxicity of multiple chemical mixtures to Vibrio fischeri: mixtures composed of similarly acting chemicals. Environ. Toxicol. Chem. 2000; 19, 2341_/2347.

[4] Sigworth, E.A. Control of odor and taste in water supplies. J.Amer. Water Wks. Assn. 1957; 49: 1507-1521.

[5] Adams, B.A. The role of actinomycetes in producing earthy tastes and smells in potable water. Dept. Of Public Wks., Roads and Transport Congress 1933. Paper No.4. London, England.

[6] Silvey, J.K. and Roach, A.W. Actinomycetes may cause tastes and odors in water supplies. Public Wks. Mag. 1956; 87: 103-106, 210, 212.

---

[7] Brook, A.J. Planktonic algae as indicators of lake types, with special reference to the Desmidiaceae. Limnol and Oceannog 1965;10; 403-411.

[8] Braarud, T.,. A phytoplankton survey of the polluted waters of Inner Oslo Fjord. Hvalraadets Skrifter, Scientific Results of Marine Biological Research 1945. No.28: 1-142.

[9] Butcher, R.W. Pollution and re-purification as indicated by the algae. Fourth Internat. Congress for Microbiology, 1949. Rept. of Proc. P. 149-150.

[10] Archibald, R.E.M. Diversity in some South Africation diatom associations and its relation to water quality 1972. Water Research 6: 1229-1238.

[11] Patrick, R. Factors effecting the distribution of diatoms. Bot. Rev. 1948; 14: 473-524.

[12] Palmer, C.M. A composite rating of algae tolerating organic pollution. J. Phycology. 1969; 5: 78-82.

[13] Palmer, C.M. Algae and water pollution. Castle House Publications Ltd. 1980; 110 p.

[14] Şen, B., Topkaya, B., Alp, M.T., Özrenk,F." Organik Madde ile Kirlenen Bir Çay (Selli Çayı, Elazığ) İçindeki Kirlilik ve Algler Üzerine Bir Araştırma." II. Ulusal Ekoloji ve Çevre Kongresi Bildiriler Kitabı, 1995; s.599-610, Ankara,

[15] Kolkwitz, R. Oekologie der Saprobien. Uber die Bezelhungen der Wasser-organismen zur Umwelt. Schriftenreihe des Vereins für Wasser-, Border-, und Lufthygiene 1950; No:4, 64 p.

[16] Lackey, J.B. Stream and richment microbiota. Public Health Report 1956; 1:708-718.

[17] Liebmann, H. Handbuch der Frischwasser-, und Abwasserbiologie. R. Oldenburg, München, Germany 1951; 539 p.

[18] Lackey, J.B. Two groups of flagellated algae serving as indicators of clean water. J. Am. Water Wks. Assn. 1941; 33: 1099-1110.

[19] Rafter, G.W. The microscopical examination of potable water. Van Nostrand Co., N.Y. 1900.

[20] Kolkwitz, R. and Marsson, M.,. Okologıe der Pflanzichen Saprobien. Ber. Deutsch, Bot. Ges 1908; 26: 505-519.

[21] Werner D.,. The Biology of Diatoms. Botanical monographs. California pres. 1977; Vol 13. 498 pp.

[22] Kalesh NS, Nair SM The Accumulation Levels of Heavy Metals (Ni, Cr, Sr, & Ag) in Marine Algae from Southwest Coast of İndia. Toxicological & Environmental Chemistry 2005; 87(2): 135-146.

[23] Jothinayagi N, Anbazhagan C. Heavy Metal Monitoring of Rameswaram Coast by Some *Sargassum* species. American-Eurasian Journal of Scientific Research 2009; 4 (2): 73-80.

[24] Alp MT, Sen B, Ozbay O. Heavy Metal Levels in *Cladophora glomerata* which Seasonally Occur in the Lake Hazar. Ekoloji, 20 (78): 13-17. doi: 10.5053/ekoloji.2011.783

[25] Alp MT, Ozbay O, Sungur M.A. Determination of Heavy Metal Levels in Sediment and Macroalgae (*Ulva* sp. and *Enteromorpha* sp.) on the Mersin Coast 2011. Ekoloji 21, 82, 47-55 (2012).

[26] MacKenthun, K.M. Radioactive wastes. Chapt 8. İn The Practice of Water Pollution Biology. U.S. Dept. Interior, Fed. Water Pol. Contr. Admin., Div. of Tech. Support. U.S. Printing Office 1969.

[27] Droste, R.L. Theory and Practice of water and wastewater treatment, John Wiley and Sons, New York 1997.

[28] Oswald,W.J. Ponds in twenty first century. Water Science and Technology 1995; 31(12):1-8.

[29] Şen, B. ve Nacar, V. Su Kirliliği ve Algler. Fırat Havzası I. Çevre Sempozyumu Bildiriler Kitabı. 1988; 405-21.

[30] Reddy, K.R. Fate of Nitrogen and Phosphorus in a Wastewater Retention Reservoir Containing Aquatic Macrophytes. Journal of Environmental Quality, 1983;12(1):137-41.

[31] Craggs, R.J., Adey, W.H., Jenson K.R., St. John, M.S., Green, F.B. and Oswald, W.J. Phosphorus removal from wastewater using an algal turf scrubber, Water Science and Technology 1996; 33(7):191-98.

[32] Rose, P.D., Boshoff, G.A., van Hille, R.P., Wallace, L.C., Dunn, K.M., Duncan, J.R. An integrated algal sulphate reducing high rate ponding process for the treatment of acid mine drainage wastewater, Biodeg. 1998; 9:247-57.

[33] Guha, H., Jayachandran, K. and Mauresse, F. Kinetics of chromium (VI) reduction by atype strain Shewanella alga under different growth conditions, Environmental Pollution 2001; 115(2):209-18.

[34] Kaewsarn, P. and Yu, Q. Cadmium (II) removal from aqueous solutions by pretreated biomass of marine alga Padina sp., Environmental Pollution 2001; 112(2):209-13.

[35] Tam, N.F.Y., Wong, J.P.K. and Wong, Y.S. Repeated use of two Chlorella species, C. vulgaris and WW1 for cyclic nichel biosorption. Environ. Pol. 2001; 114(1):85-92.

[36] Weber, K., Probes, B., Lyvansky, K., Kredl, F. and Beryl, I. Removal of biogenic elements, polychlorinated diphenyls and heavy metals during the biogical final treatment of wastewaters. Acta Microbiol. Pol. 1981; 30:255-58.

[37] Shashirekha, S., Uma, L. and Subramanian, G. Phenol degradation by marine cyanobacterium Phormodium valderianum, J. Indust.Microbiol. Biotechnol. 1997; 19(2):130-33.

[38] Lloyd, B.J. and Frederick, G.L. Parasite removal by waste stabilisation pond systems and the relationship between concentrations in sewage and prevalence in the community, Water Science and Technology 2000; 42(10):375-86.

[39] Oswald, W.J. The role of microalgae in liquid waste treatment and reclamation. In: C.A. Lembi and J.R. Waalnd (eds). Algae and Human Affairs, Cambridge University Press 1988a; 403-31.

[40] Oswald, W.J. Microalgae and Wastewater Treatment. In: Microalgal Biotechnology, M.A. Borowitzka and L.J. Borowitzka (eds). Cambridge University Press, New York 1988b; pp.357-94.

[41] Grobbelaar, J.U., Soeder, D.J. and Stengel, E.,. Modelling algal production in large outdoor cultures and waste treatment systems, Biomass 1990; 21:297-314.

[42] Arceivala, S.J. Simple waste treatment methods. Metu Eng. Fac. Pub. 1973 No 44, Ankara.

[43] Lovaie, A. and De La Noüe, J. Hyperconcentrated cultures of Scenedesmus obliquus: A new approach for wastewater biological tertiary treatment, Water Res 1985; 19:1437-42.

[44] Laliberte, G., Proulx, D., De Pauw, N. and De La Noüe, J.,. Algal Technology in Wastewater Treatment. In: H. Kausch and W. Lampert (eds.), Advances in Limnology. E. Schweizerbart'sche Verlagsbuchhandlung, Stuttgart 1994; 283-382.

[45] Filip, D.S., Peters, T., Adams, V.D. and Middlebrooks, E.J. Residual heavy metal removal by an algae-intermittent sand filtration system 1979. Water Res. 13:305-313.

[46] Nakajima, A., Horikoshi, T., and Sakaguchi, T. Studies on the accumulation heavy metal elements in biological system XVII. Selective accumilation of heavy metal ions by Chlorella vulgaris. Eur. J. App. Microbiol. Biotechnol. 1981; 12:76-83.

[47] Ting, Y.P., Lawson, E. and Prince, I.G. Uptake of cadmium and zinc by alga Chlorella vulgaris: Part I. İndividual ion species. Biotechnol. Bioeng. 1989; 34:990-99.

[48] Hassett, J.M., Jennett, J.C. and Smith, J.E.,. Microplate technique for determining accumulation of metals by algae. Appli. Environ. Microbiol 1981; 41:1097-106.

[49] Sakaguchi, T., Nakajima A. and Horikoshi, T. Studies on the accumulation heavy metal elements in biological system XVIII. Accumilation of molybdenum by green microalgae. Eur. J. App. Microbiol. Biotechnol. 1981; 12:84-89.

[50] Wikfors, G.H. and Ukeles, R. Growth and adaptation of estaurine unicellular algae in media with excess copper, cadmium and zink and effect of metal contaminated algal food on Crassostrea virginica larvae. Mar. Ecol. Prog. Ser. 1982; 7:191-206.

[51] Rich, L.G. Low Maintenance Mechanically Simple Wastewater Treatment Systems. McGraw-Hill, New York, 1980; 211.

[52] Tebbutt, T.H.Y. Principles of Water Quality Control. 5th ed, Butterworth-Heinemann, Oxford 1998.

[53] Mara, D.D., Mills, S.W., Person, H.W. and Alabaster, G.P. Waste stabilization ponds: A viable alternatives for small community treatment systems, Journal of Water and Environmental Management 1992; 6(1):72-78.

[54] NRC,. Microbial Processes: Promising Technologies for Developing Countries. National Academy Press, Washington D.C. 1979; 198p.

[55] Oswald, W.J. and Golueke, C.G. The high rate pond in waste disposal. Devel. Indust. Microb. 1963; 4:112-19.

[56] Shelef, G., Moraine, R. and Oron, G. Photosynthetic Biomass Production from Sewage. Arch. Hydrobiol. Beih. 11:3-14.

[57] Esen, I. and Puskas, K.,. Algae removal by sand filtration and reuse of filter material, Waste Management 1991; 11:59-65.

[58] Oswald, W.J. Introduction to Advanced Integrated Ponding Systems. Water Science and Technology 1991; 24(5):1-7.

[59] DOE,. Alternative Wastewater Treatment: Advanced Integrated Pond Systems, US Departmant of Energy, Office of Energy Efficiency and Renewable Energy 1993, Technical Information Program Document No: DOE/CH100093-246, Washington. 8p.

Permissions

The contributors of this book come from diverse backgrounds, making this book a truly international effort. This book will bring forth new frontiers with its revolutionizing research information and detailed analysis of the nascent developments around the world.

We would like to thank Dr. Walid Elshorbagy and Dr. Rezaul Kabir Chowdhury, for lending their expertise to make the book truly unique. They have played a crucial role in the development of this book. Without their invaluable contribution this book wouldn't have been possible. They have made vital efforts to compile up to date information on the varied aspects of this subject to make this book a valuable addition to the collection of many professionals and students.

This book was conceptualized with the vision of imparting up-to-date information and advanced data in this field. To ensure the same, a matchless editorial board was set up. Every individual on the board went through rigorous rounds of assessment to prove their worth. After which they invested a large part of their time researching and compiling the most relevant data for our readers. Conferences and sessions were held from time to time between the editorial board and the contributing authors to present the data in the most comprehensible form. The editorial team has worked tirelessly to provide valuable and valid information to help people across the globe.

Every chapter published in this book has been scrutinized by our experts. Their significance has been extensively debated. The topics covered herein carry significant findings which will fuel the growth of the discipline. They may even be implemented as practical applications or may be referred to as a beginning point for another development. Chapters in this book were first published by InTech; hereby published with permission under the Creative Commons Attribution License or equivalent.

The editorial board has been involved in producing this book since its inception. They have spent rigorous hours researching and exploring the diverse topics which have resulted in the successful publishing of this book. They have passed on their knowledge of decades through this book. To expedite this challenging task, the publisher supported the team at every step. A small team of assistant editors was also appointed to further simplify the editing procedure and attain best results for the readers.

Our editorial team has been hand-picked from every corner of the world. Their multi-ethnicity adds dynamic inputs to the discussions which result in innovative outcomes. These outcomes are then further discussed with the researchers and contributors who give their valuable feedback and opinion regarding the same. The feedback is then collaborated with the researches and they are edited in a comprehensive manner to aid the understanding of the subject.

Apart from the editorial board, the designing team has also invested a significant amount of their time in understanding the subject and creating the most relevant covers. They scrutinized every image to scout for the most suitable representation of the subject and create an appropriate cover for the book.

The publishing team has been involved in this book since its early stages. They were actively engaged in every process, be it collecting the data, connecting with the contributors or procuring relevant information. The team has been an ardent support to the editorial, designing and production team. Their endless efforts to recruit the best for this project, has resulted in the accomplishment of this book. They are a veteran in the field of academics and their pool of knowledge is as vast as their experience in printing. Their expertise and guidance has proved useful at every step. Their uncompromising quality standards have made this book an exceptional effort. Their encouragement from time to time has been an inspiration for everyone.

The publisher and the editorial board hope that this book will prove to be a valuable piece of knowledge for researchers, students, practitioners and scholars across the globe.

List of Contributors

Rezaul K. Chowdhury and Walid El-Shorbagy
Department of Civil and Environmental Engineering, United Arab Emirates University, Al Ain, United Arab Emirates

Jiří Šajer
T. G. Masaryk Water Research Institute, p. r. i., Ostrava Branch Department Macharova 5, Ostrava, Czech Republic

Edward Ming-Yang Wu
I-Shou University, Kaohsiung, Taiwan

Karmen Margeta
University of Zagreb, Faculty of Chemical Engineering and Technology, Croatia

Nataša Zabukovec Logar
National Institute of Chemistry, Ljubljana, Slovenia

Mario Šiljeg
Croatian Environment Agency, Zagreb, Croatia

Anamarija Farkaš
Institute for International Relations, Zagreb, Croatia

Ivan X. Zhu and Brian J. Bates
Xylem Water Solutions Zelienople LLC, Zelienople, PA, USA

Letícia Nishi, Angélica Marquetotti Salcedo Vieira, Ana Lúcia Falavigna Guilherme, Milene Carvalho Bongiovani and Rosângela Bergamasco
Universidade Estadual de Maringá, Brazil

Gabriel Francisco da Silva
Universidade Federal de Sergipe, Brazil

Adina Elena Segneanu, Carmen Lazau, Paula Sfirloaga, Paulina Vlazan, Cornelia Bandas and Ioan Grozescu
National Institute for Research and development in Electrochemistry and Condensed Matter –INCEMC Timisoara, Romania

Cristina Orbeci
Politehnica University Bucuresti, Romania

Erik Gydesen Søgaard and Henrik Tækker Madsen
Section of Chemical Engineering, Aalborg University Esbjerg, Denmark

Ramiro Escudero, Francisco J. Tavera and Eunice Espinoza
Instituto de Investigaciones Metalúrgicas, Universidad Michoacana de San Nicolás de Hidalgo, Morelia, Michoacán, México

Rangarajan T. Duraisamy, Ali Heydari Beni and Amr Henni
Industrial/Process Systems Engineering , Faculty of Engineering and Applied Sciences, University of Regina, Regina, Saskatchewan, Canada

Florica Manea and Aniela Pop
Faculty of Industrial Chemistry and Environmental Engineering "Politehnica" University of Timisoara, Romania

Chunli Zheng
School of Energy and Power Engineering, Xi'an Jiaotong University, China

Ling Zhao and Zhimin Fu
College of Environment & Resources of Inner Mongolia University, China

Xiaobai Zhou
The Environmental Monitoring Center of Jiangsu Province, Nanjing, China

An Li
School of Petrochemical Engineering, Lanzhou University of Technology, China

Agnes Barbério
Institute of Bioscience, University of Taubaté, Brazil

Vera Raicevic and Blazo Lalevic
University of Belgrade, Faculty of Agriculture, Department for Microbial Ecology, Zemun-Belgrade, Serbia

Mile Bozic and Zeljka Rudic
Institute for the Development of Water Resouorces "Jaroslav Cerni", Belgrade, Serbia

Dragan Kikovic
Faculty of Natural Science, Kosovska Mitrovica, Serbia

Ljubinko Jovanovic
Educons University, Sremska Kamenica, Serbia

Bulent Sen, Feray Sonmez and Ozgur Canpolat
University of Firat, Faculty of Fisheries Department of Aquatic Basic Sciences Elazığ, Turkey

Mehmet Tahir Alp
University of Mersin, Faculty of Fisheries Department of Aquatic Basic Sciences Mersin, Turkey

Mehmet Ali Turan Kocer
Mediterranean Fisheries Research Production and Training Institue, Antalya, Turkey

www.ingramcontent.com/pod-product-compliance
Lightning Source LLC
Chambersburg PA
CBHW070713190326
41458CB00004B/965